The Selten School of Behavioral Economics

Axel Ockenfels • Abdolkarim Sadrieh
Editors

The Selten School of Behavioral Economics

A Collection of Essays in Honor of Reinhard Selten

Springer

Editors
Prof. Dr. Axel Ockenfels
University of Cologne
Staatswissenschaftliches Seminar
Albertus-Magnus-Platz
50923 Cologne
Germany
ockenfels@uni-koeln.de

Prof. Dr. Abdolkarim Sadrieh
University of Magdeburg
Department of Economics
Universitätsplatz 2
39106 Magdeburg
Germany
sadrieh@ovgu.de

ISBN 978-3-642-44300-8 ISBN 978-3-642-13983-3 (eBook)
DOI 10.1007/978-3-642-13983-3
Springer Heidelberg Dordrecht London New York

Cover design: WMXDesign GmbH, Heidelberg, Germany

Printed on acid-free paper

Springer is part of Springer Science+Business Media (www.springer.com)

Preface

Nevertheless I still nourish the hope that some of the students who now work on experimental economics under my guidance will become university teachers.

Reinhard Selten (1995): Autobiography. Tore Frängsmyr (ed.): *Les Prix Nobel. The Nobel Prizes 1994*, Stockholm: Nobel Foundation.

Reinhard Selten, to date the only German Nobel Prize laureate in economics, celebrates his 80th birthday on October 5, 2010. For more than half a century he has contributed to economic research, inseminating various areas in microeconomics, especially game theory and behavioral economics, as well as various neighboring disciplines. While his contributions to game theory are well-known, the behavioral side of his scientific work has received less public exposure, even though he has been committed to experimental research during his entire career, publishing more experimental than theoretical papers in top-tier journals. This Festschrift is dedicated to Reinhard Selten's exceptional influence on behavioral and experimental economics.

Throughout his academic career, Reinhard Selten advised and supported many young economists, both in game theory and experimental economics. While in the early years of his career, many of the game theorists he supported proceeded to be successful academics, his Ph.D. students in experimental economics often left academia for successful business careers. One reason probably was that the experimental method was not widely accepted in the discipline and academic jobs were rare for experimenters. It was only after the onset of the "behavioral revolution" in the mid 1980s that a discernible number of experimental papers started appearing in economic journals, opening doors for academic careers. At about this time, Reinhard Selten relocated from the University of Bielefeld to the University of Bonn. Supported by the University, the State, and the Deutsche Forschungsgemeinschaft (German National Science Foundation) in 1985/1986 Reinhard Selten started the first fully computerized laboratory in Europe that was exclusively devoted to research in experimental economics.

The opportunity to do academic work with Reinhard Selten at the Bonn Laboratory of Experimental Economics – in combination with a top-tier Ph.D.-program in economics – created an excellent research atmosphere that attracted a number of highly motivated young scientists and brought them in contact with renowned researchers from all over the world. The experimental economics lab gradually became a well-established hot-spot for behavioral research in Europe, cooperating with distinguished researchers across Europe and the USA. A number of the graduate students and post-docs, who worked with Reinhard Selten at the Bonn Laboratory for Experimental Economics – especially during the 1990s and early 2000s – advanced in academia and are now holding positions and running their own labs. These academic scholars, Selten's students devoted to experimental and behavioral economics, constitute *The Selten School of Behavioral Economics*. They have brought in some of the leading international scholars in experimental research to jointly contribute to this Volume in honor of the "Meister" (which is Reinhard Selten's nickname in the lab and refers to the old German term for accomplished arts and crafts teachers, expressing an especially high degree of esteem).

The collection of papers presented in this book documents the historical role of Reinhard Selten in the development of the research methodology in experimental economics and his contributions to several sub-fields of behavioral economics. The topic of each paper was selected by one of the Meister's students, who also took the active role of coordinating the team of distinguished co-authors. (We have marked each coordinator's name with an envelope symbol on the opening page of his or her paper. The coordinators will act as the corresponding authors). Next to the academic insights that the papers bring into sub-fields of experimental and behavioral research, they also provide a glance at Reinhard Selten's academic and personal interaction with his students and peers. Because we firmly believe that good research needs inspired and critical personal interaction as much it needs formal documentation, we decided to mingle these two incredibly rewarding aspects together.

In some of the papers, you will find personal surveys of a sub-field and you can trace the influence that Reinhard Selten has had on the development of that subject matter, not only by publishing scientific papers, but also with his input in personal communication and his unique personality. Other papers report specific research projects that have been suggested or decisively influenced by the Meister. Yet other papers simply report the interaction with Reinhard Selten under specific circumstances. All of the papers taken together, we hope, will transmit the image of a great scientist and remarkable humanist, who has used his extraordinary talent to push science forward, without pushing compassion away.

October 2010

Axel Ockenfels
Abdolkarim Sadrieh

Abstract

Reinhard Selten, to date the only German Nobel Prize laureate in economics, celebrates his 80th birthday in 2010. While his contributions to game theory are well-known, the behavioral side of his scientific work has received less public exposure, even though he has been committed to experimental research during his entire career, publishing more experimental than theoretical papers in top-tier journals. This Festschrift is dedicated to Reinhard Selten's exceptional influence on behavioral and experimental economics. In this collection of academic highlight papers, a number of his students are joined by leading scholars in experimental research to document the historical role of the "Meister" in the development of the research methodology and of several sub-fields of behavioral economics. Next to the academic insight in these highly active fields of experimental research, the papers also provide a glance at Reinhard Selten's academic and personal interaction with his students and peers.

Acknowledgements

We are greatly indebted to Kristina John (Faculty of Economics and Management, University of Magdeburg) for her support in preparing and formatting this book. The timely realization of this book would have not been possible without her marvelous and dedicated effort. We also thank all other members of the Chair in E-Business at the Faculty of Economics and Management, University of Magdeburg, who supported us in the production phase.

Furthermore we wish to express our deep gratitude to Dr. Werner A. Müller (Executive Vice President Business/Economics and Statistics, Springer, Heidelberg) for his sustained support of the project at every stage, even when we were – once again – lagging behind the time schedule. Without his support and the help of the team at Springer this project would not have been possible.

Contents

Contributors

Klaus Abbink CBESS, School of Economics, University of East Anglia, Norwich, NR4 7TJ, UK, k.abbink@ueq.ac.uk

Dieter Balkenborg Department of Economics, University of Exeter, Streatham Court, Rennes Drive, Exeter, EX4 4PU, UK

Otwin Becker University of Heidelberg, Tannenweg 21 A, 69190 Walldorf, Baden, Germany, O.Becker.Walldorf@t-online.de

Siegfried K. Berninghaus Karlsruher Institut für Technologie (KIT), Institut für Wirtschaftstheorie und Statistik, Lehrstuhl für Wirtschaftstheorie (VWL III), Postfach 69 80, D-76128, Karlsruhe

Gary E. Bolton Laboratory for Economic Management and Auctions, Smeal College of Business, Pennsylvania State University, 334 Business Building, University Park, PA 16802, USA

Jordi Brandts Institut d'Anàlisi Econòmica CSIC, Campus UAB, 08193 Bellaterra, Barcelona, Spain, jordi.brandts@uab.cat

Giorgio Coricelli Institut des Sciences Cognitives CNRS, 67 Boulevard Pinel, 69675 Bron, France

Ernst Fehr Institute for Empirical Research in Economics, University of Zurich, Blümlisalpstrasse 10, CH-8006 Zürich, Switzerland

Simon Gächter University of Nottingham, Sir Clive Granger Building, University Park, Nottingham, NG7 2RD, UK

Sebastian J. Goerg MPI for Research on Collective Goods, Kurt-Schumacher-Str. 10, D-53113 Bonn, Germany

Werner Güth Strategic Interaction Group, Max Planck Institute of Economics, Kahlaische Straße 10, D-07745 Jena, Germany

Christine Harbring Lehrstuhl für Human Resource Management, Fakultät für Wirtschaftswissenschaften, Karlsruhe Institute of Technology, Waldhornstr. 27, 76131 Karlsruhe, Germany, christine.harbring@kit.edu

Heike Hennig-Schmidt *B*onn*E*con*L*ab, Laboratory for Experimental Economics, University of Bonn, Adenauerallee 24-42, 53113 Bonn, Germany

John D. Hey Department of Economics and Related Studies, University of York, Heslington, York YO10 5DD, UK

Bernd Irlenbusch Seminar für Allgemeine Betriebswirtschaftslehre, Unternehmensentwicklung und Wirtschaftsethik, Universität zu Köln, Herbert-Lewin-Str. 2, 50931 Köln, Germany, brend.irlenbusch@uni-koeln.de

Claudia Keser Georg-August-Universität Göttingen, Platz der Göttinger Sieben 3, 37073 Göttingen, Germany, claudia.keser@uni-goettingen.de

Sebastian Kube Department of Economics, Institute for Empirical Research in Economics, University of Bonn, Adenauerallee 24-42, 53113 Bonn, Germany, kube@uni-bonn.de

Ulrike Leopold-Wildburger Institut für Statistik und Operations Research, University of Graz, Universitätsstraße 15, A-8010 Graz, Austria

Manfred Milinski Max-Planck-Institute for Evolutionary Biology, August-Thienemann-Str. 2, 24306 Plön, Germany

Rosemarie Nagel Department of Economics and ICREA, Universitat Pompeu Fabra, Ramón Trias Fargas, 25-2708005 Barcelona, Spain

Tibor M. Neugebauer Faculty of Law, Economics and Finance (LSF), University of Luxembourg, 4- rue Albert Borschette, L-1246 Luxembourg, Luxembourg, Tibor.Neugebauer@uni.ln

Axel Ockenfels Universität zu Köln, Albertus-Magnus-Platz, D-50923 Köln, Germany, ockenfels@uni_koeln.de

Axel Ostmann Universität des Saarlandes, Building FZU (50.40), D-66123 Saarbrücken, Germany

Carmen M. Pasca LUISS Guido Carli, Viale Pola, 12 - 00198 Rome, Italy

Bettina Rockenbach Universität Erfurt, Postfach 900 221, 99105 Erfurt, Germany, bettina.rockenbach@uni-erfurt.de

Alvin E. Roth Harvard Business School, Baker Library 441, Soldier's Field Road, Boston, MA 02163, USA

Abdolkarim Sadrieh Fakultät für Wirtschaftswissenschaft, Otto-von-Guericke-Universität Magdeburg, Postfach 41 20, Magdeburg 39016, Germany, sadrieh@ovgu.de

Dirk Sliwka Seminar für Allgemeine Betriebswirtschaftslehre und Personalwirtschaftslehre, Universität zu Köln, Herbert-Lewin-Str. 2, 50931 Köln, Germany, dirk.sliwka@uni-koeln.de

Matthias Sutter Institut für Finanzwissenschaft, University of Innsbruck, Universitätsstrasse 15, A-6020 Innsbruck, Austria, matthias.sutter@uibk.ac.at

Saraswati Talloo Department of Economics, University of Exeter, Streatham Court, Rennes Drive, Exeter EX4 4PU, UK

Fang-Fang Tang Southwestern University of Finance and Economics, Chengdu, and National School of Development at Peking University, Beijing, China, fftang@accer.edu.cn

Reinhard Tietz Fachbereich Wirtschaftswissenschaften, Goethe-Universität, Steinhausenstr. 23, D 60599 Frankfurt am Main, Germany, tietz@wiwi.uni-frankfurt.de

Horst Todt Fakultät für Wirtschafts- und Sozialwissenschaften, Universität Hamburg, von-Melle-Park 9, 20146 Hamburg, Deutschland, horst.todt@gmx.de

Bodo Vogt Fakultät für Wirtschaftswissenschaft, Lehrstuhl für Empirische Wirtschaftsforschung, Otto-von-Guericke-Universität Magdeburg, Postfach 41 20, 39016 Magdeburg, Germany

Frans van Winden CREED, Faculty of Economics and Business, University of Amsterdam, Roetersstraat 11, Amsterdam, 1018 WB, The Netherlands

Part I
Exceptional Academic Behavior

Chapter 1
Reinhard Selten a Wanderer

Abdolkarim Sadrieh

When I first met Reinhard Selten, I was a sleepless young student, double-degreeing in economics and computer science (which, studying in Bonn, basically meant that I was spending my days and nights studying applied mathematics). I knew Herbert Simon's work well (but not Reinhard Selten's) and I had planned to shake-up economic research by introducing macroeconomic models based on boundedly rational decision-makers.[1] That was why I had to study computer science. I needed to learn how to write simulation software for my virtual economies. Things were going OK. The university had just received about 40 first generation IBM PCs and – being a second year computer science student – I was privileged to have courses learning to program in Pascal, Modula, ProLog, and other cool languages that I cannot recall. Then, looking for an interesting course in economics, I came across a seminar given by Reinhard Selten on "experimental industrial organization." Frankly, I had absolutely no idea what this seminar would be like, but I was pleased by the announcement that "warned" participants that in this seminar they would have to "program a strategy" to run on a PC. The seminar started with a 2-week crash course to teach the participants how to program a duopoly strategy in Turbo Pascal. Clearly, there was no course in economics that was closer than this one to programming the boundedly rational agent simulations that I was so interested in. It was immediately clear to me that I must sign-up for this course.

The next thing I knew, I was listening to Reinhard Selten give the introductory lecture at the first class meeting. He first explained the game (a simple Cournot Game with asymmetric cost) and then described the procedure. I did not know then that I was in the middle of a strategy method seminar (Selten 1967). Since I had

[1]Agent-based simulation economics actually has gained some ground in the mean time (see Tesfatsion 2001).

A. Sadrieh (✉)
Fakultät für Wirtschaftswissenschaft, Otto-von-Guericke-Universität Magdeburg, Postfach 41 20, Magdeburg 39016, Germany
e-mail: sadrieh@ovgu.de

A. Ockenfels and A. Sadrieh (eds.), *The Selten School of Behavioral Economics*,
DOI 10.1007/978-3-642-13983-3_1, © Springer-Verlag Berlin Heidelberg 2010

always preferred studying on my own to visiting lectures, I had never actually seen Reinhard Selten give a lecture before. What he said was so clear, so precise, and so much committed to his vision of bounded rationality that it took him only those first 90 minutes to convince me that economic research without experimentation makes no sense at all. When I left that class, I still wanted to run agent-based macroeconomic simulations with boundedly rational agents. But, now I had learned that the only sensible way of constructing these agents, was to run experiments that provide the necessary information on behavior-based decision heuristics.

What followed in the strategy seminar were spontaneous decision experiments, in which my fellow students made "silly" decisions, instead of simply choosing to play the Cournot equilibrium. In class, we talked about these results. I was dumbfounded to find that my fellow students, who were making "silly" decisions, were actually earning much higher payoffs than I was. So I reconsidered and tried to give a best response to their choices, instead of giving a best response to what I thought they should have chosen. This helped. Once we started programming, I wrote a complicated strategy that used a stochastic procedure to estimate the response function of my opponent, before giving a dynamic best-response to the estimated strategy. Meanwhile the others were using "rules of thumb" to find ways cooperate in the asymmetric Cournot Game. My "optimization" strategy turned out to be wildly suboptimal against the large number of conditional cooperators. But, while my strategy lost in every tournament, I won much more than I had expected. After the course, I was asked to join Reinhard Selten's experimental group as a student assistant. I stayed for the next 12 years, moving up the "career ladder," from student assistant, to research assistant (Ph.D.-student), and finally to the position of the managing director of the laboratory (post-doc). And to this very day, over a quarter century after the strategy seminar that brought us together, Reinhard Selten enjoys reminding me of the bold and youthful stupidity of my "optimization" strategy.[2]

But, the academic career and a deeper understanding of strategic and boundedly rational behavior were not the only things that I took away from that seminar. I also learned about the dignity of taking and holding onto a scientific point of view that you believe in. Reinhard Selten is doubtlessly one of the most highly respected game theorists. He is a distinguished member of numerous renowned scientific academies, a Nobel Memorial Prize laureate, and – as many scholars in economics will agree – a genius mind. But, despite all the game theory successes that he was piling up in the 1980s, he invested almost all his resources at that time into starting a computerized laboratory for experimental economics – a dream that he had developed ever since he had worked at Austin C. Hoggatt's computerized lab in the 1960s at the University of California at Berkeley. Some purely theory oriented

[2]The results of that strategy seminar were eventually published in Selten, Mitzkewitz, and Uhlich (1997). On page 541, the authors' comment on my strategy as follows: "In fact, the participant who wrote a strategy with success rank 20 firmly believes that this approximative dynamic programming approach based on an estimated response function of the opponent can be improved to a degree which will make it superior to all final strategies in a tournament against them. We doubt that this is the case".

game theorists seem to believe that Reinhard Selten is a turncoat, who has lost or (even worse) has left the path to the "pure and true" cause of game theory. But, those who have this perception, neither seem to share Reinhard Selten's notion of game theory,[3] nor have they carefully read any of his early work, in which he sketches out his research agenda more than half a century ago (Sauermann and Selten 1959). For him – as also for numerous other early experimenters – game theory and behavioral economics have always been two sides of the same coin, i.e., the two paths to understanding behavior in strategic interaction. Asking Reinhard Selten to give one up, would be like asking the parent of twins to abandon one for the sake of the other.

Let us stick to that picture for a minute: Like the parents of twins should not scold one and spoil the other, Reinhard Selten has not only urged game theorists to face the facts of observed behavior, but has also often been the scientific conscience of experimental economics. I clearly remember numerous instances – long ago – when experimental research with very weak statistics was presented at conferences (especially if random-matching treatments were played in only one or two sessions, yielding only one or two independent observations), Reinhard Selten would not hesitate to speak up in an upset voice: "If you publish this with so few observations, you will be polluting science! It may take years – if not decades – until somebody else can reexamine the question and check for the reliability of your data!" In later years, he could sit back and watch how the young experimental economists from all across the globe (including a number of the contributors to this volume) spoke up whenever weak statistics were presented. The standards have gone up tremendously and I have absolutely no doubt that Reinhard Selten – to a large extent – takes the credit for the high standards that we maintain in experimental economic research today.

As the example above shows, Reinhard Selten has always taken responsibility for the area of research in which he is active. One of the most important issues, he believes, is to keep the credibility of research as high up as possible. Once the credibility is lost, he likes to point out, there is no sense in doing research, because there is a pooling at the worst possible state. The market for science is not much different from Akerlof's market for lemons. Hence, no matter into which field Reinhard Selten takes a scientific "field trip," he is always extremely cautious and well-prepared. For example, when Reinhard Selten wrote a game theoretic paper on the foraging behavior of solitary bees together with Ronen Kadmon and Avi Schmida (1991), he not only learned all there was to learn about solitary bees, but also invested in a number of plant identification books that he would carry with him on any of his numerous hiking tours. For the next couple of years, hiking with Reinhard Selten was a stop-and-go process, because he enjoyed stopping at any unknown flower and flipping the pages of his plant identification books, trying to identify the flower. Most of the time, I recall, he was successful finding an

[3]Reinhard Selten has been famously cited for saying: "Game theory is for proving theorems, not for playing games". (Goeree and Holt 1999).

illustration that resembled what we saw. But, sometimes when we were frustrated by the futile page flipping, he suddenly looked up and claimed with the type of certainty only a seasoned forest guide could possibly have: "Aha! This is a foreign plant that has either been blown into this forest or – even more likely – has been brought in by tourists stuck to their car or their shoes!" Or on another occasion: "Aha! This seems to be the yellow variant of the . . ."

Hiking with Reinhard Selten is indeed important, because it is one of the two possible states of the world, in which you can discuss both scientific and non-scientific topics extensively and receive excellent advice. The other option is spending the afternoon at a café with coffee and cake. I must admit that I usually preferred the latter option, because I am lazy and dislike hiking in the rain.[4] Obviously, the *non plus ultra* is hiking first and spending the rest of the afternoon in a café, which is probably also his favorite option. My assumption is that the hiking fatigue opens the mind for unusual and creative ideas that can then be thoroughly analyzed and evaluated with an intake of coffee and calories.

The metaphor of the wanderer also holds in Reinhard Selten's academic life. Like a wanderer dislikes being stuck in a forest that he has seen before, Reinhard Selten dislikes staying on a research topic longer than necessary. This does not mean that he switches quickly. On the contrary, he sticks to the topic until he has a satisfactory answer, no matter how many years go by.[5] But, writing several papers on the same topic always seems boring to him. When we suggested a new, presumably "exciting" variant of an experimental game that he already knew, Reinhard Selten would shake his head in discomfort and say: "Well, you can do that. It sounds interesting and it certainly is not forbidden. But, you know, I would prefer if you come up with a new experimental paradigm." And often, while explaining a design, we noticed that his mind had wandered away. He was designing a new paradigm, while we were still trying to understand the old.

Another important aspect of the hiking for Reinhard Selten has always been keeping in touch with the people. Listening to their stories and transforming the nature of the relationship into friendship. Reinhard Selten always maintains a cordial, sincere, and almost caring relationship to those with whom he works closely. His doors are always open both for those, who are close by, and for those, who

[4]Light rain is no issue for Reinhard Selten – not even, if it happens to fall in the Negev – simply because he always carries an umbrella. This obviously implies an extremely high risk-aversion parameter, which in fact may be in line with the extent of his early arrival time at train stations and airports. But, the extremely high risk-aversion parameter sharply contradicts Reinhard Selten's choice of research topics. These choices seem often strongly risk-seeking and sometimes almost contrarian.

[5]Once, working on the last part of my Ph.D.-thesis, I was frustrated and I felt my time is running out with only a year left on the contract. So I went to get some advice from Reinhard Selten. I was optimistic, because usually talking to him solved all problems immediately. When I arrived at the meeting, the first thing that came to my mind was to mention that my time is running out. He was puzzled by my assessment of the situation and answered: "Well, you know, some things take time. My paper '. . . where 4 are Few and 6 are Many' took me almost 10 years to write. So, in comparison, you still have some time." (See Selten 1973.)

have been long gone. His memory – in general – is long-term and rather detailed, especially when supported by his wonderful wife Elisabeth. The couple is known for their magnificent hospitality (that has become more challenging, due to the increased physiological difficulties in the last years). They are also praised for their open-mindedness, which may be positively correlated to their interest in Esperanto and the peaceful global society that is part of the vision of the Esperantist community. I will never forget the superb dinners, to which we were invited at the family home. Knowing the bounded rationality of her guests, Elisabeth Selten sometimes started the dinner by staining the tablecloth with a small drop of gravy and proclaiming: "Now, that the tablecloth is stained, you can eat comfortably and do not need to worry about spilling anything". We would toast and start dining, with the family cats sitting quietly in different corners of the room, observing the feast and waiting for the delicious leftovers. The cats, we learned, were there, because they had "tested" the services in all the houses of the neighborhood and had decided that the Selten family home was a premium (or should I say "puur-mium") location.

It seems that the only competitors the cats have in that household are books. For German standards this is a very spacious house, but I remember that at one point the number of books had reached such a seriously threatening high number that Reinhard Selten consulted us, saying that his house had been totally "over-booked." The books had taken up every angle, including the space under the beds and the corners in some of the rooms. Our crisis plan first took out almost 40 years of journal issues, donating them to one of the new universities in the eastern part of Germany. An additional library built in the basement of the house finally stabilized the situation for the next couple of years. We hope that adding a copy of this book to the huge collection at the Selten family home will be a pleasure that does not cause any new space problems, threatening the family's or the cats' habitat.

References

Goeree JK, Holt CA (1999) Stochastic game theory: for playing games, not just for doing theory. Proc Natl Acad Sci U S A 96:10564–10567

Kadmon R, Selten R, Shmida A (1991) Within-plant foraging behavior of bees and its relationship to nectar distribution in Anchusa strigosa. Isr J Bot 40:283–294

Sauermann H, Selten R (1959) Ein oligopolexperiment. Z Gesamte Staatswiss 115:427–471

Selten R (1967) Die Strategiemethode zur Erforschung des eingeschränkt rationalen Verhaltens im Rahmen eines Oligopolexperiments. In: Sauermann H (ed) Beiträge zur experimentellen Wirtschaftsforschung. J.C.B. Mohr, Tübingen, pp 136–168

Selten R (1973) A simple model of imperfect competition, where 4 are few and 6 are many. Int J Games Theory 2(3):141–201

Selten R, Mitzkewitz M, Uhlich GR(1997) Duopoly strategies programmed by experienced players. Econometrica 65(3):517–555

Tesfatsion L (2001) Introduction to the special issue on agent-based computational economics. J Econ Dyn Control 25(3/4):281–293

Chapter 2
Encounters with Reinhard Selten: An Office Mate's Report

Otwin Becker

Background Story

In winter term 1960/1961 about 100 students applied for the economic seminar held by Professor Heinz Sauermann. As an economics student at the University of Frankfurt, I took part in that seminar. Among the students it had the reputation of being the most challenging course in the field of economic theory.

After passing a test, about 20 of the best students were allowed to participate. Sauermann always organized this seminar in economic theory together with all his research assistants, Reinhard Selten being one of them, and with some guests (former assistants, future colleagues).

At the end of the term I decided to talk to Sauermann during his office hour and ask for a position as a student assistant. To take that step was not easy for me and I was all the more surprised and happy by his answer: "You can start tomorrow".

During the seminar I did make some critical remarks but more with the intention that I, with a major in Business Administration, could forget about economics later on and would not have any more contact with it at my already planned and partly started second study program in mathematics after finishing my degree. But it should all turn out differently.

First Encounters with Reinhard Selten

A very important step for me was that Professor Sauermann, who then already had more than ten assistants, had Reinhard Selten and me to share an office. It stayed like that for many years and looking back I can say that nothing better could have happened to me.

O. Becker
University of Heidelberg, Tannenweg 21 A, 69190 Walldorf, Baden, Germany
e-mail: O.Becker.Walldorf@t-online.de

A. Ockenfels and A. Sadrieh (eds.), *The Selten School of Behavioral Economics*,
DOI 10.1007/978-3-642-13983-3_2, © Springer-Verlag Berlin Heidelberg 2010

The Sauermann-assistants soon recognized that we both used to discuss all different kinds of problems for hours and soon the saying spread: "Even if you do not see the two of them, you will hear them".

We never attended the same lectures but shared interests for the same research fields, such as psychology. From my almost 10 years of professional practice before studying (6 years as insurance broker, 4 years as certified financial accountant), I was able to share all different aspects of my practical experience. On the other side, Reinhard Selten as mathematician and especially as game theorist was an excellent teacher. We did not only concern ourselves with economic questions; we also enjoyed solving all kinds of riddles. Reinhard Selten was a very successful problem solver.

Common Fundamental Beliefs

Regarding economics as an academic discipline we shared a number of common fundamental beliefs from a methodological point of view. For us it seemed very doubtful and unrealistic that human behavior (regardless of the decision-maker being a household or a firm) could be explained by the rationality postulate. It appeared to be very far from reality for a number of reasons. To justify this opinion is easy. One need only think of the large set of goods desired by the household and of the quantities of producible goods by the firms to have reasonable doubts about the practical solvability of the household maximizing utility and the firms maximizing profits. In addition, the rationality principle requires a temporal interdependency of economic decisions (always meaning long-term planning). With the complexity of the decision tasks alone it seemed impossible, even with a short-term time horizon.

We both agreed that for a introductory lecture it is sufficient to demonstrate the fundamental tasks of economic decisions in a two-goods-world and to illustrate rational behavior in this two dimensional cosmos; and why not, it all seems very reasonable at first sight.

Certainly this is not a realistic consideration of the real economic decisions tasks. This can be proven normatively and exploratory. If one is not content with an "as-if-explanation," something new had to be conceived. In this connection Reinhard Selten, like Herbert A. Simon, whom he knew in person, thought about developing an own concept in the direction of "limited rationality," that has to be applicable and that must be able to provide experimentally verifiable results. This laid the foundation for the later by Reinhard Selten developed "Aspiration Adaptation Theory". But one life task persisted: the experimental consideration (i.e. verification) of this theory, which Reinhard Selten is still preoccupied with until today.

Naturally Reinhard Selten's academic interests were not only focused on this topic. A lot of our discussions aimed at topics that were not represented when building up the departments at the Frankfurt Faculty of Economics and Social

Sciences. Apart from the field of game theory, this concerns: operations research, business informatics including computer applications and finally econometrics.

Led by Reinhard Selten, soon a group was formed that according to their interests dealt with questions from the just mentioned areas. An accompanying study of literature for our increase of knowledge played an important role.

Within the group there was more than one person capable of programming so there were no significant problems with the numerical analysis. There was only seldom the need to buy software, the more so as for numerical analysis Selten could fall back on me (FORTRAN) and on Reinhard Tietz (ALGOL).

A New Seminar Develops at the Sauermann Chair

Thanks to Reinhard Selten Heinz Sauermann could be convinced to offer another seminar at his chair, especially because of the current and future planned DFG projects and to give the research in this area an institutional frame. It was his seminar for "Mathematical Economic Research and Econometrics".

First Forecast Experiments with the Time Course HISTO

The first experimental collaboration with Reinhard Selten were preceded by discussions on so called "technical stock price analysis" that was also popular in the 1960s in Germany. We asked ourselves a just as obvious as provoking question: What is it like, if someone looks at a price trend graph of the recent past of a share and knows neither anything about the company, nor anything about the price movement of other corporations? Are there concrete general rules about price movements of shares that forecasts could be based on them?

A well-directed experimental approach was easy to realize. After a preliminary study with some empirical data, I attended the task as follows: I created a long random series of realizations of a linear second order difference equation with damped oscillations and a "white noise term," so that graphically a picture of a more or less irregular sine wave was produced.

From a longer series of more than 200 periods, 42 periods were chosen for the actual experiment. The test subjects had no information on where the data came from, nor on what it meant. For us this seemed to be best to compare it to a newspaper reader, who finds a time course whose meaning he does not know, but thinks about how this time course would continue if it was a stock price.

In individual tests, the test subject had the task to predict the next number in the row. Thereupon they were told the actual number, which they had to write in the chart, to go on predicting the next number.

Our first aim was simply to explain the mean predictive value of the test subjects.

In the Gestalt Psychology there have been similar experiments a lot earlier (already in the 1930s). The test subjects were showed a pile of dots at the beginning and had to guess, what object they would form. The number of dots was intentionally that small, so the test subjects were not able to guess right at the first try. The relevant object was a cup e.g., of which only a few dots or rather fragments were shown. Then, successively more dots were added until the test subjects were able to identify the object. It took some test subjects longer to identify the object than others. Our experiment is of course a lot more complicated but there are some analogies. By adding more dots, the sine wave could not be recognized wholly but the local extrema should have been considered for their prognosis as "dots of special value" (Selten) by the test subjects. On this a first explanatory model for the mean projection was built.

The time course HISTO, as we named it, is still today subject of academic research. Already in the 1960s the number of participants was extensively increased. In cooperation with Ulrike Leopold-Wildburger, a "Bounds and Likelihood-Theory" was developed that explained the mean projections comparatively well. The difference to the first, predominantly wholistic explanatory model lays mainly in the reversal probabilities of the time course being estimated from the reversed cases observed up to then. In later versions of the experiment, additional information in terms of leading series with a lead of a period and more or less high deviations compared to the base course were added. In the next step experiments were conducted in which structural breaks are built in the base course.

Some of these experiments were conducted during my time at the Karl Franzens-University in Graz. Ulrike Leopold-Wildburger, who under Selten already habilitated several years ago, played an important role by carrying on the experiments initiated by me.

The current state of the experimental research of the HISTO-data in Graz is that the test subjects are attached to an electronic indicator, the so called "eye tracker", with which in nearly all of the 42 periods all movements of the eye in viewing direction, location and time period can be recorded. All in all it has to be noted that meanwhile there are several publications on the HISTO experiments. Further publications are in preparation.

OR-Problems, Ultima Ratio and Good Heuristics

The field of operations research is another chapter of the collaboration with Reinhard Selten. Though from mathematical view seemingly trivial, yet n-factorial sequences have to be compared and the value n by no means is small, otherwise this problem would not exist. At this class of problems there is always the question if all possible cases can be calculated (the ultima ratio), or if only the search for a preferably efficient heuristic is promising.

Firstly, one of these n-factorial complete problems is described, for which a certain practical meaning cannot be denied. It is the *travelling salesman problem*:

A travelling salesman has to visit n people (or stations) as fast as possible. All information is restricted to a type (n,n) matrix, from which emanates what time it takes to get from station i to station j. For small n this task is trivial, as the solution is only to compare all n! periods and to look for a period with the shortest time requirement.

First of All a Side Note on Available Computer Technology Now and Then

Is n, the number of stations, notably smaller than 20, already in the 1960s this task could be solved in adequate time with the main frame computers (qua ultima ratio) available. But despite all technical progress, for $n > 200$ the ultima ratio would still take too much computing time. Besides, this entire problem with deterministic running times is an exercise at the most, because realistically, these running times often have quasi random character.

From the view of a practitioner it is solely about finding a good heuristic that has a tolerably acceptable efficiency relation (running time of the optimal solution in relation to the heuristic).

Reinhard Selten suggested an approach for the production of all periods, which he recalled from a combinatorics lecture and that was easy to program for any n. This was the trick: To every number from 1 to n a running direction (left or right) was assigned. At the beginning e.g. all numbers run to the right. All permutations only result from exchanging one number with its neighboring number. There is an exactly defined running rule. If no number can be exchanged according to this rule any more, the procedure stops automatically and all permutations have been played. There is one basic rule: Every number can only be exchanged in its momentarily running direction and only by a higher number.

Step 1: You start with the elemental period $+1, +2, +3, \ldots, +n$, giving all numbers the algebraic sign $+$. It means their exchange direction is to the right.

Step 2: Every try to exchange starts with $K = 1$, wherever this number is located, followed by testing the basic rule:

(a) If allowed, the exchange takes place and you go back to step 2.
(b) If not allowed, a change of direction of the number 1 takes place and a skip to step 3 with $K = 2$.

Step 3: The try to exchange the number K in its momentary direction is tested.

(a) If allowed by the basic rule, the exchange takes place and you go back to step 2.
(b) If not allowed, only a change of direction of the number K takes place and $K + 1$ continues with step 3 as long as $K < n$.

In case of $K = n$ all permutations are produced.

The simple example for n = 3 produces the six permutations in the following row:

+1+2+3	+2+1+3	+2+3+1	+3+2−1	+3−1+2	−1+3+2

In the last position, no number can be exchanged.

There is no need to test whether all n! periods are checked using this procedure, because at the end in the case of n = 6, for example, the last position is: −1−2−3−4+6+5. The first four numbers cannot be exchanged to the left, the number 6 can never be exchanged and the number 5 cannot be exchanged to the right.

The mathematical proof for the possibility of producing all n! sequences is waived here. A draft of the proof in the direction of the above mentioned procedure should suffice: Imagine for the numbers of 1 to (n−1) any arrangement of the (n−1)! permutations. In every one of these (n−1) factorial permutations you put the number n in front and pull this number diagonally through all of the (n−1)!-permutations until the end. Therewith you produce exactly n(n−1)! = n! permutations.

In a program it is recommendable to fill the positions 0 and n + 1 with the number 0 to avoid exceeding the boundary. Regarding the target function, the new value can be calculated by exchanging three numbers from the old value of the objective function as one passage is reversed and both the new tie-in sections have to be replaced, too. Anyway, this "elegant" solution of Selten with n as input parameter of the program is better than all other programming solutions that work with an n-times nested Do-nest.

Now what about a feasible heuristic for the concrete task of combinatory optimization? It is known for a long time as "nearest neighbor". The name is strictly speaking actually self explaining. To determine the efficiency of this heuristic you can, for example, produce a higher number of purely random topologies for a yet acceptable n. For the time needed from i to j, you take a value proportional to the Euclidean distance of the random target points. For the constructed time matrices you generally find a mean efficiency of more than 90%. Admittedly, to be correct, the efficiency can drop to nearly 50% occasionally. But these cleverly constructed extreme counter examples are based on neighborhood relations that you can get under control by using easy to observe modifications.

Another Example: Helmstädter's Linearity Measure

The sector alignment of an input–output-matrix in the broader sense follows the typical alignment of a production process, starting with the primary industry, over the manufacturing industry in different sectors, to the consumer oriented output stages. The allocation of the production sums above the main diagonal indicates on how the production process develops straight (in the sense of "linear") towards the output stages. Helmstädter's question is: At which allocation of the sectors is the sum of the delivery chains above the main diagonal at a maximum? If all industries would only deliver downstream, or technically speaking, if all sectors were aligned in way that below the main diagonal of a input–output table no regressive delivery chains would exist, than maximum linearity exists according to Helmstädter. As I/O tables often consist of more than 30 sectors, even here it is not possible to calculate all possible n! rows in an adequate computing time.

A reasonable heuristic to use is a ratio criterion (sum of all delivery chains above/sum of all delivery chains below the main diagonal) to stepwise decide whether a sector can be integrated in the still available free spot ahead (maximum ratio) or back (minimum ratio). The deliveries of the already assigned sectors play no role anymore at the calculation of the ratio. A control calculation shows that a reassignment of the middle sectors often results in slightly improved solutions but this reassignment is often accompanied by small robustness. The conclusion is: for smaller n an exact efficiency value is calculable and the described heuristic reaches a relatively high efficiency.

The SINTO Market

There exists a company game, called SINTO market, which Selten and I invented in 1967 and we played it for the first time at the Control Data Company in Frankfurt/M. This off-campus location was chosen because the University of Frankfurt/M. did not have the adequate computer equipment needed for our experiment. We needed a main frame computer system working for several hours. Nowadays you only need a mediocre computer.

The development of the game was preceded by numerous considerations about corporate strategies, e.g. oligopolistic price policy, brand products produced from a uniform raw material, possibilities of quality variation and product-related advertising with aftereffects of it in later periods. This was all supposed to be realized after the company foundation. The products to be introduced on the market are subjected to a typical product life cycle and the course of the game should reach exactly 15 periods to the maximum sale stage.

In previous years we closely examined a series of practical, managerial questions, also covering the field of quality variation, which now suggested that we should experimentally examine a practical example ourselves. It was supposed to be an example for a market, on which a certain good is offered in different quality variations and by a small number of companies. Therefore, we invented an oligopoly market in which three companies could offer and sell the same product (called SINTO) in different quality variations on one market.

Each of the companies was allowed to produce and sell up to ten qualitatively different brands of SINTO. For each brand, three quality criteria had to be guaranteed that each could be varied in 10 steps from 0 (low level) to 9 (maximum level). To avoid any previous knowledge on the product itself and its production, we invented with SINTO an albuminous "dietary sublement substance, that was not yet invented but could be at any time," i.e. a new product to be introduced on the market.

In order not to distract the decision-makers in the three companies with financing problems and questions of profit distribution, we invented parent organizations to which a certain amount from the earned profit had to be paid. So these parent organizations were responsible for all financial duties and responsibilities of the companies.

The company's scope of action: In each of the 15 periods they were able to cancel an already existing brand, introduce new brands and at the same time carry on the old ones and, if desired, change their quality criteria. In every period they had to decide for each brand for the production output, the offer price and advertising expense.

At the beginning of each subsequent period the companies were presented the following documents:

1. The closing balance
2. The profit and loss statement
3. A market overview on all products on the market including their prices and quality criteria
4. A breakeven analysis structured by brands

Thus, up to 30 different brands can be introduced to the market. The first two quality criteria (fine graindness and tartness) are cost neutral and only serve for differentiation of taste. To present the participants with an indicator for the preferences of the customers, they were told that there is a buyer for each criteria combination but the middle criteria stages are preferred. In contrast, the third criterion is not cost neutral but a quality criterion that causes constantly increasing additional costs.

As an ideal casting for the companies we imagined teams of up to six people. The team should decided on a division of labor, consult together on the situation of the company, and then reach a decision. So far for the short description.

Regarding the course of the game, we were interested from a theoretical point of view in various questions:

1. Which price, investment, and advertising policy is chosen?
2. How many different brands are introduced?
3. Where are the brands placed in the market landscape?
4. How do the companies organize their decisions?

The underlying mathematical model was designed by Reinhard Selten in a way that it could be aggregated regardless of the numerous decisions of the companies and it therefore allowed detailed analyses and benchmark tests with theoretical solutions.

From 1967 until today, this game has not lost any of its attractiveness. Even for experienced Business Administration Majors it is not easy to play; not to mention students that are often not even able to understand all the documents correctly.

This also explains why a lot of companies could not generate even half as much profit as with an experienced play.

For this experiment there are several result oriented questions. It allows e.g. an answer to the question if and in what ways firms with female or male staff differ. Decades later, we could convince Ulrike Leopold-Wildburger, University of Graz, and Oliver Heil, University of Mainz, to repeat the game. The data from Graz presented similar results to our former observations.

One interesting result from the Graz experiment should be pointed out. A company with solely female staff – on average – earns less profit than with mixed staff or solely male staff. The women comparatively often had too low starting prices. Male teams spend a lot on advertising. A typical mistake was obviously not to act whole-heartedly in case a brand completely sold out. In this case, one expects that the obviously too low starting price is drastically increased and the additional cost of stockpiling is accepted, especially since the sales are planned for the following period.

A Resume of Reinhard Selten Back Then Until Today

To draw a resume of the years spent together with Reinhard Selten is easy for me: He was an excellent provider of ideas, a mathematician quickly adjusting to the special problems in Economics, an all-rounder who liked talking about psychology as much as about science and eventually one of the best game theorists worldwide already at an early age.

And nothing of that has changed ever since.

Chapter 3
Reinhard Selten's Frankfurt Years from the Perspective of a Co-player

Reinhard Tietz

The international reputation of Reinhard Selten is based on three pillars of his research activities that – similarly as in his "Three-Level-Theory"[1] – form an interplay. Those are namely: first Game Theory working with the assumption of strict rationality; second Experimental Economics studying real behavior in laboratory situations and third the theory of bounded rationality bridging the resulting contrast. He already laid the foundation for these three directions in his time in Frankfurt while studying mathematics, economics and psychology.

After his Abitur in Melsungen, Reinhard Selten enrolled in summer term 1951 at the J.-W.-Goethe University in Frankfurt for mathematics.[2] Encouraged by the book "Theory of Games and Economic Behavior" by von Neumann and Morgenstern[3] published in 1944, he took part in a game theory seminar for economists organized by Ewald Burger. Referring to his Master's Thesis "Bewertung strategischer Spiele", the supervisor Burger in 1957 said that Selten "was the first to pick up the question about the valuation of games in an extensive form" and that this Master's Thesis can be compared to a Ph.D. Thesis. Already in 1960, his thesis was published in the Zeitschrift für die gesamte Staatswissenschaft.[4]

At first, the ten axioms required by Selten were only compatible for two-person-games. When he added the "monotony axiom" in his Ph.D. Thesis, he achieved

---

[1]R. Selten (1978), The chain store paradox, Theory and Decision 9, pp. 127–159.

[2]Who is interested to be informed about the manifold work and the vita of Reinhard Selten from his own pen, may look to: R. Selten (1994), In search of a better understanding of economic behaviour, in: Arnold Heertje (ed.), The Makers of Modern Economics, New York, pp. 115–139.

[3]J. von Neumann and O. Morgenstern (1944), Theory of Games and Economic Behavior, Princeton.

[4]R. Selten (1960), Bewertung strategischer Spiele, ZgS 116, pp. 221–281.

R. Tietz
Fachbereich Wirtschaftswissenschaften, Goethe-Universität, Steinhausenstr. 23, D 60599 Frankfurt am Main, Germany
e-mail: tietz@wiwi.uni-frankfurt.de

A. Ockenfels and A. Sadrieh (eds.), *The Selten School of Behavioral Economics*,
DOI 10.1007/978-3-642-13983-3_3, © Springer-Verlag Berlin Heidelberg 2010

a "valuation of n-person games".[5] The proof of the characterization theorem claiming the existence and the unambiguousness of the value function the second supervisor Wolfgang Franz appreciated by saying that the "author is able to see through a complicated matter mathematically and to reveal complex coherences with exhausting and determined work". This characterization of Selten's working method is still valid today. An English version of his Ph.D. Thesis was published in the U.S. in 1964.

In 1957, at the Faculty of Economics and Social Sciences, Reinhard Selten became an assistant of Heinz Sauermann, who – as Selten wrote himself – was one of the few German economists that predicted the increasing mathematization of the economic theory.

I first met Reinhard Selten during my study period in winter term 1961/1962 at the lecture "Theory of Economic Cycles and Growth" held by Heinz Sauermann. Reinhard Selten then, in his role as assistant, very impressively presented the solutions of second order difference equations to explain the economic cycle model of Samuelson.[6]

During my studies, my focus laid on fiscal policy and income distribution and my Master's Thesis was supervised by Heinz Sauermann. After finishing my studies in 1963, Sauermann offered me a DFG-position as research assistant, which therefore assigned me to Reinhard Selten and to the Seminar for Mathematical Economics and Econometrics – the origin of Experimental Economics.[7]

As test subject and student with a wide field of interests, Reinhard Selten had learned experimental work with the well-known Frankfurt Gestalt psychologist Edwin Rausch. Despite the ruling opinion that experiments in economics are not possible, Sauermann and Selten boldly started experimental research in 1958. The existence of many different competing approaches to the oligopoly problem was probably partly responsible that the experimental research began right there.[8] Already in 1959, "An Oligopoly Experiment" by Sauermann and Selten was

[5] R. Selten (1961) Bewertung von n-Personenspielen, Diss. Frankfurt, published as: Valuation of n-Person Games, in: M. Dresher, L.S. Shapley, and A.W. Tucker (eds.), Advances in Game Theory, Princeton 1964, pp. 577–626.

[6] P. Samuelson (1939), Interactions between the multiplier analysis and the principle of acceleration, Review of Economic Statistics, 21, 75–78.

[7] On the beginnings of experimentel economics cf. e.g. also: H. Sauermann and R. Selten (1967), Zur Entwicklung der experimentellen Wirtschaftsforschung, in: H. Sauermann (ed.) (1967), Beiträge zur experimentellen Wirtschaftsforschung (Vol. 1), Tübingen, pp. 1–8. On the experiments in Frankfurt cf.: H. Sauermann (1970a), Die experimentelle Wirtschaftsforschung an der Universität Frankfurt am Main, in: H. Sauermann (ed.) (1970b), Contributions to Experimental Economics, Vol. 2, Tübingen, pp. 1–18; R. Selten and R. Tietz (1980), Zum Selbstverständnis der experimentellen Wirtschaftsforschung im Umkreis von Heinz Sauermann, ZgS 136, pp. 12–27; R. Tietz (1990), On Bounded Rationality: Experimental Work at the University of Frankfurt/Main, JITE, Vol. 146, pp. 659–672.

[8] Cf. Also the hints to corresponding research in the U.S.: R. Selten and R. Tietz (1980), esp. p. 15.

released.[9] In a quantity tripoly the Cournot behavior assumption of the short term profit maximization can be widely observed, which could be proofed by a motive analysis using the protocol method showing that the test subjects had exactly the intention assumed.

Even with this first experiment, a concern of Selten is standing out, which is so characteristic for his experiments conducted in Frankfurt. In the conflicting fields of – usually strictly rational – theory and bounded rational behavior, the work in the experiment is double tracked. For the experiment there should at one point be a theoretical solution – often in the sense of an equilibrium – that can be taken into consideration as a standard of comparison for the observed behavior. Either the experiment is to be designed in a way that it supports an already known theory, or a theory according to the experiment has to be developed. This theory will be compared with the experimental behavior to test how close the numerical approximation is.

On the other side, there are attempts to exploratively gain at least components for bounded rational approaches out of the observed behavior and if necessary, of the underlying considerations. Therefore, subjective considerations of the test subjects are used (e.g. protocol method, motive analysis, strategy method) in addition to the "hard" experimental data collection.[10]

In many cases it can be shown that the decisions are guided by discrete criteria, which can be interpreted as aspiration levels. For the examined oligopoly situations, basic considerations of the "Aspiration Adaptation Theory of the Firm" by Sauermann and Selten (1962) give useful explanations.[11] In this dynamic theory, the decision behavior of a firm is guided by the satisfaction of aspiration levels. It therefore needs not an utility or profit maximization approach.

The theory differentiates between stages of the decision process in which the expectations about the viability of the goals lead to the adjustment of the aspiration. Modified basic considerations of this "Aspiration Adaptation Theory" later form my starting point for a comprehensive macroeconomic model which is solved with alternating aspiration adaptation processes.[12] This model provides experimental environments for labor market negotiations, which can be explained with an afterwards

[9]H. Sauermann and R. Selten (1959), Ein Oligopolexperiment, ZgS 115, pp. 427–471, reprinted in: H. Sauermann (ed.) (1967), pp. 9–59.

[10]The planning report method, later developed by the author, can be seen as an extension of the protocol method or as "ex ante protocoll method". With their help the subjects are asked before the decision also expectations and aspirations. Cf.: R. Tietz (1996), Experimentelle Wirtschaftsforschung – Wege zur Modellierung eingeschränkter Rationalität, in: de Gijsel et al. (eds.), Ökonomie und Gesellschaft, Jahrbuch 13, Experiments in Economics – Experimente in der Ökonomie, Frankfurt – New York, pp. 120–155, esp. pp. 127f.

[11]H. Sauermann and R. Selten (1962), Anspruchsanpassungstheorie der Unternehmung, ZgS 118, pp. 577–597.

[12]R. Tietz (1973), Ein anspruchsanpassungsorientiertes Wachstums- und Konjunkturmodel (KRESKO), Tübingen.

developed negotiation theory. This theory works with the principles of aspiration balance and aspiration securing.[13]

Also the second Frankfurt experiment "Investment Behavior in an Oligopoly Experiment"[14] was already completed as a DFG-report in 1962. In this quantity oligopoly the capacity as well as the production methods with different capital intensities and unit costs can be adjusted. The motive analysis presents a classification of the behavior in four possible balanced, i.e. non contradictory, causal diagrams.

Although the third oligopoly experiment was already completed when I joined in 1963, the DFG-report was not finished until December 1963.[15] The motive analysis showed that the test subjects were guided by different decision criteria. It was my task to investigate with the help of computer simulations reasonable strategies in the sense of complete behavior plans.[16] Nine strategies of mean aggressiveness were used to explain the experimental observations. From the simulations recommendable strategies could be picked for most game types.

I participated also in the investigation with the "Strategy Method", which aims at an analog direction.[17] Here the test subjects had to decide in primary and secondary games. Afterward, their task was to develop a strategy for the game situation in the course of the semester, which represents a complete behavior plan for all differentiated situations. It turned out that often the prices were orientated by competing prices.

In Selten (1967b) a model for the evaluation of theoretical values for the total profit (final capital) was already used operating with an approximation formula assuming pure competition and constant prices (not given in the experiment). Selten himself even then pointed out the weaknesses of this model.[18] These critical points

[13]R. Tietz (1976), Der Anspruchsausgleich in experimentellen Zwei-Personen-Verhandlungen mit verbaler Kommunikation, in: H. Brandstätter and H. Schuler (eds.), Entscheidungsprozesse in Gruppen, Beiheft 2, Zeitschrift für Sozialpsychologie, Bern – Stuttgart – Wien, pp. 123–141. Cf. on the three mentioned aspiration theories: R. Tietz (1997), Adaptation of Aspiration Levels – Theory and Experiment, in: W. Albers, W. Güth, P. Hammerstein, B. Moldovanu, and E. van Damme (eds.), Understanding Strategic Interaction – Essays in Honor of Reinhard Selten, Berlin – Heidelberg – New York – Tokyo, pp. 345–364. For the description of the experiment cf. R. Tietz (1972), The Macroeconomic Experimental Game KRESKO – Experimental Design and the Influence of Economic Knowledge on Decision Behavior, in: Heinz Sauermann (ed.) (1972), Contributions to Experimental Economics, Vol. 3, Tübingen, pp. 267–288.

[14]R. Selten (1967a), Investitionsverhalten im Oligopolexperiment, in: H. Sauermann, (ed.), (1967), pp. 60–102.

[15]R. Selten (1967b), Ein Oligopolexperiment mit Preisvariation und Investition., in: H. Sauermann, (ed.), (1967), pp. 103–135.

[16]R. Tietz, Simulation eingeschränkt rationaler Investitionsstrategien in einer dynamischen Oligopolsituation, in: H. Sauermann (ed.) (1967), pp. 169–225.

[17]R. Selten (1967c), Die Strategiemethode zur Erforschung des eingeschränkt rationalen Verhaltens im Rahmen eines Oligopolexperimentes, in: H. Sauermann (ed.) (1967), pp. 136–168. This unstructured method has to be distinguished from the structured strategy method, in which all possible decisions are given in a questionnaire. Cf. on the strategy method: R. Tietz (1996), pp. 127f.

[18]R. Selten (1967b), pp. 116f.

always stayed on his mind. By leaving out the investment possibilities, a simplified model of the experimental situation, an "oligopoly model with demand inertia", made a "game theoretical approach" possible with the concept that he later called "subgame-perfectness", which has great importance for the reliability of equilibria.[19]

Ten years later, lecturing in Bielefeld, he improved his approach. By introducing the "principle of the trembling hand" (allowing small mistakes), he redefined the "perfect equilibrium point".[20]

Selten's improvement of the Nash Equilibrium with the perfectness-concept resulted in reviving the non-cooperative game theory. By this concept the game theoretical modeling and solution of numerous economic problems has become possible. Experimental economics without his approach is even today unthinkable. For his Perfection Concept he was awarded the Nobel Prize in 1994.

The idea to include also irrational actions in the game theoretical analysis already occurred during his time in Frankfurt. In a model, developed with my cooperation in 1967, for irreversible games in arms control policy, the evaluation of armament situations is not only based on rationally expected results. Also such equilibria were considered that could only be achieved with irrational behavior in the beginning, which resulted in a differentiated evaluation of "arms races".[21]

In a market experiment,[22] in which 12 buyers faced 4 sellers, a door for experimental negotiation was opened in 1965 being quite significant for my own experiments later. On either side of the market, the participants were distinguishable by four types of cost and redemption functions. Contracts on selling prices and quantities of a homogenous good could be closed during negotiations. Numerous phenomena of the market could be isolated, e.g. the prominence of numbers at contract prices and quantities, demand inertia, quantity competition, and customer loyalty. In follow up experiments[23] customer loyalty was identified as a result of aspiration satisfaction and a fair aspiration balance in the sense of a relation equilibrium.[24]

---

[19]R. Selten (1965), Spieltheoretische Behandlung eines Oligopolmodells mit Nachfrageträgheit, ZgS 121, pp. 301–324 and 667–689.

[20]R. Selten (1975), Reexamination of the perfectness concept for equilibrium points in extensive games, International Journal of Game Theory 2(3), pp. 25–55.

[21]R. Selten and R. Tietz (1972a), Security Equilibria, in: R. Rosecrance (ed.), The Future of the International Strategic System, San Francisco – Scranton – London – Toronto, pp. 103–122 and R. Selten and R.Tietz (1972b), A Formal Theory of Security Equilibria, ibid, pp. 185–202.

[22]R. Selten (1970b), Ein Marktexperiment, in: H. Sauermann, (ed.) (1970b), pp. 33–98.

[23]H.J. Crössmann (1982), Entscheidungsverhalten auf unvollkommenen Märkten, Frankfurt am Main, and U.W. Vossebein (1990), Eingeschränkt rationales Marktverhalten, Frankfurt am Main.

[24]H.J. Crössmann and R. Tietz (1983), Market Behavior based on Aspiration Levels, in: R. Tietz (ed.) (1983), Aspiration Levels in Bargaining and Economic Decision Making, Lecture Notes in Economics and Mathematical Systems, – Experimental Economics, Vol. 213, Berlin – Heidelberg – New York – Tokyo, pp. 170–185. and R. Tietz (1992), Semi-normative theories based on bounded rationality, Journal of Economic Psychology 13, pp. 297–314, reprinted in: A. Sadrieh and J. Weimann (2008), Experimental Economics in Germany Austria and Switzerland, – A collection of papers in honor of Reinhard Tietz, Marburg, pp. 1–15.

In 1966 an experiment for coalition negotiations was conducted explicitly examining psychological influences.[25] The game was in form of a characteristic function with one big and four small players. Different winning coalitions were possible, dividing the total profit of DM 40 among themselves.

To test what factors are important for the success of a player, Selten developed the rank sum test, for which tables of significance were calculated by me.[26] The test shows that those stable final coalitions (with minimal number of participants) prevail, in which the mean levels of demands of the players (in the sense of defended aspiration levels) complement best to the coalition profit of DM 40 ("coordination phenomenon").

For the psychological influences the following is found: The weaker the trust between the four small players and the smaller the variance of the offers of the big player, the bigger is the negotiation success of the big player.[27] The success of the small player in coalitions of two is positively influenced by risk-averse behavior, conformism, and higher mean level of demands.

This experiment brought Selten again to developing a theory of his own – the equal share analysis, in which the three hypotheses of the order of strength, exhaustivity, and equal division core concur.[28] It presents a good explanation and seems to exceed the Aumann-Maschler-theory when concerning the size of the prediction-area.[29]

Selten's "chain store paradox" published in 1978 has its roots also in Frankfurt.[30] At a clam-dinner at the "Fressgasse" together with Armin Gutowski, the two competitive approaches were discussed. The rational game-theoretically correct Induction Theory and the intuitively convincing Deterrence Theory. According to him not the clams but the contradicting approaches made him physically uncomfortable. That is why this work may be especially characteristic for Reinhard Selten, as it clearly reflects his inner tension in the triangle of rational game theory, experimental reality and comprehensible bounded rationality. This tripartite research program never quite let go of him.

In 1965, Reinhard Selten officially started his successful work as an academic teacher in Frankfurt with the lecture "Mathematics for Economists". Apart from numerous trips to international academic conferences, his time in Frankfurt was only interrupted by visiting professorships at the Institute for Advanced Studies in Vienna 1967 and at the University of California, Berkeley 1967/1968. It was there

[25]R. Selten and K.G. Schuster (1970), Psychologische Faktoren bei Koalitionsverhandlungen, in: H. Sauermann (ed.) (1970b), pp. 99–135.

[26]R. Selten and R. Tietz (1967), Der Rangsummentest – Beschreibung und Signifikanztafeln, in: Rudolf Henn (ed.), Operations Research-Verfahren III, Meisenheim am Glan, pp. 353–375.

[27]R. Selten and K.G. Schuster (1970), p. 135.

[28]R. Selten (1972), Equal Share Analysis of Characteristic Function Experiments, in: H. Sauermann (ed.) (1972), pp.130–165.

[29]R. Selten and W. Krischker (1983), Comparison of Two Theories for Characteristic Function Experiments, in: R. Tietz (ed) (1983), pp. 258–264.

[30]R. Selten (1978).

where the long lasting and fruitful collaboration started with one of the other 1994 Nobel Prize winners, John C. Harsanyi, regarding the problem of incomplete information[31] and equilibrium selection.[32]

In 1968, Selten habilitated in Economics at the Faculty for Economics and Social Sciences with his work "Price policy of the multi-product firm in the static theory".[33] The supervisor Heinz Sauermann had "the opinion that this professorial writing was the most mature and important analysis" that the faculty was presented since him being there. The supervisor Waldemar Wittmann even considered the work "as one of the most important German-speaking contributions to economic theory in the postwar period". Further mentioning of his work while being in Frankfurt[34] or as professor in Berlin (1969), Bielefeld (1972), and Bonn (1984) is not possible here.

The 7 years spent together in Frankfurt have left quite an impression on me and my work. The intense discussions not only took place in the office but while exercising the "peripatetic method". Numerous problem areas were discussed of which a few aspects are to be mentioned here. The name KRESKO for my macro-model was suggested by Reinhard Selten and is Esperanto. For the complex model with many variables and behavioral equations some consistency problems had to be solved at the beginning with the help of dimensional analysis. With the advice for the necessity of reproducibility of constant growth rates,[35] Selten made a valuable suggestion.

The discussions with Selten more than surely pushed the further development of my experimentally obtained aspiration-oriented negotiation theories. He more than often insistently demanded that a good negotiation theory must explain why it would not lead to the recommendation to start the negotiation with extremely high requests. It is a preferable semi-normative property of a descriptive theory that the observance of behavior recommendations, derived from the theory, are not destabilizing as to favor equilibria.[36] The firstly developed planning difference theory determines the player that has some more room left for his bidding and therefore has

[31]J.C. Harsanyi and R. Selten (1972), A Generalized Nash Solution for Two Person Bargaining Games with Incomplete Information, Management Science, 18, pp. 809–106.

[32]J.C. Harsanyi and R. Selten (1988), A General Theory of Equilibrium Selection in Games, Cambridge.

[33]R. Selten (1970b), Preispolitik der Mehrproduktenunternehmung in der statischen Theorie, Berlin-Heidelberg-New York.

[34]Of the other experiments during Seltens years in Frankfurt I will mention the relatively complex SINTO-market, on which I participated only as experimental subject. Cf. the contribution of Otwin Becker in this Volume or e.g. O. Becker and R. Selten (1970), Experiences with the Management Game SINTO-Market, in H. Sauermann (ed.) (1970b), pp. 136–150.

[35]R. Tietz (1973), pp. 86–89.

[36]R. Tietz, W. Daus, J. Lautsch, and P. Lotz (1988), Semi-Normative Properties of Bounded Rational Bargaining Theories, in: R. Tietz, W. Albers, and R. Selten (eds.) (1988), Bounded Rational Behavior in Experimental Games and Markets, Lecture Notes in Economics and Mathematical Systems, – Experimental Economics, Vol. 314, Berlin – Heidelberg – New York – London – Paris – Tokyo, pp. 142–159.

to concede first.[37] But according to this theory, higher opening bids are no disadvantage. The dynamic aspiration balancing theory then considers the bargaining strength with a system of decision filters. The aspiration securing principle than gives reason for the disadvantage of excessive requests and therefore the demanded attribute of stability against itself.[38]

The meetings with Reinhard Selten have always been very inspiring and fruitful. As he is open-minded for all kinds of different questions, discussions often lead to new ideas of one's own. One wishes for far more opportunities to do so. Ad multos annos!

References

Becker O, Selten R (1970) In: Sauermann H (ed) Experiences with the management game SINTO-market, pp 136–150

Crössmann HJ, Tietz R (1983) In: Tietz R (ed) Market behavior based on aspiration levels, pp 170–185

Crössmann HJ (1982) Entscheidungsverhalten auf unvollkommenen Märkten. Barudio & Hess, Frankfurt am Main

Harsanyi JC, Selten R (1972) A generalized nash solution for two person bargaining games with incomplete information. Manage Sci 18:80–106

Harsanyi JC, Selten R (1988) A general theory of equilibrium selection in games. MIT Press, Cambridge

Rosecrance R (ed) (1972) The future of the international strategic system. Chandler, San Francisco

Samuelson P (1939) Interactions between the multiplier analysis and the principle of acceleration. Rev Econ Stat 21:75–78

Sauermann H (ed) (1967) Beiträge zur experimentellen Wirtschaftsforschung, (Vol. 1). JCB Mohr, Tübingen

Sauermann H. (1970a) In: Sauermann H. (ed) (1970b) Die experimentelle Wirtschaftsforschung an der Universität Frankfurt am Main, pp 1–18

Sauermann H (ed) (1970b) Beiträge zur experimentellen Wirtschaftsforschung – Contributions to experimental economics, vol. 2. JCB Mohr, Tübingen

Sauermann H (ed) (1972) Beiträge zur experimentellen Wirtschaftsforschung – Contributions to experimental economics, vol. 3. JCB Mohr, Tübingen

Sauermann H, Selten R (1959) Ein Oligopolexperiment. Z. Gesamte Staatswiss 115, 427–471. Reprinted in: Sauermann H (ed) (1967), 9–59

Sauermann H, Selten R (1962) Anspruchsanpassungstheorie der Unternehmung. Z Gesamte Staatswiss 118:577–597

---

[37]R. Tietz and H.-J. Weber (1972), On the Nature of the Bargaining Process in the KRESKO-Game, in: H. Sauermann (ed.) (1972), pp. 305–334. The measure used there originally for the willingness to concede, the difference between the opening offer (first demand) and the value regarded as attainable was named "tactical reserve". In later writings we prefer for he same measure the name "concession reserve", whereas the name "tactical reserve" is used for the difference between the opening offer and the planned value (goal) only.

[38]R. Tietz (1976), Der Anspruchsausgleich in experimentellen Zwei-Personen-Verhandlungen mit verbaler Kommunikation, in: H. Brandstätter and H. Schuler (eds.), Entscheidungsprozesse in Gruppen, Beiheft 2 der Zeitschrift für Sozialpsychologie, Bern – Stuttgart – Wien, pp. 123–141.

Sauermann H, Selten R (1967) In: Sauermann H (ed) Zur Entwicklung der experimentellen Wirtschaftsforschung, pp 1–8
Selten R (1960) Bewertung strategischer Spiele. Z Gesamte Staatswiss 116:221–281
Selten R (1961) Bewertung von n-Personenspielen. Diss, Frankfurt, published as: valuation of n-person games. In: Dresher M, Shapley LS, Tucker AW (eds) Advances in game theory. Princeton University Press, Princeton 1964, pp 577–626
Selten R (1965) Spieltheoretische Behandlung eines Oligopolmodells mit Nachfrageträgheit. Z Gesamte Staatswiss 121:301–324, 667–689
Selten R (1967a) In: Sauermann H (ed) Investitionsverhalten im Oligopolexperiment., pp 60–102
Selten R (1967b) In: Sauermann H (ed) Ein Oligopolexperiment mit Preisvariation und Investition, pp 103–135
Selten R (1967c) In: Sauermann H (ed) Die Strategiemethode zur Erforschung des eingeschränkt rationalen Verhaltens im Rahmen eines Oligopolexperimentes, pp 136–168
Selten R (1970a) In: Sauermann H (ed) Ein Marktexperiment, pp 33–98
Selten R (1970b) Preispolitik der Mehrproduktenunternehmung in der statischen Theorie. Springer, Berlin
Selten R (1972) In: Sauermann H (ed) Equal share analysis of characteristic function experiments, pp 130–165
Selten R (1975) Reexamination of the perfectness concept for equilibrium points in extensive games. Int J Game Theory 2(3):25–55
Selten R (1978) The chain store paradox. Theory Decis 9:127–159
Selten R (1994) In search of a better understanding of economic behaviour. In: Heertje A (ed) The makers of modern economics. New York, Harvester Wheatsheaf, pp 115–139
Selten R, Krischker W (1983) In: Tietz R (ed) Comparison of two theories for characteristic function experiments, pp 258–264
Selten R, Schuster K (1970) In: Sauermann H (ed) Psychologische Faktoren bei Koalitionsverhandlungen, pp 99–135
Selten R, Tietz R (1967) Der Rangsummentest – Beschreibung und Signifikanztafeln. In: Henn R (ed) Operations Research-Verfahren A Hain, III. Meisenheim am Glan, pp 353–375
Selten R, Tietz R (1972a) In: Rosecrance R (ed) Security equilibria, pp 103–122
Selten R, Tietz R (1972b) In: Rosecrance R (ed) A formal theory of security equilibria, pp 185–202
Selten R, Tietz R (1980) Zum Selbstverständnis der experimentellen Wirtschaftsforschung im Umkreis von Heinz Sauermann. Z Gesamte Staatswiss 136:12–27
Tietz R (1967) In: Sauermann H (ed) Simulation eingeschränkt rationaler Investitionsstrategien in einer dynamischen Oligopolsituation, pp 169–225
Tietz R (1972) In: Sauermann H (ed) The macroeconomic experimental game KRESKO – experimental design and the influence of economic knowledge on decision behavior, pp 267–288
Tietz R (1973) Ein anspruchsanpassungsorientiertes Wachstums- und Konjunkturmodell (KRESKO). JCB Mohr, Tübingen
Tietz R (1976) Der Anspruchsausgleich in experimentellen Zwei-Personen-Verhandlungen mit verbaler Kommunikation. In: Brandstätter H, Schuler H (eds) Entscheidungsprozesse in Gruppen, Beiheft 2, Zeitschrift für Sozialpsychologie. Huber, Bern, pp 123–141
Tietz R (ed) (1983) Aspiration levels in bargaining and economic decision making, Lecture notes in economics and mathematical systems. Experimental economics, vol 213. Springer, Berlin
Tietz R (1990) On bounded rationality: experimental work at the University of Frankfurt/Main. J Inst Theor Econ 146:659–672
Tietz R (1992) Semi-normative theories based on bounded rationality. J Econ Psychol 13:297–314. Reprinted in: Sadrieh A, Weimann J (2008) Experimental economics in Germany Austria and Switzerland – a collection of papers in honor of Reinhard Tietz. Metropolis, Marburg, pp 1–15
Tietz R (1996) Experimentelle Wirtschaftsforschung – Wege zur Modellierung eingeschränkter Rationalität. In: de Gijsel et al. (eds) Ökonomie und Gesellschaft. Jahrbuch 13, Experiments in Economics – Experimente in der Ökonomie. Campus Verlag, Frankfurt, pp 120–155

Tietz R (1997) Adaptation of aspiration levels – theory and experiment. In: Albers W, Güth W, Hammerstein P, Moldovanu B, van Damme E (eds) Understanding strategic interaction – essays in honor of Reinhard Selten. Springer, Berlin, pp 345–364

Tietz R, Weber H-J (1972) In: Sauermann H (ed) On the nature of the bargaining process in the KRESKO-game, pp 305–334

Tietz R, Daus W, Lautsch J, Lotz P (1988) Semi-normative properties of bounded rational bargaining theories. In: Tietz R, Albers W, Selten R (eds) Bounded rational behavior in experimental games and markets. Lecture notes in economics and mathematical systems. Experimental economics, vol. 314. Springer, Berlin, pp 142–159

von Neumann J, Morgenstern O (1944) Theory of games and economic behavior. Princeton University Press, Princeton

Vossebein UW (1990) Eingeschränkt rationales Marktverhalten. Lang, Frankfurt am Main

Chapter 4
Reinhard Selten and the Scientific Climate in Frankfurt During the Fifties

Horst Todt

It is worthwhile to consider the scientific climate at the Economic Faculty of the Johann-Wolfgang-Goethe-University in Frankfurt/Main during the 1950s of the twentieth century. I will do this admittedly from a subjective point of view as I see the situation today after 50 years have gone. My personal valuation has been changing already during the fifties, and even more during the decades afterwards.

The Second World War with the defeat of Germany had been only a few years ago. Germany was a poor country at least compared with the present state of affairs. The economic situation of most people was bad and could become only better in the course of time. Hence younger people were carried away by a wave of needs and wishes and by optimism.

Most students strived just for higher qualification in order to climb up the ladder of a professional career in the economy. Some students, however, were more concerned with the question in how far economics in Germany was in line with the international discussion. I was among these students. We recognized that a gap had opened between the German way to tackle economic problems and the style of the international discussion especially within the Anglo-American world. Germany had been at least partly cut off the scientific development by the Nazi Regime. In the field of economics the gap seemed to be even more serious than in other sciences because of the tradition of the German Historical School of Economics. This school showed considerable animosity against theoretical thinking. This attitude was still persistent and averse to quickly keeping up with the international standards of the time that had become more theoretical.

Those students, who were critical about the German tradition, certainly doing some injustice to the merits of the Historical School, were keen in becoming acquainted with modern theory. They felt very much attracted by the chair of Heinz Sauermann that was open to new developments. Heinz Sauermann proved

H. Todt
Fakultät für Wirtschafts- und Sozialwissenschaften, Universität Hamburg, von-Melle-Park 9, 20146 Hamburg, Deutschland
e-mail: horst.todt@gmx.de

A. Ockenfels and A. Sadrieh (eds.), *The Selten School of Behavioral Economics*,
DOI 10.1007/978-3-642-13983-3_4, © Springer-Verlag Berlin Heidelberg 2010

to be the central figure of all approaches to fill in the gap between international and German standards in economic theory.

Openness to theoretical thinking means also openness to mathematical methods. So it was quite natural to use the knowledge and help of mathematicians in order to capture modern theoretical economics. There was a young lecturer (Privatdozent) in the mathematical department of the Frankfurt University, Ewald Burger, who had beyond his duties in the mathematical department a teaching assignment for "mathematics for economists". He usually taught standard methods of mathematics nowadays being a matter of course for theoretical economists, but not in those times. Could he be brought to introduce students of economics into such exciting new fields of research like game theory? He could!

On the initiative of a group of students of economics (again myself among them) Ewald Burger started his seminar in mathematical economics in 1954, open for economists and mathematicians. It became a permanent institution. The first topic was game theory.

In the very beginning the members of the Burger Seminar were mainly students of economics. Only one student of mathematics took part: Reinhard Selten. He presented a proof of the existence of equilibrium points (saddle points) in mixed strategies of two-person-zero-sum-games. With regard to the economists in the class an elementary proof was chosen that was long but without any advanced methods. The proof was brilliantly presented and needed about 1 h time.

Ewald Burger, normally a very mild, modest, and friendly man used to become completely intolerant whenever he encountered the slightest imprecise formulation or even mistake in a mathematical presentation – and this used to happen very often in his seminars. Then he became really angry. Reinhard Selten's presentation passed without any critical remarks.

The seminar on game theory was the starting point of a development with quite a few consequences of some importance. Ewald Burger became the supervisor of the diploma thesis and the doctor thesis of Reinhard Selten. The Burger Seminar became a highly reputed institution where mathematicians and economists met. It continued over several years and had the strong moral support of Heinz Sauermann. When Ewald Burger left Frankfurt in order to take over a chair of mathematics in Cologne the seminar ended. It was a loss for Frankfurt.

In later years there were hardly any students in the seminar. It was an event for the scientific staff. Even professors of economics including Heinz Sauermann appeared there occasionally. The Burger seminar changed into a meeting centre for mathematicians and economists. Last not least economists – students, staff, and above all Heinz Sauermann – took notice of Reinhard Selten.

For all students and staff members who happened to come into closer contact with Reinhard Selten this meant stimulating discussions. His widely spread interests were impressive. The leading German textbook on economic theory in those times was that of Erich Schneider. There was hardly a student of mathematics who took interest in such matters, but Reinhard Selten did. He even fought with the "Capital" of Karl Marx a book usually found on book shelves – but not in the hand of readers.

It was, however, not so much the common base of knowledge in economics that made conversations with Reinhard Selten inspiring. By far more important was his background in other scientific fields, notably in psychology. Even in those early times he adopted a critical position against the economic theory of behaviour. The hypothesis of human beings guided by purely rational considerations appeared doubtful to him. The way psychologists investigated the problem of human behaviour, i.e. by experiments, seemed more reliable to him.

All this might represent Reinhard Selten as a person who knows to use a broad overview over various disciplines to combine different traits of thinking. He certainly did this, yet this description falls short in fully depicting his original mind. An episode may highlight the appreciation he received even in young years.

I happened to have quite a good personal relationship to Ewald Burger. He fascinated me – and not only me – by his extremely good memory; his reputation was, of course with some exaggeration, that "he knows mathematics by heart". He pretended to have no other interests but mathematics especially mathematical logic. Nevertheless he had a pretty good command of several languages that he needed in order to read mathematical contributions. Confronted with his reputation he answered me: "I would rather have such brilliant ideas like Selten".

My observations of the situation were interrupted in 1957 for 1 year which I spent abroad. When I returned 1958 Reinhard Selten had become a staff member of Heinz Sauermann and this was no surprise for me. I, too, became a staff member of Heinz Sauermann after my return in 1958.

Our time as students was over. The new role of staff members changed the perspectives, and opened new views. Reinhard Selten managed to convince Heinz Sauermann immediately that economic behaviour had to be probed by experimental research and received all support for doing so. Heinz Sauermann created an atmosphere of tolerance and freedom. All of us took much advantage out of his attitude. Nevertheless, it was clear that he as the head of the chair was the boss.

Reinhard Selten used to have a somewhat extraordinary rhythm of life. He rarely was present in the morning. It was not decent to ring him up before noon. Yet it was quite possible to contact him at 2 o'clock in the morning. He was anyway present whenever it was important. Soon he acquired the reputation of being well informed and deeply scrutinizing the problems he encounters. His influence within the chair was tremendous. Occasional presentations gave some hint of his thinking. By far more important were his individual discussions with members of the chair. He very much preferred this way of scientific contact. All members of the chair were convinced that he is a brilliant mind und would certainly become a famous man in future years. Moreover they tried to draw advantage of the conversations with him for their own research.

Lots of different topics were brought up at the chair in the course of time, many of them by Reinhard Selten. The concepts of rational thinking and Herbert A. Simon's criticism of these concepts as an explanation of real behaviour were a core matter of discussion. Experimental economics did not exist as a subject of its own in these years; it was about to be created.

The Sauermann Chair had a specific structure in these years. There was the traditional task of teaching a huge number of students in economic theory, supervising their theses, seminar papers, and so on. There also was the Institute for Tourism (where I was an assistant) doing research work in this field and finally "Experimental Economics" (not officially called so) a new department, the realm of Reinhard Selten. There was cooperation between the "departments" and no rivalry. The department of Experimental Economics was expanding and soon had the biggest number of assistants.

Looking back at these times it seems to me that the output rate in experimental economics was accelerating faster than the size of the staff. They actually practiced "learning by doing". Starting from zero they had a long way to go. Was there in the beginning some doubt whether the chosen way was the right one? If there was some doubt, it could not be recognized by the behaviour of Reinhard Selten. Nor could it be read from his face. He always appeared convinced and convincing. I do not know whether he really never was bothered by doubts.

Heinz Sauermann supported the program decidedly, and became one of the most ardent adherents. Moreover he engaged himself in the scientific discussion. Nevertheless the department of Experimental Economics stood alone in the research field for a long time.

For those who had known Reinhard Selten for a long time all these activities did not really fit to him. His appearance has always been that of a theoretically oriented personality far away from the practical needs of this world. This man, however, did such a practical thing like experiments, and moreover initiated the movement of experimental economics. In so far his experimental career was astonishing. Looking back at all this from the present state of affairs we must say: Reinhard Selten was a Schumpeterian entrepreneur in the field of science.

After all it was an inspiring time. We all were young and ready to take up suggestions. Quite a few we owe to Reinhard Selten. Many thanks!

Chapter 5
Reinhard Selten Labs, Bounded Rationality and China

Fang-Fang Tang

Oct. 5th is Doktorvater's 80th birthday. As one of his disciples, the best way to congratulate is a paper. As a scholar, he has won world-wide respect by his pioneering work in game theory, experimental economics and bounded rationality. His articles on evolutionary biology, psychology and political science have been widely cited. The esteem of his peers is vividly reflected in a well-circulated article recently. Professor Vernon Smith (2005) said that Professor Selten should be awarded the Nobel Memorial Prize in Economic Science for the second time. Therefore I decided, after e-mail communications with Professor Karim Sadrieh, to write an article from another perspective, in a more personal way. I will write about what I know about Professor Selten, and his influence on me, and his academic impact in a fast-developing country, my home country, China.

Two Decades Ago

Going through the old documents, I feel sentimental that it is such a "survey" of my past 20 years' life with Professor Selten. His influence on me is fundamental, if I am allowed to say so. Almost all my progress made in experimental economic research in China is related to Professor Selten'. Story with Professor Selten starting from a letter by late Professor John C. Harsanyi on June 10, 1991 (see, also, Tang 2002): "Dear Fang-Fang, thanks for your letter of May 24. ... Another possibility for you would be to write to my German friend, Professor Reinhard Selten. ... Unfortunately, Prof. Selten had a mild heart attack last February, but now he feels much better. I talked to him yesterday to inquire about his health. I also mentioned your name, telling him that you would like to be admitted to a good Ph.D. Program.

F.-F. Tang (✉)
Southwestern University of Finance and Economics, Chengdu, and National School of Development at Peking University, Beijing, China
e-mail: fftang@ccer.edu.cn

A. Ockenfels and A. Sadrieh (eds.), *The Selten School of Behavioral Economics*,
DOI 10.1007/978-3-642-13983-3_5, © Springer-Verlag Berlin Heidelberg 2010

He said that he may be willing to admit you as his Ph.D. student. If he does, you will find that he is a first-rate teacher." Yes indeed, our Master is. I have learnt a lot from him since then, not only in game theory and experimental economics, but also about life, for which I will expand more later. More than that, he was willing to take me to study under his guidance, without knowing anything about me at that moment – certainly not having seen me at all. That is, one has to say, risky. In this sense, he acted more like an academic "venture capitalist", and if I may say this myself, not unsuccessfully. In every sense, I have been deeply touched and grateful.

I wrote to Professor Selten about my work in fuzzy preferences and social choice. He replied promptly about his view on July 25, 1991. The letter has such a profound influence on me that I think that it is appropriate to cite it full:

> Dear Mr. Fang-Fang Tang,
>
> Your interest in vague preferences could be the basis of fruitful research work. However I feel that your efforts should take a different direction. Experimental work done by psychologists and economists has revealed many "anomalies" like preference reversal, the common ratio effect, etc. which are really strong empirical regularities. The evidence suggests that people do not have preferences but must construct them on the basis of the task. One needs a theory of preference construction rather than a theory of vague preferences. Preference judgments are the result of information processing and this information processing must be described. Moreover any theory in this area must be guided by empirical evidence. The work done by my group mainly concerns experimental economics. We have a computerized laboratory where we perform experiments on games and markets. The exploration of human behavior cannot work. However this does not mean the end of theory but the beginning of a different kind of theory.
>
> Yours sincerely, Reinhard Selten

It was a revolutionary day as I can still vividly recall. Actually as a science and technology student (my Bachelor degree was in applied mathematics and operations research while my master degree was in systems engineering and transportation studies), I had done a personal "experiment" among my classmates and friends about the transitivity of preferences through binary choices of various combinations of commodities, when I was still in Shanghai Jiao Tong University around the mid-1980s. I found that the real choices by my Chinese friends were not fully transitive in fact. Of course I did not pay them at all, it was more like an exploration simply based on the scientific spirit combined with the engineering method to check something I felt dubious. I did not know that there was a field called experimental economics: How could economics be experimented with? That was unimaginable in those days unheard of! It was only a naive impulse to test the theory of preferences in the real world. Nevertheless it did tell me something. This experience in Shanghai gave me some comprehension of Professor Selten's letter which opened up my mind to a new window since such a feedback came from "one of the leading *game theorists* in the world" as in Professor Harsanyi's letter to me.

Interesting enough, after almost 20 years when I have initiated and helped to establish several economic research labs in China, I'm still frequently asked to answer the same questions: What is experimental economics? How can economics be experimented with? How can you use a few student subjects with a small amount of monetary payment to derive implications for "real" life? It reminds me of the old days I myself experienced. It is understandable therefore that I never lost patience to explain what it is and how it could be, for hundreds of times. I also cite Mr. Deng Xiaoping's words "Experimentation is the only key to test truth", since almost every modern Chinese has heard about this famous sentence. Of course, some typical Chinese questions also arise, such as "Can you predict the financial crisis in the lab?" For such, I could only smile, and cite what a friend told me to do: "As the philosopher Hegel pointed out, the seemingly occasional point is the junction of two necessary paths." Science never pretends to be able to do everything, but it does not mean that science cannot help us to solve practical issues. Unduly too much attention has been paid to practical issues while insufficient efforts on fundamental research, for which I hope that some balance will be struck some day.

The letter from Professor Selten on July 25, 1991 started my journey to Bonn when I did not even know the German alphabets. It also started my intellectual journey into experimental economics and the theory of bounded rationality. I arrived in Bonn in October 1991 and met Professor Selten in person for the first time in his small office on the ground floor of Juridicum of Bonn University, a street away from the Rhein River, opposite to the then German Foreign Ministry. He spoke very softly, and from time to time went astray into somewhere in his spiritual universe nobody knew when he paused speaking for some minutes. I got used to it in the next 5 years being with him. At first, I did not know whether I should say something or simply wait for his mind coming back to our conversation. The silence caused some nervousness in my mind since the other teachers I knew well by that time were all fast speakers. I tried to say something after waiting for a while. Then, I realized that it was totally unnecessary, because I was simply not at the same intellectual level as he was before I managed to read sufficient research literature about the topics we were talking about. From that time, I learnt that the speed in speech was not equivalent to the depth of thinking. I began to appreciate the slowness in talking, another side effect! Our Master may have never realized how he influenced us unconsciously. Competitions among the students were invisible but always there. Everybody tried to show his or her intelligence by prompt reactions during seminars or discussions. I was no exception but gradually I tried to slow down my reaction pace towards deeper thought. Now, after my Ph.D. students have graduated in China, the habit of answering important questions in a slow manner has become a usual norm in the research teams I lead. "No need to show off, please, but rather, explain to us more carefully, youngsters", I like to say. "You do not need to agree with me," which I emphasize so often that some of my former students even made a statue for me, in which my finger was pointing upward (my naughty daughter broke that finger during her play). I must attribute the source of such a cultural trace back to my days in Bonn, from Professor Selten. The influence is profound, probably beyond what he might have thought, in the Far East, the then mysterious country of China for him. One never knows what comes out from a student, is it not so?

From the first conversation with Professor Selten, he has been emphasizing the importance of bounded rationality and how experimental economics can help us to construct a framework of descriptive human behavior structure. He joked all the time about the "corrupt utility theory" based on perfect rationality hypothesis where people maximize their utility and firms maximize the profit. Such things do not exist in the real world, as he labelled himself a "radical left". This needs courage, because it means that his past decades' work is exactly criticizing the hypothesis of his game theory work where everything is based on maximization. I frequently cite this example to my own students that a true scholar needs the courage to face his or her own soul, no matter how established one might be. If the facts do not support your beliefs, you have to respect the facts rather than your beliefs. Surely, game theory itself is a normative subject about how people should be in a framework of perfect rationality. It is important to explore such a world to see how things may turn out to be if people follow the axioms. On the other hand, we also need to explore how people really behave, since we cannot build our economic theory on the unrealistic hypothesis of perfect rationality forever, when the modern information technology enables us to experiment in laboratory to search for the empirical regularities of human behavior structure with massive data collection and processing capabilities. I mostly agree with Professor Selten's view, and indeed have been following his thoughts in the past two decades.

One of the research papers he recommended is "Anticipatory Learning in Two-Person Games"(Selten 1991a), in which he proposed a framework with much less rationality requirement on players to learn in a game theoretical setting. It was a fascinating paper, though mathematically somewhat involved. I enjoyed this piece of work and took the experimental task to test this theory in Bonn Laboratory. I still vividly remember the computing environment in our old Bonn Lab in the early 1990s, where the terminals were monographic and we needed to program our experiments ourselves. Professor Selten was clear about what we needed to achieve, but interestingly enough he almost never used any computer himself. In fact, he always wrote amazingly neat by pencil on folders of plain paper, each piece of paper neatly organized into a plastic folder. In the pockets of his suit and his briefcase, there were pencils and an eraser. In a very old-fashioned style, he wrote clearly on the paper by pencils, erased errors by rubber, and rewrote until everything was right. From the first day I saw him in Bonn to the "unexpectedly many trips" (in his words) I later accompanied him in China, the paper-pencil way never changed. Technically those days, Karim Sadrieh and Klaus Abbink were organizing things in Bonn Lab. At that time, these two capable young men sat in the compact working room of the Juridicum underground floor, smoking a lot and drinking a lot of black coffee. They worked very hard to program a software tool called "RatImage" by Turbo Pascal programming language. It is not widely use any more, but at that time, they had managed to program one experiment by only 99 sentences using RatImage, which was so amazing that it was used in advertisements for this tool. In some sense, it is a great pity that this tool was not more widely used. My students use mainly Z-Tree if the experiments are not too complicated to program. For my doctoral dissertation, Karim and Klaus guided me to program the whole experiment by Turbo Pascal myself since the experiment was somehow complicated in data collection. I tried to collect every information that I could, and presented the subjects with all possible ways of payoff calculation on the screen.

It took almost half a year to program and to conduct the experiments. This sort of work is unimaginable for the youngsters since my own students can enjoy all the progress in computing facilities now. They only need several weeks at most to program an experiment. Extraordinarily tedious work at that time, but I did learn a lot about how to manage things. The computer I was equipped to work in Bonn was an IBM-386, without a mouse. My daughter would laugh at such a museum machine. Anyway it was quite a nice equipment at that time, although it took a lot of work to move the cursor on the screen. I programmed all the simulation programs and calculation programs by Turbo Pascal on that machine. The speed was not great, and I needed to run the simulation programs manually by setting the parameters of simulation by fine grids step by step to search for the best possible fit with the data. Day and night, I tested all the possible learning models I could find upon the suggestion of Professor Selten and printed out the calculation outcomes and compared them with piles of papers. It was not very environmentally friendly I have to admit, but there were not many choices at that time. I did use the printing-outs in multiple ways to minimize the waste, and of course sent the used papers for recycling. It turned out that the stability criteria proposed by Selten (1991a) worked pretty good (see Tang 2001) and the simple Harley model from theoretical biology performed very well to fit our learning data (Tang 2003). I also tested the exponential version and a modified version in a polynomial way in my own experiment, and found that the later was performing somewhat better (with only two parameters). This was surprising but also pleasant since such a simple structure of Harley model did so well. The point by Professor Selten on bounded rational way of learning was well reflected in the calculation results we were exploring, which we did not expect (see, also Chen and Tang 1998, Nagel and Tang 1998, and Abbink et al. 2001).

The limitation of calculation capability in Bonn Lab was obvious when I heard about the Quantal Response Equilibrium framework by Professors Richard McKelvey and Thomas Palfray, then in Caltech. I tried to contact them to ask for help. They promptly responded and very kindly offered their help and the use their supercomputer facility in Caltech to calculate the QRE for our data. I guessed that such calculation would only take a few seconds in the powerful facility there in Caltech, but transferring the data to them from Bonn took me almost 2 h by FTP. It was a single color (green) terminal in the Bonn Lab which was connected to satellite channel for data transfer by FTP. I had to sit by it to perform manual steps and transferred the files one by one carefully. Such a file can now be easily sent by a single click through the modern Internet. But at that time, Karim seriously instructed me "Do not walk away, wait and watch until everything is done". Good advice. I like to tell this story because the youngsters now take the huge technical advances for granted. If there is anything the computer cannot readily perform, some kind of panic mood could emerge, since the technical tools may be too ready for them. I see a possible loss of trial-and-error spirit since too many things are too ready for this generation of Chinese. Things becoming so easy is both good and bad. The good side is that we can do things much faster now. The bad side is that it develops a kind of laziness. Scientific exploration is working on new uncertain things. Nobody knows what will come out. One needs to have patience to try things out. From that time, I developed a new interest towards the Internet related things, which led me to teach e-commerce in a marketing department in Hong Kong. An unexpected outcome indeed!

Professor Selten guided me to work on the learning models' testing independently those days, we did not know that Professor Alvin Roth (then in Pittsburgh) was also working on the learning issues in a similar way, until the Nobel Symposium on Game Theory in Stockholm in 1993 (see Roth and Erev 1995). Professor Selten came back to Bonn and told me about the work by Professor Roth, which led to the communications between us. Professor Roth immediately invited me to visit him in Pittsburgh for an academic exchange. That was my only trip to the United States so far. The visit was very fruitful that we fully exchanged what we had been doing independently. After I came back to Bonn, I reported to Professor Selten what I had learnt in Pittsburgh and our data analysis into individual behavior.

The 5 years in Bonn with Professor Selten had a profound impact on me, not only academically. From the numerous conversions we had, which usually he asked me to come to his office as the last person on Thursdays (when he came to office) so we could have more time to discuss various issues, the scope of experimental economics and more than that what bounded rationality could change our economics profession became clearer to me. I'm now advocating and spreading the ideas we were talking about all over China. The first lesson of my teaching courses started with an overall interactive discussion with students on the fundamental assumptions in economics and possible ways to do economics without maximization hypothesis. The way I teach is to encourage exploration rather than simply reciting books and papers.

The Selten School had the German experimental economic tradition, which was to explore things first, rather than to try to confirm theories, consciously or unconsciously. This deep impact on me has helped me to open up my mind, and through my teaching, the minds of younger generation in China. The youngsters are smart, but they do need encouragement to explore new things. They have adopted the habit of obedience from Chinese culture. The rigorous training under Professor Selten equipped me with solid tools for problem-solving. And this is what Chinese young generation badly needs: to do things, not just for "cheap talks". The second feature of my classes is to require my students to do a research project themselves. It does not matter how small it is. It just needs to be on a concrete problem. This is also what I have learnt in Bonn, especially from Professor Selten. He always told me not to learn too many things, because it wastes a lot of valuable time. "Learn the things when you need to use them in a research project, and you learn them well." Yes indeed, I more and more come to realize that nobody can know everything if not crazy.

The best way to learn things is to do it on your own, then the knowledge becomes systematically organized and unforgettable. This does not mean that one does not need to take courses in fundamental subjects. But in the Chinese education system, unfortunately, there are too many examination-oriented courses. They are aiming at good grades rather than at developing a research-oriented problem solving mindset. Chinese students are well-known all over the world for their skills in written exams.

There is no need to further enhance that skill I should think. In fact, like Professor Selten, I like teaching but not to set exam paper questions. I prefer to encourage students to conduct research projects which will help them to integrate what they learn in really doing something. The way I learnt from Professor Selten is to try, to explore, to think freely or even "eccentrically" as he himself has done some times. His method of thinking has been extremely helpful, far more than the specific points he taught me here and there. "Just do it", as a Nike slogan says. Actually in March 2010, after I came out of hospital due to a small operation, a research student who worked with me on an experiment came to me and told me that he did not have the necessary knowledge of statistical tools to conduct his research. It seemed that he expected me to conduct these statistical tests. For his own good, I simply told him what Professor Selten told me: "There is no need to read the whole huge volumes of statistical handbooks, you are a smart boy, go to learn the relevant pages of the needed tests and do them, and believe in yourself." I did that myself and gained confidence and problem-solving capabilities. I believe that the youngsters now can also manage and do well, even better hopefully, as long as they are encouraged to develop their confidence and equipped with the proper working methods. If they continue the way to read all volumes of various subjects before using anything, they can read until their hair becomes white without any research progress. I had that tendency to read too much but do too little, Professor Selten taught me how to do good research (if I had done any), I am deeply grateful and wish to transfer this efficient working method to the youngsters now. A Master is indeed a Master, the things I learnt from him have been beneficial to my whole life ever since.

Years in Bonn were such a golden time. There were many eminent economists giving us seminars almost every week. I met the people whom I only read from books before I went to Bonn. They talked about various subjects and interacted with us in restaurants and bars after the seminars. Many of them came to visit Professor Selten. From this experience, I came to realize how important it is for a school to have a Master. Without the presence of Professor Selten in Bonn, it would be questionable whether some of those scholars would come to the "small boring town" for a serious visit. People make the place indeed, and in fact a Master makes a School!

Among the young visitors, Dr. Yan Chen from University of Michigan talked to me about her paper with Professor Charlie Plott on the experiment of public goods mechanism design. The paper was published in the Journal of Public Economics, which was initially the term paper for the "Experimental Economics" course by Professor Plott. The design was nice and interesting, but unfortunately they mixed the testing of different mechanisms within the same experimental session. The experiment was on one mechanism in the first half of the sessions and then followed by another mechanism in the latter half. When Yan described what they had done, I immediately recognized the problem since Professor Selten always focused upon the cleanness of statistical tests: He did not compromise on the fundamental nature of a "clean" test. This simple but profound influence on us the "pupils" in Bonn Lab was lasting. We followed it strictly all the time, although we may need to run more experimental sessions with higher costs. At that time, I pointed out the flaw in

Chen-Plott design and also the possibility of testing learning effects in the sequential data to Yan. She honourably took the suggestions and we did the project of a clear test on various mechanism designs of public goods provision, especially the Ledyard mechanism versus others. The results were clear-cut and the Ledyard design was functionable although the theoretical model seemed complicated compared to other designs (see Chen and Tang 1998). A seemingly tiny habit of insisting on the proper handling of statistical tools turned out to be very important. Professor Selten actually paid attention to technical details carefully, not only on "grand" ideas. Even for my English writing, he clearly corrected my spelling of "can not", telling me that there is no such a thing in English because "cannot" is put together as one single word. I was thankful for this and later I found that many of my students in Singapore and Hong Kong always spelled in the wrong "can not" way. Thus I always tell my classes in the very first lesson about the term paper writing requirement that "cannot" is one word and that one should write properly although we are not native English speakers.

The restaurants and bars in Bonn were a lot of fun. After the seminars which were usually scheduled in late afternoons, senior and young scholars gathered together for dinners and drinks. Professor Selten joined us often. Sparkling conversations and interesting mini talks occurred often in such places. We sometimes wrote joking short notes about neo-classic economics in the bars with good laughter. For the Bonn Lab guys, birthday parties were another occasion for informal gatherings, when the birthday person needed to buy cakes for everyone else. Klaus Abbink happened to be born on the same day as I was, though 3 years later, thus we split the bill. Klaus's mother sometimes left the dog to him to take care of and that dog was fond of cakes, interestingly enough. We always joked about how fat the dog was. It indeed needed more exercise but he preferred to sleep under Klaus' desk. Professor Selten gathered quite some characters in the Bonn Lab, a 2-m tall and thin Klaus, a much shorter Berndt, a Chinese (the only Asian hanging around there those days), the very neat and capable Bettina Rockenbach and the equally capable but much messier Karim, and so on. Under the benign leadership of Professor Selten, the atmosphere in Bonn Lab was simply superb and warm. I absorbed this spirit and later I tried to develop a family-like atmosphere in the labs I helped to build, by taking students to dinners and hiking together, etc.

Yes, hiking is very important. This was the only exercise of Professor Selten and probably the only hobby he has. He took us for hiking on weekends from time to time. In the hiking occasions I went with him, we chatted widely from economics to everyday life. I could always remember the highly enjoyable moments all the time. In his car trunk, there were amazingly hundreds of hiking maps of German mountains and hiking trails. These maps were so good that we could find almost everything we needed for hiking there. Professor Selten always carried a small bag of a plant book with him as well. This kind of book contained descriptions of the plants, pictures and their Latin names. He taught me with this book, how to recognize a plant, first from its leaves then its classification of categories. I asked him whether he liked animals (in addition to his well-known cats), he said that the animals in the forest run too fast while he was rather slow to follow their movement.

So he preferred plants. It was such a simple answer from a then-not-very-old naughty gentleman. He was quite childishly cute from time to time. I still vividly remember that once he tried to put on my coat a plant's fruit which was like a ball with lots of plush-like tiny sticks. I ran a step away from this ball thing, then he giggled to put the sticky ball on his own tie (he always wears a tie anyway) and proudly showed me that the ball would not fall down. That picture was just like a naughty boy in my memory. In another hike, we saw a church down at the foot of a beautiful mountain. I asked him to guess the age of the church, he said a few hundred years. When we walked downhill to the church, I found the statue which inscribed the age and a description of the church which was built in early 1900s. He saw it as well, but walked away quietly without saying a word. After a few minutes, he suddenly spoke softly, "I will never guess the age of a church any more". I will never forget the beautifully happy moments with him. He is rare, like his surname ("Selten" in German means seldom), an unusually intelligent and cute man. How precious and enjoyable to have the wonderful opportunities to have worked with him, and hiked with him. Every time he was in China, we went hiking whenever we got a chance. Every time he comes to China (sometimes Mrs. Selten also comes with him), my soul has been full of joy while saddened every time when I accompanied them to Chinese airports to see them back to Germany. Such a love has been deeply rooted back to the Bonn days. They are always in my heart.

Experimental Economics Labs in China

After my graduation from Bonn, I gradually moved back to the Far East, 4 years in Singapore and 7 years in Hong Kong before the now Chief Economist of World Bank, Professor Justin Yifu Lin, persuaded me to join China Center for Economic Research (now National School of Development as Premier Wen Jiabao supported its expansion) in Peking University. I began my journey to bring what I have learnt from Professor Selten in my Bonn years to help Chinese academic institutions establish experimental economic labs and conduct relevant research in economics experiments and bounded rationality.

The very first trip I helped to organize for Professor Selten to the Far East was jointly sponsored by City University of Hong Kong and Sichuan Province "Enterprise Competition Strategy Forum" back in the spring of 2001. Professor Selten was the key-note speaker in that Forum, where the Seltens were also received by an official banquet of the then Sichuan provincial governor. I accompanied them in the entire journey as the "body guard" and interpreter. It was a tough trip for the wheelchair but very fruitful indeed, especially for Sichuan Province. Since there were so few internationally well-known scholars to visit the very inland southwest province at that time, Professor Selten's tour was so warmly welcomed, it caused a "Nobel heat" in Chengdu, the capital city of Sichuan. The event caused such a stir that a taxi driver who carried one of the local assistants to the Forum site asked whether he had at least a master degree, since "without at least a master degree you might not be able to understand such a Master's speech." Sichuan local organizers even used one of the four provincial police cars to clear roads to ensure our timely arrival since the traffic conditions there were bad. This high-level forum was held in Gold Bull Guesthouse which was the province official guesthouse (the then Premier Zhu Rongji also stayed in that guesthouse during that week although he was inspecting something else for a very short period). Hundreds of senior academics, officials and executives from the region attended the Forum. Media coverage of Professor Selten's visit was overwhelming, side-by-side with news of Premier Zhu in Sichuan on the cover page of almost every newspaper. During the official reception banquet in Sichuan Guesthouse, Professor Selten discussed various issues with the then Provincial Governor Zhang and emphasized the importance of professional education system for economic development, citing the example of the German "Hochschule" which means "high schools" for practical professional training. Sichuan was a very populous province with one of the largest populations among the provinces in China. It was a very important point for the Governor, but due to some reason, he might not fully understand the implication. Nowadays, Chinese education system reform is putting high priority on developing a nation-wide professional training education system, years later.

Professor Selten was also invited to visit two local universities, Southwest Jiao Tong University (the top research university in China on railway system and especially the high-speed trains) and Southwest University of Finance and Economics (which was under the Chinese central bank before it was switched to the Chinese Ministry of Education). Professor Selten met the university presidents and council chairmen. Professor Selten made excellent "propaganda" for experimental economic research during his visit in China. He patiently explained what experimental economic labs were and why research in bounded rationality of human behavior was so important. He planted the academic seeds, though he might not realize what a wonderful job he had done. A modern research lab was successfully established – with strong support of Dean Jamie Jia of the School of Economics and Management – in Southwest Jiao Tong University in 2005 (named after late Professor Herbert A. Simon and himself – Herbert A. Simon and Reinhard Selten Research Lab in Behavioural Decision-Making, as Professor Selten honourably recommended that Professor Simon' name should be placed in front of his since Professor Simon's works significantly influenced him). Southwest University of Finance and Economics is now building its experimental economics lab, which is 9 years later after Selten's visit.

Developing extraordinarily quickly, Chengdu is a rather modern city now. But back to 2001, the trip was not easy for a person who was in a wheel chair like Mrs. Selten. For instance, the old Chengdu airport was rather small and the day we departed, the elevator was out of service. We, four strong young men, had to carry the wheelchair and Mrs. Selten upstairs. There were few facilities for handicapped people by that time, such as toilet, and stairs in the buildings and hotels. Even so, we managed to show them around as much as we could, and we had a good time. Some of the ancient historical sites they visited were destroyed by the terrible earthquake on May 12, 2008. Fortunately, the construction and reconstruction progress in

China were amazing. Chengdu airport is now very comfortable and modernized (designed by the same designer of the Frankfurt airport). The urbanization process in China is so fast that most of the rural areas we visited are now a part of the greater Chengdu city. Towns are connected by highways and wide roads which were rice fields. Professor Selten came to visit Chengdu several times afterwards, because the Sino-German Center and China Natural Science Foundation has been supporting the Sino-German Summer School on Experimental Management and Applications which places students 1 year in Chengdu and another year in Bonn. So far three summer schools have been held. For each summer school, 15 Ph.D. students each from China and Germany gathered together for seminars and interactions with senior researchers. With the strong support from Professor Selten, Professor Zhuyu Li of Sichuan University, Dr. Heike Schmidt and other local organizers worked strenuously to make these academic exchanges come true. The profound impact on relevant research in economics and related management fields, especially from a cross-cultural perspective, and in bounded rationality will show up later (see, e.g., Chen and Tang 2009).

When I was teaching in Hong Kong, Nankai University's Business School approached me through a friend in Hong Kong, Professor Junxi Zhang, to explore the possibility of inviting Professor Selten to visit Nankai University and establish some experimental research. This visit was arranged during the Nankai Corporate Governance Forum in 2003. Again, Professor Selten advocated the importance to build a modern research lab and the local host took this suggestion. The first economic research lab named after a Nobel Prize winner in China was established in November 2003, in Nankai Business School. Professor Selten also designed two experiments for this lab, one on corporate governance from a non-cooperative game-theoretical perspective, the other on enterprise formation from a cooperative game-theoretical perspective. We have been working on the two projects since then.

In December of the same year, Professor Selten was awarded the Honorary Doctoral Degree in Social Sciences from Chinese University of Hong Kong, where I had been a professor, at the 60th University Degree Award Ceremony and the 40th Anniversary of the University. This trip was initiated by an earlier trip by Professor Selten to Hong Kong in which Professor Leslie Young of their Finance Department asked me to help invite Professor Selten to deliver the Wei Lung Lecture in the University.. The lecture was so well received that the University senior members asked me to approach Professor Selten whether he was willing to accept the Honorary Doctoral title. I still remember the opening speech by Professor Selten, since it was exactly at the same time that the as the announcement that as Professor Vernon Smith was awarded the Nobel Memorial Prize. Professor Selten highly praised the pioneering work by Professor Smith and the importance of experimental economic research. He mentioned that the first experimental economics paper published by Professor Smith was in 1961 and "my first experimental economics paper was published in 1959". Everybody laughed friendly. Another story I must relate is about one hiking tour I accompanied him not far from the campus. We hiked in the beautiful area, in high spirits in spite of the humidity until sunset.

All the sudden, we realized that we were lost. I tried very hard to explore several directions to look for the right way to go out to the mountain road where the university car driver would be waiting for us. It took time since several efforts were in vain. The "young old" for the first time seemed a little panicked by this situation. I asked him to stay put by the hiking trail junction to wait for me. I was anxious to look for the right path, and heard his voice "Fang-Fang, where are you?" I came back to him and told him that I was looking for the right way. Finally I found it and we went out to the car parking place. That moment and the calling voice "Fang-Fang" have been in my heart. I would never let you down, my Doktorvater, for my love in heart. He has taken me into a life path which I cherish. I would do whatever I could to find the way for us to go out, needless to say. Later on during a hiking tour into Hong Kong's most beautiful bay mountain area together with Professor Sir James Mirrlees and Professor Lu Yuan of Strategic Management, I mentioned and joked about this experience, and got the reaction "How exciting." Perceptions are different from different perspectives. There is no standard of rationality, even among close friends.

And fun memory: On the Shenzhen hiking trail, an old European Gentleman (in suit and tie as always) hiking with two long sticks like summer skiing in the warm and humid forest. It was quite a scene. Professor Selten might have never imagined that he would have hiked so many Chinese mountains. During one of his visits to Hong Kong, Guangdong Academy of Social Sciences invited him to pay a visit to Guangzhou and we hiked in the famous Bai Yun (White Cloud) Mountain. Shenzhen, the border city next to Hong Kong, is on the way. It was during the famous annual "Hi-Tech Exhibition Fair" period. I asked him whether he would be interested to attend the "Hi-Tech" event. He smiled with his childlike way that he preferred a "low-tech" fair. Clear enough, let's go hiking then. I took him to hike two mountains in the city, and even walked through some dark under-bridge tunnels in between. We took the small taxi cars, not easy, but he managed. He walked around and ate in local restaurants with my family. He loved it. As usual, he always brought "Play Mobil" toys for my equally naughty daughter, the elegant German sets where even figure's heads are movable. He truly spoiled the little girl to the extent that she began to "order" from the in-box catalogue what she would want next time. She eventually accumulated a whole cabin of German "Play Mobil" and with beautiful furry cats and rabbits etc.

Bounded Rationality and E-Commerce

During the data transmission days from Bonn to the United States, I became curious about the Internet issues. When I was teaching in Singapore in the late 1990s a capable young student, Ho Hoong Pok, writing his bachelor thesis came to me asking me to serve as his thesis supervisor. Hoong Pok noticed a peculiar phenomenon that a few online branches of traditional bookstores set the online book prices significantly higher on average than some pure Internet retailers he surveyed.

This observation captured my attention although most people at that time were not much interested in the so-called "e-commerce" stuff. Hoong Pok and I began to explore this issue in online book pricing, and later joined by Dr. Xiaolin Xing (now in Fannie Mae) in online DVD pricing. We took great pains to collect weekly data online by thoroughly checking the major online retailers' pricing quotes of tens of popular titles and tens of random titles for statistical representativeness. We found that indeed, for most situations this phenomenon exists: for example, Tang and Ho (2003) who collected data by three online branches of multi-channel retailers (BarnesandNoble.com, Borders.com and Wordsworth.com) versus two pure play e-tailers (Amazon.com and Books.com) on 50 titles of books, from February 6 to March 6, 1999 once every 2 days (thus 3,750 price observations). A puzzling price difference pattern similar to Tang and Xing (2001) on DVDs was discovered (average price of US$16.07 of six online branches of multi-channel retailers versus US$17.22 of six pure play e-tailers, or 73% versus 78% in terms of percentage prices), thus we continued to collect price data by five online branches of multi-channel retailers versus five pure play e-tailers from April 29 to June 3, 2000 (once every week), with a total of 2,900 price observations during this second stage (updating only the bestseller titles while keeping the random titles unchanged). The price difference patterns that were discovered in the first stage remained robust in the second stage, in fact, amplified: average price of US$15.06 versus US$16.98, or 66.62% versus 75.1% in terms of percentage prices and price dispersions around 20–30% less among the pure play e-tailers than the price dispersions among the online branches of multi-channel retailers, in every sense from dollar price ranges or standard deviations to percentage price ranges or standard deviations. A series of completed studies and ongoing studies have followed and continue to follow this methodology proposed by Tang and Ho (2003) since then, to explore various categories of products (e.g., books, CDs, DVDs, pre-recorded videotapes, electronics, toys, hotel rooms) and across different geographic regions (e.g. the United States, Mainland China, Australia and South Korea). The phenomenon of the significant differences between average pricing levels and even more importantly between the price dispersion levels of the two types of online retailers (online branches of multiple-channel retailers versus pure online rètailers) continues (Li et al. 2008; Liu and Tang 2005; Lu et al. 2008; Tang 2004, 2008; Tang and Gan 2004; Tang and Lu 2001; Tang and Xing 2003; Tang and Zong 2008; Xing and Tang 2004; Xing et al. 2004, 2006; Zong et al. 2008). A simple but fundamental issue is touched: Why is the law of one price still violated in the online market where the search cost is so low or even "negligible" as Mr. Bill Gates et al. heralded so optimistically for "frictionless commerce"?

To be honest, we have no satisfactory theory to explain such a striking difference after a decade empirical work. Interestingly, from the feedback of some review, we had this sentence "two key limitations of their study are that they analyzed only one category and they did not offer any theoretical explanation for their findings". For the first comment, we have spent years expanding the scope of the categories of goods we researched on thus we need not repeat. For the frequently required demand for a theory, I really have only this to say, "The second limitation is due to the careful research attitudes of this group of researchers, because they do not want to offer any theory before the empirical patterns have stabilized as a robust phenomenon." This is what I have learnt from Professor Selten in my Bonn days, in his words (Selten 1991b): "It makes no sense to speculate on the evolution of unicorns unless unicorns have been found in nature" (p. 21); "A thorough knowledge of nature cannot be replaced by abstract principles" (p. 9); "Look at human anatomy and physiology: bones, muscles, nerves, and so on. Human anatomy and physiology cannot be derived from a few general principles" (p. 18) and "It is better to make many empirically supported ad hoc assumptions, than to rely on a few unrealistic principles of great generality and elegance" (p. 19). Even for the theory construction per se, there could be a profound influence. See below for how immaturely published results can cause a stir in the research community.

In a well-cited survey paper, Pan et al. (2004), like many others, have quoted Bakos (1997) with significant attention. For instance, in the introduction part of Pan et al. (2004, p. 117), it is already pointed out that: "Finally, and most importantly, research on online price dispersion can help us better understand whether this new retail format really does provide the gains in informational efficiency that many have predicted (e.g., Bakos 1997)." Then, on the same page, but in the next section: "Is Price Dispersion Narrower Online than Offline?" Pan et al. (2004) quoted again: "There are many reasons to expect price dispersion to be lower online than offline. Search costs are typically lower on the Internet than off-line, suggesting reduced price dispersion among e-tailers than among conventional retailers (Bakos 1997)." This is not an occasional matter, but unfortunately is a widespread problem, in that quite some studies (see, e.g., Kuksov 2004, Wu et al. 2004, Zhu 2004, etc.) have continued to quote Bakos (1997) as a kind of cornerstone study without knowing that the major results in Bakos (1997) had been proved critically wrong or unreasonable. It is a pressing urgency to call sufficient attention to this misleading problem, because we ourselves were also misled by those wrong results for quite some time.

Harrington (2001) proved that Bakos (1997) Result 1 "When the cost of product information is positive but the cost of price information is close to zero, near perfect competition prevails when there are sufficiently many firms, that is, the equilibrium price is close to marginal cost" is mathematically wrong. Harrington (2001) pointed out that, "in contrast to Bakos' statement, approximately competitive pricing does not prevail when the cost of acquiring price information is arbitrarily small. If a pure-strategy equilibrium exists, a little cost to searching over price results in prices being bounded above competitive prices" (p. 1729). Further, on the same page, "this proves that there is no symmetric pure-strategy equilibrium in which consumers search. Whether an equilibrium with searching exists remains an open question."

Another striking result in Bakos (1997), also quite new to the literature on pricing in markets with search, is the Result 2: When the cost of price information is positive and the cost of product information is zero, the equilibrium price is decreasing in the cost of the price of information. Unfortunately, upon careful examination, Harrington (2001) pointed out: "Hence, the strategy profile that Bakos puts forth as an equilibrium is an equilibrium in this more general model if and only if $\alpha = 1$ and/or $c_2^{\,1} = 0$. Since Bakos clearly assumes $c_2^{\,1} > 0$, it follows that he is implicitly assuming $\alpha = 1$. This assumption is never stated and, more to the point, is not very plausible. By what mechanism would a consumer be refunded for his search if he did not purchase the good?" (p. 1731)

Harrington (2001) thus concluded that: "In conclusion, the most interesting results of the analysis of Bakos do not survive closer scrutiny. On a positive note, Bakos' paper raises some important issues related to electronic commerce that, hopefully, will encourage further research into this emerging area." I have nothing more to add to Harrington's comments in this aspect, except that it is probably important to be open-minded and more exploratory in empirical studies, as I learnt from Professor Selten, rather than to insist on having a theory all the time before empirical studies or as a requirement for purely empirical studies. Time flies and facts exist, whether one has a theory or not. How many wonderful theories in human history have emerged and gone with the wind? For instance, one of the ancient Chinese theories believed that the world was composed by five elements of gold, wood, water, fire and earth. For more recent theories, one may like to mention about the once extremely popular chaos theory, which was also applied to the financial market, which Professor Selten mentioned to me during conversation. Nevertheless, it does not seem that our financial markets function as that kind of theory tried to explain, except that our financial markets do seem more chaotic now than before. From this perspective, it is a misleading requirement to put too much emphasis on empirical studies to have a theory in mind before or during the research process. In natural sciences, purely exploratory investigations are at least as important as theoretical constructions. In the natural world, empirical findings are about how the facts per se stand on their own. Isn't it important to find a new gene? I suspect that the natural science journals would insist on having a theory for a paper that reports the discovery of a new gene and the structure of the new gene itself. Even for social science and business study disciplines, it may be a resources-distorting policy to force those investigations that are empirical in their fundamental nature to construct a theory for constructing a theoretical explanation artificially. And, indeed, when the purely theoretical efforts such as Bakos (1997) could have gone so wrong, why should the purely empirical exploration be required to provide a theoretical explanation during the time when there are in fact no reasonable (not even to mention satisfactory) theories? In this regard, the German school of experimental economics led by Professor Selten, one of the greatest game *theory* Masters, always encourages exploratory empirical studies, rather than insisting on having a theory first to explore the facts, and its philosophy may still have a point to be thought over.

Any conclusion, either empirical or theoretical, has to be careful in it's interpretation. Indeed, theoretical constructs must be built based on robust empirical findings. Any theory without empirical foundation would be dubious. To build a theory for establishing a theory is not a proper concern from this view. It falls squarely on the criticism by Selten (1991a, b) as quoted above. Thanks to the disciplined training in Bonn by Professor Selten, and even more to the open-mindedness, he showed to us to explore, to probe, to experiment without fear. I will continue to carry this spirit in the rest of my life. As I always whole-heartily say to

Professor Selten, I wish him another two decades to work on the bounded rationality theory, and to see our research progress.

Happy Birthday on October 5, 2010, our dearest Master. Here is my little wish that this article sends you some joy and fun from the Far East, a student of yours, and always a student of yours. In a Chinese saying, “One day a teacher, one life a father.”

References

Abbink K, Bolton G, Sadrieh K, Tang F-F (2001) Adaptive learning versus punishment in ultimatum bargaining. Games Econ Behav 37(1):1–25

Bakos Y (1997) Reducing buyer search costs: implications for electronic marketplaces. Manage Sci 43(12):1676–1692

Chen Y, Tang F-F (1998) Learning and incentive compatible mechanisms for public goods provision: an experimental study. J Polit Econ 106(3):633–662

Chen K, Tang F-F (2009) Cultural differences between Tibetan and ethnic Han Chinese in ultimatum bargaining experiments. Eur J Polit Econ 25:78–84

Harrington JE Jr (2001) Comment on “reducing buyer search costs: implications for electronic marketplaces”. Manage Sci 47(12):1727–1732

Kuksov D (2004) Buyer search costs and endogenous product design. Mark Sci 23(4):490–499

Li H, Tang F-F, Huang Y, Song F (2008) A longitudinal study on pricing, price dispersion and market dynamics in the Australian online DVD market. J Prod Brand Manage 18(1):60–66

Liu Y, Tang F-F (2005) An empirical analysis on pricing patterns in China’s online book market. J Int Consum Mark 18(1/2):117–136

Lu Y, Xing X, Tang F-F (2008) Retailers’ incentive to sell through a new channel and pricing behavior in a multi-channel environment. Ann Econ Finance 9(2):315–343

Nagel R, Tang F-F (1998) An experimental study on the centipede game in normal form – an investigation on learning. J Math Psychol 42(2):256–284

Pan X, Ratchford BT, Shankar V (2004) Price dispersion on the internet: a review and directions for future research. J Interact Mark 18(4):116–135

Roth AE, Erev I (1995) Learning in extensive games: experimental data and simple dynamic models in the intermediate term. Games Econ Behav 8:164–212

Selten R (1991a) Anticipatory learning in two-person games. In: Selten R (ed) Game equilibrium models I. Springer, Berlin, pp 98–154

Selten R (1991b) Evolution, learning, and economic behavior. Games Econ Behav 3(1):3–24, This is the entire text of the 1989 Nancy L. Schwartz Lecture delivered at the J. L. Kellogg Graduate School of Management, Northwestern University, Evanston, Illinois

Smith V (2005) A nobel for human betterment. Wall St J, Monday, 17 Oct 2005, 17

Tang F-F (2001) Anticipatory learning in two-person games: some experimental results. J Econ Behav Organ 44(2):221–232

Tang F-F (2002) John C. Harsanyi: memory from China. Games Econ Behav 40(1):150–152

Tang F-F (2003) A comparative study on learning in normal form games: some simulation results. J Econ Behav Organ 50(3):385–390

Tang F-F (2004) Hybrids vs. doctors: some online pricing patterns in the South Korean Book Market. Int J Mark Advert 1(3):316–328

Tang FF (2008) Pricing differences between pure internet retailers and multichannel retailers Online. Invited chapter for Encyclopedia of e-business development and management in the global economy, forthcoming in 2010, (eds. In Lee), by IGI Global

Tang F-F, Gan L (2004) Pricing convergence between DocComs and hybrids: empirical evidence from the online toy market. J Target Meas Anal Mark 12(4):340–352

Tang F-F, Ho HP (2003) A puzzle in online pricing: early evidence from the book market. Rev Bus Res 1(2):1–11
Tang F-F, Lu D (2001) Pricing patterns in the online CD market: an empirical study. Electron Mark 11(3):171–185
Tang F-F, Xing X (2001) Will the growth of multi-channel retailing diminish the pricing efficiency of the web? J Retailing 77(3):319–333
Tang F-F, Xing X (2003) Pricing differences between DocComs and multi-channel retailers in the online video market. J Acad of Bus Econ 2(1):152–161
Tang FF, Zong J (2008) Hotel electronic distribution and online price dispersion in mainland China. China Econ J 1(3):303–315
Wu D, Ray G, Geng X, Whinston A (2004) Implications of reduced search cost and free riding in e-commerce. Mark Sci 23(2):255–262
Xing X, Tang F-F (2004) Pricing behavior in the online DVD market. J Retailing Consum Serv 11 (3):141–147
Xing X, Tang F-F, Yang Z (2004) Pricing dynamics in the online consumer electronics market. J Prod Brand Manage 13(6):429–441
Xing X, Yang Z, Tang F-F (2006) A comparison of time-varying online price and price-dispersion between multichannel and dotcom DVD retailers. J Interact Mark 20(2):3–20
Zhu K (2004) Information transparency of business-to business electronic markets: a game-theoretic analysis. Manage Sci 50(5):670–685
Zong J, Tang FF, Huang W, Ma J (2008) Online pricing dispersion and dynamics in mainland Chinese hotels. J China Tourism Res 4(3/4):248–260

Part II
Strategic Behavior

Chapter 6
Drei Oligopolexperimente

Klaus Abbink and Jordi Brandts

Introduction

Reinhard Selten's work marks the beginning of the experimental analysis of oligopoly. Indeed, his research includes both one of the first experiments on quantity competition and the first one on price competition. In the aftermath a very rich literature has emerged, dealing both with theory-based issues linked to the many oligopoly models that appeared as the result of the application of game-theory to industrial organization and with more applied questions that came out of the economic analysis of regulation and anti-trust issues.

Our own work has been very much influenced by the experimental work of Selten by motivating us to view human interaction in terms of boundedly rational behavior. In this short piece we first describe two of Selten's seminal experimental papers on oligopoly and briefly discuss elements of this work that influenced us. After that we discuss three experimental papers of ours about price competition. An important difference between Selten's oligopoly experiment and ours is that his involve a considerably higher degree of complexity than ours. The trade-off between the advantages of simple and complex experiments is precisely one of the themes of this paper.

Reinhard Selten's classic oligopoly experiments were mostly exploratory exercises. They were not focused on theories and the testing of hypotheses. The researcher is simply interested to set up an interesting situation, observe behavioural patterns and then find theories to explain the observations. Arguably, this approach, while still widely used, has become a little less fashionable as the field of

K. Abbink (✉)
CBESS, School of Economics, University of East Anglia, Norwich NR4 7TJ, UK
e-mail: k.abbink@uea.ac.uk

J. Brandts
Institut d'Anàlisi Econòmica CSIC, Campus UAB, 08193 Bellaterra Barcelona, Spain
e-mail: jordi.brandts@uab.cat

A. Ockenfels and A. Sadrieh (eds.), *The Selten School of Behavioral Economics*,
DOI 10.1007/978-3-642-13983-3_6,

experimental economics grew. Exploratory experiments have been replaced by hypothesis-based studies, which ask a well-defined research question in advance and tailor an experimental design that is particularly suitable to give a clear answer to the research question. This suitability is then the main priority, not whether the situation that is studied is particularly natural or even interesting. In fact, to design an experiment that sharply separates hypotheses one has to simplify so much that the designs often become quite detached from any realism.

Of course, part of the reason why Selten's early experiments were as they were is that the field of experimental economics was not yet invented, so there was no vast body of literature that could have generated ex-ante predictions. Similarly, the pioneers could not know what is the appropriate level of complexity for an experimental design. From today's point of view the richness of the early designs is overwhelming, Selten's later experiments have become much simpler as well. But it would be wrong, in our view, to call the early experiments outdated and only historically interesting. Selten has always passionately promoted the exploratory approach to experimental economics, preferring complex analysis of natural situations over simple answers of straightforward questions.

In recent years, the authors of this paper conducted three studies on oligopoly experiments, more precisely experiments involving price competition. In one sense, they are diametrically opposite to Selten's early papers, since we follow the approach of radically simplified designs, sometimes barely recognisable as market environments, and clear straightforward research questions. In another sense, our own experiments owe a lot to Selten's pioneering efforts, since many features we use in our settings have been introduced in these early pieces. We use the other approach to experiments not because we believe that it is superior. Perhaps it is just that we are lazier and more impatient than Selten, and therefore lean towards experimental designs that promise quick and simple answers. Perhaps it is that the research ideas that we had at the time called for this type of experiment. In any case, when we were invited to contribute to this festschrift it seemed interesting to relate our own oligopoly trilogy to Reinhard Selten's classics. Of course, this cannot mean comparing them to find out which are better – the only chance our own experiments could compare favourably would be that we had the unfair advantage of more than four decades of experimental research we could build on. Rather, we wish to draw the readers' attention to two points. First, we would like to make researchers in oligopoly experiments aware of how much they owe to Selten's pioneering work – often without knowing it. Second, we wish to emphasise that Selten's exploratory approach is a strong tool that should receive more attention and be used more often – even though (or because) we did not follow it in our own studies. In fact, in one of our papers we were narrowly fooled by the results of a seemingly clever design to sharply test two hypotheses. Natural, richer designs are needed to give us an insight into what matters in the real world.

We will now summarise the two path-breaking oligopoly studies by Reinhard Selten. Although these studies are classics, they are nevertheless much less read than they would deserve. The most obvious reason for this under attention is that they are written in German, so they are not accessible to many readers in the experimental economics community. We therefore describe the two studies in a little bit of detail.

Genesis: Sauermann and Selten's "Ein Oligopolexperiment" (1959)

This seminal article is a first in many ways: The first published article by Reinhard Selten, the first oligopoly experiment in history, it even marks the beginning of modern experimental economics altogether. It was published in 1959 in the *Zeitschrift für die gesamte Staatswissenschaft* and translated into English 1 year later (though in a collective volume that is hard to obtain today). This paper is, together with Hoggatt (1959), the first experimental work on quantity competition. The aim of this research is to contribute to the elaboration of a decision theory useful for economic contexts. We will discuss the paper by Sauermann and Selten to illustrate its rich complexity and to highlight how the most early work already contained many elements that still today are of great interest.

In their introduction to "Ein Oligopolexperiment" the authors refer to two pieces of work as motivations for their own work. One of them are the management games developed by the American Economic Association for the training of higher executives as in Ricciardi (1957) and the other is the pure research market experiment paper by Chamberlin (1948). They add that the design of their experiment is closer to those of the AMA, but that their own objective is primarily to make a research contribution.

Indeed, their experiment is a rich quantity competition experiment with many particular features, which appears to be inspired both by the sparse models of economic theory and by the desire to incorporate – albeit in a simplified manner – relevant features of the business environment they are interested in.

Three firms produce a homogeneous good and compete in quantities over 30 periods. The firms differ in their capacities and in their cost conditions. There is a smaller firm, a middle-size firm and a larger firm. Cost conditions are such that the larger the firm the higher are fixed costs but the lower the marginal costs. Capacities are known by all firms while cost conditions are private information of the firms. Firms produce on demand, so that there are no inventories. There is also no investment. Firms can go into debt at a fixed interest rate. The debt positions of firms are, like the cost conditions, private information, but information about these two aspects of other firms' behaviour can be acquired. At the beginning of each period, firms can spend resources to find out the cost conditions and the debt positions of their competitors.

The demand is constant over time and simulated, i.e. there are no humans making purchasing decisions. The demand function is not known to firms, but since it is stationary its shape can be learnt over time. However, the demand function has a quite particular shape, which makes learning from experience through repeated play not easy.

Price determination works in the standard quantity competition fashion. In each period the three firms independently set a quantity and the sum of the quantities determines the uniform price which firms receive for all produced units. The resulting costs can be paid either with revenue from sales or with debt.

An important feature of the design is that each firm is formed by a group of – on average – five participants. At the beginning of the 30 rounds one of them is randomly assigned the role of entrepreneur. Firms' decisions are discussed by the whole group, but the entrepreneur has the power to make the final decision. One of the other members of the group has another important role to which we come back below.

The theoretical analysis of the experimental situation is based on the one shot game with complete information, although both demand and cost conditions of others are not known. It turns out that there are four different equilibria, involving different kinds of asymmetries between the firms. In two of the equilibria the three firms produce different output levels, with magnitudes in the same order as the capacity levels of the three firms. In the two other equilibria either the small and the medium-sized firm or the medium-size and the large firm produce the same output levels.[1] In addition the authors discuss the different Pareto-optimal quantity combinations, which could be reached by some kind of tacit collusion.

One, from today's perspective, quite remarkable feature of the experiment are the motive protocols. This feature of the experiments reflects Selten's life-long interest in not only studying behaviour, but also understanding humans' reasoning behind behaviour. During the time in which the members of a firm discuss the decisions they are going to make, one of the members takes notes of the motives that lead the group to making a particular decision. These notes are later studied and the motives are classified into a number of categories.

The results show a number of interesting features. With respect to the quantities chosen by participants, three results are central. First, play does not tend to settle at any of the Pareto-optimal points, i.e. tacit collusion did not emerge. The finding is that collusion without communication is hard to achieve. Subsequent experiments with a simpler quantity competition design with a unique equilibrium by Huck et al. (2004) have confirmed this result; they find that only in the case of two firms can collusion arise. Second, behaviour is heterogeneous and still quite unstable towards the end of the experiment. This is perhaps due to the fact that, given the rather complex environment, 30 periods is not enough for allowing behaviour to settle down. However, the environment is probably too complex to allow for some kind of formal analysis using learning models. Third, two of the four equilibria are more frequent in the data. These are those for which the ordering of payoffs is consistent with what a priori seems to be the ordering of

[1]The four equilibrium quantity triplets are (4, 5, 7), (5, 5, 6), (5, 6, 7) and (4, 6, 6), with the third of these only being a weak equilibrium.

the strength of the firms in the market. This is a simple empirical equilibrium selection criterion.

A second part of the results correspond to what comes out of the analysis of the motive protocols. The different motives for making decisions were grouped into 11 groups of substantial motives and six groups of other motives. The 11 groups are then ordered according to the frequency with which they were used. The three most frequent groups together capture 50% of the frequency. The most frequent group of motives is labelled "good experience" and corresponds to decisions that were made because they were successful in the past. Observe that this group of motives is basically reinforcement thinking. The second most frequent group corresponds to best-response to competitors' choices in previous rounds, i.e. what is sometimes referred to as the Cournot-dynamic. The third group of motives is labelled "chains of motives" and captures all those cases in which participants typically based their reasoning on some experienced facts, some expectation about others' behaviour and a quantity decision. An example is good experience with a particular quantity in the past, the information that the competition has high debt and therefore will not produce and the decision to, therefore, maintain the same quantity as before.[2]

At this point it should be clear that the experiment by Sauermann and Selten is considerably more complex than the more modern studies on quantity competition, which starting with Holt (1985) mostly deal with the simplest static game with symmetric firms and complete information about costs and demand. The Sauermann and Selten (1959) approach appears to be more natural, since it nicely incorporates many interesting elements. The simplifications they use are based on common sense and are not straight-jacketed by the need to end up with a design that corresponds to a well-defined game for which it is easy to find the equilibria. The approach that is used in the modern literature is more appropriate for the testing of simple hypothesis and for finding simple results that are easily communicated.

The First Price Competition Experiment: Selten (1967)

Selten later continued to work on imperfect competition and published various several other articles involving oligopoly experiments. In particular, Selten (1967) appears to be the first published experiment on price competition. This paper is much in the same vein as the paper on quantity competition discussed above. The experimental set-up used by Selten is arguably even more complex than the one on

[2]The other substantial motive groups are fighting motives, attempts at influencing the price to increase profits, exploration around previous decisions, output limitation due to liquidity constraints, attempts at fixing a price equal to average price, retreat and optimization according to accounting. For details about these motives please see the original article, which also contains a discussion of how different motives are used at different points in time.

quantity competition discussed. It is not at all designed to allow for a test of the simple Bertrand hypothesis that under price competition two firms would suffice to bring prices down to marginal cost. Remarkably, this simple experiment was not run until much later by Dufwenberg and Gneezy (2000). In Selten (1967) we find a rich experiment involving capacities that could change over time and rather complex demand conditions including demand inertia, demand growth and the fact that price differences do not only affect current sales but also market shares. In addition, subjects only receive some qualitative information about this demand. The analysis of the decision data focuses on pricing and investment behaviour. Selten formulates a particular theoretical model meant to be a simplified representation of the experimental environment and finds that many of the regularities of the data are consistent with the model. In the experiments subjects were asked for their decision motives and the answers are classified and studied, as for the quantity competition experiment above.

Later another strand of the experimental literature started studying markets in which each seller posted a price and also decided on the maximum of units that they would be willing to sell at that price. The focus in this line of research, which emerged in the United States, was the comparison between the double-auction and the posted-price market institutions – Plott and Smith (1978) – in relation to the convergence to the Walrasian outcome. Only later did the issue of comparing behaviour in posted-price environments to the Nash equilibria of the corresponding game, as in Brown-Kruse et al. (1994) and Davis and Holt (1994).

What is interesting in Selten's seminal work on quantity and on price competition is that it is a hybrid between the pure analysis of interaction in the kind of conflicts characteristic of oligopoly and an attempt at gaining insights into some of the complexities of oligopolistic markets and of complex organisations, albeit in simplified contexts. In subsequent research, simplicity of the design quickly became a highly valued quality of experiments and it continues to be today. However, some of the features of Selten's early work have reappeared in more modern research. Some examples of this follow.

There is recent interest in studying the effects of group decision-making like in the work of Cooper and Kagel (2005), Abbink et al. (2010), Gilet et al. (forthcoming) and Feri et al. (forthcoming). Some of the questions that are studied are to what extent group-decisions differ from decisions made by unitary players, but also how different group-decision making rules matter. Another example of a Selten-theme that is currently being studied is the analysis of decision-making protocols. Using video-technology (which was of course unavailable in the 1960s) Bosman et al. (2006) analyse how groups make decisions in a power-to-take game. A third feature of the early Selten work that has been picked up on modern studies is the multiplicity of equilibria. Some modern experiments on oligopoly are in fact empirical studies of equilibrium selection. Apart from Abbink and Brandts (2008), discussed below, another example is the paper on supply function competition in electricity markets by Brandts et al. (2008).

Our Own Price Competition Trilogy: Motivation and Background

In our own oligopoly experiments we have also been guided by various motivations. First, price competition is a more intuitively appealing model than quantity competition, since firms in most markets can choose their prices. The three experiments we discuss below involve price competition. Second, from a more theoretical point of view we have been interested in studying environments in which price competition takes place in a way that avoids the so-called Bertrand-paradox. At the same time our experiments ask questions that can be of interest from a more applied viewpoint. Third, our experiments are all rather simple and, hence, more in the line of more current oligopoly experiments than in the Selten tradition. However, the explanations for our results are in terms of simple principles of bounded rationality. See Armstrong and Huck (2010) for a survey of recent experimental and behavioural research on firms' departure from fully rational behaviour.

In Abbink and Brandts (2005, 2008) we study price competition environments in which in equilibrium two firms do not lead to a zero-profit situation. Our initial focus is on finding out whether this prediction emerges in the data and whether increasing the number of firms leads to lower, like in Cournot's well-behaved model of quantity competition. It turns out that the analysis of the data in Abbink and Brandts (2008) make it possible to gain insights about boundedly rational behaviour in coordination games, that go beyond the initial motivation for our study.

In Abbink and Brandts (2009) we use an extremely simple design to make a first approach to an issue that has emerged from the policy debate, namely whether collusion will be stronger in growing and in shrinking markets. Given that we study a policy-relevant issue a more complex design could be of interest to obtain a richer picture of behaviour. However, our very simple design does yield some intriguing insights.

Many features that characterise the experiments of our own price competition trilogy were already contained in the classics. Nevertheless, the approach we take is somewhat opposite. While Reinhard Selten created environments that incorporated many features in the same experiment, we boil the market games down to very simple settings, sometimes so simple that they are barely recognisable as oligopoly models. In fact, two of our settings were simplified to an extent that it wasn't even necessary to tell the subjects that they were playing market games – the rules were explained as simply choosing numbers.

Price Competition with Increasing Marginal Costs

This is the first experiment in our trilogy, and it is also the one that, as we will learn, showcases the power as well as the dangers of theory-testing in artificial experiments. In the case of constant marginal costs, as introduced by Joseph Bertrand in

1883, firms will compete each other down to marginal cost prices. It took more than a century until Krishnendu Ghosh Dastidar (1995) found out that things change dramatically if we consider the case of increasing marginal costs. This is the more astonishing since decreasing returns are by no means a rare and exotic setting, but arguably the most natural framework to consider for economists, and no other changes in the game are needed. The rules are still the same. There are n firms, each simultaneously posting a final price for the homogeneous good. The firm with the lowest price must serve all the demand. If more than one firm has posted the lowest price, then demand is shared equally among the firms with the lowest price. There are no fixed costs. Hence a firm whose price is not the lowest makes zero profit.

Changing the cost function from horizontal to upward sloping makes the equilibrium prediction strongly indeterminate. There is no longer one unique equilibrium price, but a continuum of equilibrium prices stated by all firms. Equilibria with high prices and profits are possible. This is illustrated in Fig. 6.1. Here all firms charge the same price p, hence they share the demand Q that corresponds to p, and each firm sells q. Since the marginal costs for each firm at q are below the price p, each firm makes a positive profit. Thus no firm has an advantage from trying to charge a price higher than p, it means it would not sell anything and hence it would not make any profit. But neither would it benefit from trying to undercut the other firms and charging a lower price. If it did, it would serve the entire demand, but that would not be a good thing in this case: The increasing marginal costs mean that the additional units would be produced at much higher costs and thus lowering the profit. Undercutting can even lead to disastrous losses if the marginal cost function is sufficiently steep.

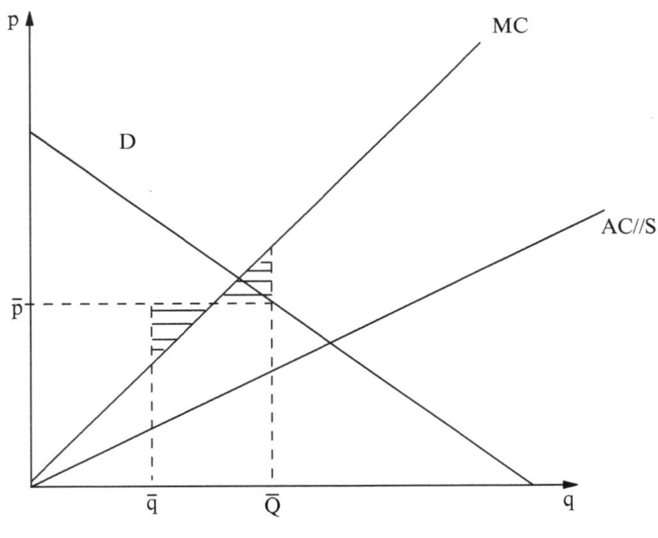

Fig. 6.1 Equilibrium price in proce competition with increasing marginal costs

The equilibrium illustrated in the figure is not unique, in fact there is a whole range of prices for which the same logic applies: If all firms charge the same price, then neither unilateral overcutting nor undercutting would improve profits. The upper bound of the range is reached when the common price is so high that the revenue from the additional units from undercutting is so large that it compensates for the increasing costs of producing them. The lower bound is constituted by the price at that the profit for the individual firm is zero in equilibrium.

Note that all but the lowest of the equilibrium prices involve positive profits for firms. Thus increasing marginal costs constitute one theoretical answer to the Bertrand paradox. It is now possible (and likely) that a small number of firms coordinate on a profitable price without having to revert to off-equilibrium collusion. The indeterminacy of the equilibrium price also changes the character of the game. It can be extremely dangerous to be the only one charging the lowest price as it can lead to ruinous losses. It will always be unprofitable to post a higher price than the competitors. To make satisfactory profits it is essential that all firms charge the same price. Thus the market game becomes a coordination game with multiple symmetric equilibria. The equilibria are pareto-ranked; within the equilibrium range higher-price equilibria involve higher profits than lower-price equilibria.

We design an experiment with the simplest model of price competition with increasing marginal costs. We are interested in (1) detecting which prices subjects will coordinate on (if they do), and (2) whether increasing competition – a greater number of firms – will lead to lower observed prices. We conduct experiments with two, three and four firms. Since the model is so simple we do not explain the underlying market environment to subjects, but only describe what the game eventually boils down to: Each player states a number, and those with the smallest number get a certain payoff. This leads to the payoff table as reproduced in Table 6.1. The table belongs to the case with four firms. For the smaller market the tables are the same, just with the columns for the larger number of lowest-price firms missing.

The game was repeated 50 times with partners matching, which was known to subjects. Our expectations were that we would see a predominance of the upper bound of the equilibrium range. These equilibria appear very attractive as they involve payoffs close to the collusive outcome, but in contrast to that they are self-enforcing. As often, our predictions turned out to be totally wrong.

Price Competition Under Cost Uncertainty

The second suggestion of how to overcome the Bertrand paradox is a variety of imperfect information. Hansen (1988) and Spulber (1995) independently introduced similar variants of a simple Bertrand game in which the marginal costs, which in this case were constant, are private information. Firms face a linear downward sloping demand function. Each firm has a marginal cost parameter c_i. Before play, the c_i are drawn randomly from a uniform distribution, independently

Table 6.1 Payoff tables in the experiment

Column 1	Column 2	Column 3	Column 4	Column 5	Column 6
Number chosen	Profit if this number is the unique lowest of the four	Profit if this number is the lowest and has been chosen by two players	Profit if this number is the lowest and has been chosen by three players	Profit if this number is the lowest and has been chosen by four players	Profit if this number is not the lowest
40	777	473	334	258	0
39	784	489	348	269	0
38	783	503	360	279	0
37	777	514	370	288	0
36	763	522	379	296	0
35	743	528	387	303	0
34	716	532	393	310	0
33	683	533	398	315	0
32	642	532	402	319	0
31	596	528	403	322	0
30	542	522	404	323	0
29	482	514	403	324	0
28	415	503	401	325	0
27	341	489	397	324	0
26	261	473	392	322	0
25	174	455	385	319	0
24	81	434	377	315	0
23	−20	411	367	310	0
22	−126	385	356	304	0
21	−240	357	344	297	0
20	−360	326	330	290	0
19	−487	293	314	281	0
18	−621	257	298	271	0
17	−761	219	279	260	0
16	−908	179	260	248	0
15	−1062	136	239	235	0
14	−1222	90	216	221	0
13	−1389	42	192	206	0
12	−1563	−8	167	189	0
11	−1743	−61	140	172	0
10	−1930	−116	112	154	0
9	−2124	−173	82	135	0
8	−2324	−234	51	115	0
7	−2532	−296	18	94	0
6	−2745	−361	−16	72	0
5	−2966	−429	−52	49	0
4	−3193	−499	−89	25	0
3	−3427	−571	−127	0	0
2	−3667	−646	−167	−26	0
1	−3914	−723	−208	−5	–

Equilibrium
Cournot
Walras
Payoff dom. Equilibrium
Carte
24

for each firm. Firms know only the realisation of their own cost parameter, but about the other firms' costs they only know the distribution from which it is drawn. The market game is then played exactly like in the classic Bertrand game. Firms state their prices, the firm with the lowest price serves the entire demand, in case of a tie the demand is shared equally.

The modification of the classic price competition game makes the market game very similar to a first-price sealed-bid auction with independent private values.

There are only two major differences. First, in this game it is not the buyers who overbid each other to obtain an asset, but the sellers who underbid each other to sell one. Second, the item is not a fixed unit, but a quantity that depends on the price. The higher the winning price, the less will be sold. The theoretical predictions are of similar nature to those of a first-price private-value seller auction. Like in an auction with fixed demand equilibrium bid functions are linear to the cost parameter. The mark-ups firms ask for are lower, though, since sellers have to consider that a high asking price not only lowers the probability of winning, but also reduces sales and thus revenue. Profits for the winner are still positive, so cost uncertainty does serve as a theoretical feature to overcome the Bertrand paradox.

Infinitely Repeated Games

The Bertrand paradox was formulated for a one-shot game. If the game is infinitely repeated, then even without any modification prices above marginal costs are sustainable. In fact, in this case any price combination can constitute an equilibrium of the supergame and thus overcome the paradox. Despite the virtually empty theoretical prediction, some qualitative assessments are possible. Game-theoretic analysis of price competition suggests that collusion will arise more easily in growing than in declining markets. Tacitly collusive agreements involve that deviations from the collusive path trigger retaliations by other firms, such that, from that point on, the deviating firm's profits will be lower than if it had stuck to the agreed behaviour. When the demand grows steadily the gains from deviating from the collusive agreement are, at any point in time, small in comparison to the future losses from retaliation. Analogously, when the demand keeps shrinking these losses will be relatively small compared to the short-term gains from deviations. Indeed, when the market is on the verge of collapsing, it will be virtually impossible to motivate firms to maintain the collusive agreement.

In the third experiment of our price competition trilogy we test this prediction in the laboratory. We use the simplest possible setup: Demand is perfectly inelastic up to a fixed price. The game is therefore very easy to understand: Both players state a number, and the player with the smaller number gets his or her number as a payoff. If both players choose the same number, they share this payoff equally. We chose a unit demand setting because it makes collusion very salient. Players do not need to compute the price that maximises joint profits, but the collusive outcome is simply the upper bound of the price range, clearly identifiable even for subjects with no economics or mathematics training.

We chose a rather extreme way in which markets would grow or shrink. In the growing treatment demand grows by 25% every round, in the shrinking treatment it decreases by 20%. Over the course of the 27 rounds of the experiment demand thus changes by the factor of 413.6. We chose this exaggerated parameterisation to ensure the best chance to detect potentially subtle effects. To keep the treatments perfectly comparable we set up the demand sequences in the two treatments as exact mirror images of one another.

In order to mimic infinitely repeated play we opted for the simplest method: Subjects were not told how many rounds the experiment would last. Though this implementation does not exactly recreate the theoretical setting, it has two advantages for our purposes. First, it is very easy to explain to subjects, and second, it helps us to keep the two treatments comparable, since all subjects play the complete sequence of demand values.

Is There More Collusion in Growing Markets?

We start the overview with the results of the third experiment. Theoretical reasoning would predict that growing markets would be more prone to collusion than shrinking markets. Table 6.2 indicates that, on average, this is not the case. The table shows average market prices, i.e. the lower of chosen percentages, for the different groups over the 27 rounds of the experiment. Indeed, the average market price is more than twice as high in shrinking as in growing markets.

This result appears counterintuitive, as it is the opposite of what the theoretical argument would lead us to expect. It seems that the prospect of great future profits in growing markets does not encourage collusion; on the contrary, high prices are much more common in shrinking markets, where we would expect greater incentives to realise a short-term gain by deviating from a collusive agreement. It seems that it is not the promise of growing profit opportunities, but rather the pressure from rapidly declining profits that exerts a disciplining effect on firms. If the prize shrinks at a dramatic rate, high profits need to be made early. The reverse effect holds for growing markets.

Since prizes are relatively small in early rounds, firms feel less pressure to coordinate quickly and can experiment with different strategies. We conjecture that these early deviations from collusion make it difficult to establish cooperation

Table 6.2 Average market prices in growing and shrinking duopolies

No.	Growing markets	Shrinking markets
1	8.96	62.52
2	1.30	82.22
3	59.04	63.30
4	15.41	10.78
5	37.22	74.00
6	6.85	89.30
7	91.07	56.67
8		75.52
9		68.15
Average	31.41	64.72

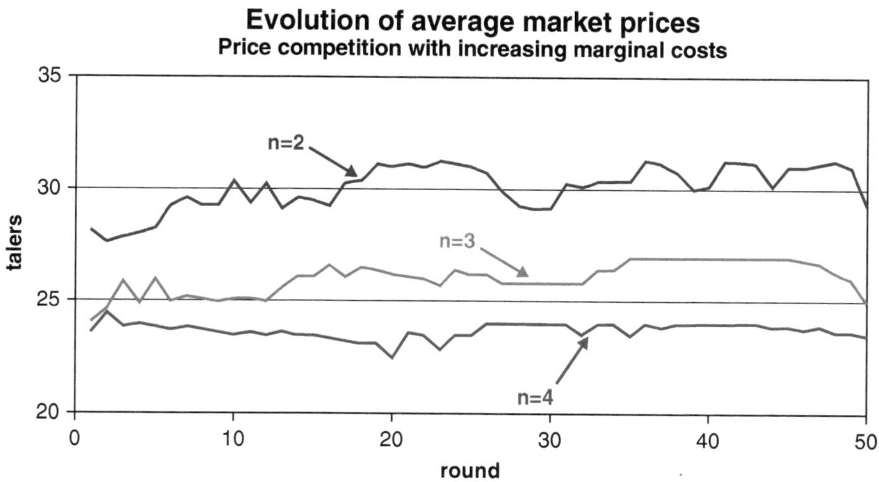

Fig. 6.2

in later rounds, when prizes become very substantial. As a result, firms in these markets fail to cooperate and realise low prices over the entire experiment.

Are More Firms Better Than Fewer?

Two of the three experiments are designed to test the effect of the number of firms. Will more competition lead to lower prices? For our two settings the answer is yes. Figure 6.2 shows the development of the average market prices in the two experiments. Clearly, more firms have an effect towards lowering the market prices.

While in the experiments with cost uncertainty prices vary wildly – perhaps not surprisingly because the cost parameter of the winning firm is drawn anew in every round – we can see that in the setting with increasing marginal costs two prices are predominant: The collusive outcome (33 with two firms, 30 with three and 28 with four), and the price of 24. The lower average prices with more firms are due to less collusion and more choices of 24 (Fig. 6.4).

A Surprise Phenomenon

As mentioned earlier, we expected the outcome to be the highest equilibrium price. It would seem to be an attractive number and, being an equilibrium, self-enforcing. The collusive outcome yields higher payoffs, but it is provides an incentive

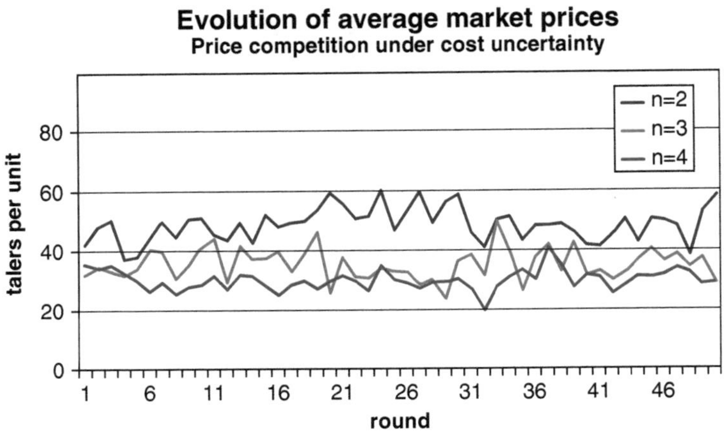

Fig. 6.3

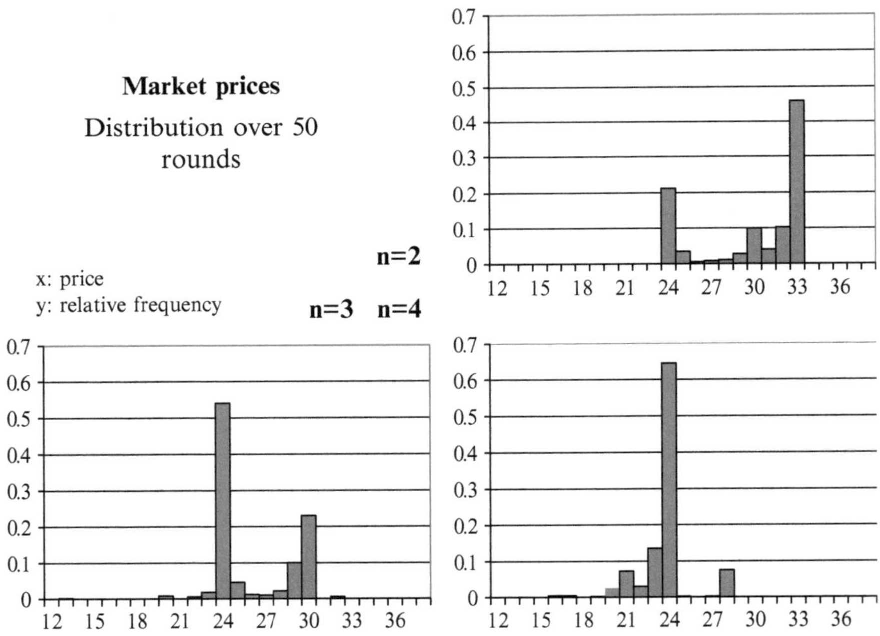

Fig. 6.4

to unilaterally undercut the collusive price. The rewards of collusion, we conjectured, would not be sufficiently high to make up for the lack of self-enforcement. Our expectations turned out to be completely wrong, as Fig. 6.5 shows. Instead, a number we had not expected at all dominates our data: the number 24. This

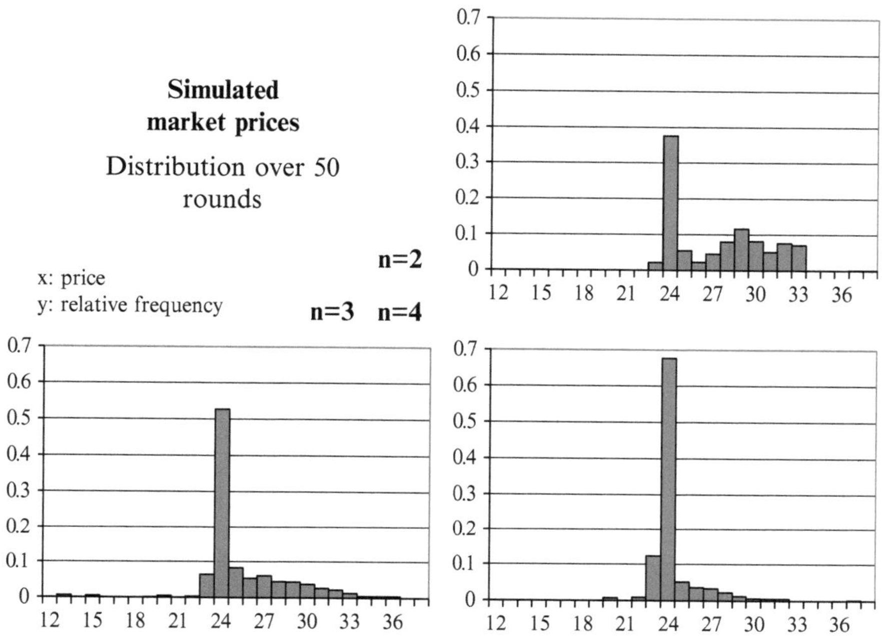

Fig. 6.5

surprised us much because 24 does not correspond to any theoretical benchmark, nor is it a round or prominent number.

What makes the number 24 such a natural choice in the present market environment? It does not match any of the theoretical benchmarks, like the payoff dominant equilibrium, nor is it a round or prominent number. The only apparent speciality of this price – which prevails regardless of the number of firms – is that it is the highest price at which unilateral underbidding is not only disadvantageous compared to equilibrium play, but also unprofitable in absolute terms (Table 6.1).

From a static perspective the feature pointed out in the previous paragraph gives this price a kind of focal quality. One might conjecture that loss-aversion could be the basis for this focality. Numerous individual decision-making experiments have found evidence for loss-averse behaviour (Kahneman and Tversky (1979)). If there is a common notion among players that individuals are loss-averse, then they might expect other players to be especially reluctant to undercut a price of 24. Therefore, at a price level of 24 players are less afraid of being undercut by others. In this way loss aversion may indirectly lead to choices of 24, as this price provides some protection against being undercut. Notice, however, that loss-aversion itself cannot directly explain the phenomenon. First, all equilibria above 24 do not involve the danger of losses either. Second, this theory does not explain why equilibrium prices of 24 and above should be undercut at all, as this would not be a best response. Nevertheless, the "undercut-proofness" of 24 may serve as a natural co-ordination device.

A different possibility is that play follows a kind of dynamic process that ends up at 24. One simple behavioural rule that leads to high degrees of co-ordination is the one of imitating the most successful behaviour in previous rounds. This is easy to apply and, as we will see, may well explain why we observe a price level of 24 so frequently. We propose a simple quantitative dynamic model of imitation in the price competition game to organise our data. In the first round, firms draw prices randomly from the distribution of choices we observe in the first round of the respective treatment. The model assumes that in general, all firms choose the most successful action they observed in the previous round with a high probability $\beta > ½$. With some smaller probability, however, they deviate from this pattern of behaviour and play some other strategy (we refer to this as *experimentation*). We assume that when firms do not imitate, they choose the next higher or next lower price level. It seems reasonable to assume that experimentation takes place locally rather than over the entire range of choices Fig. 6.5 shows the distribution of market prices in simulations we ran with the model, using $\beta = 0.76$ for duopolies, $\beta = 0.86$ for triopolies and $\beta = 0.87$ for tetrapolies. The figures show that the model captures many qualitative features of the data, especially for $n = 3$ and $n = 4$. As in the experimental data a whole range of lower equilibrium prices does not appear in the simulated results. The modal price is 24, with a tendency to slightly higher prices in $n = 3$ than in $n = 4$. Even quantitatively, the frequency of $p = 24$ choices is similar to the observed one for the treatments with more than two firms. Further, in rounds in which the market price is different from 24, these prices tend to be lower with larger n, a phenomenon also apparent in our data (see Fig. 6.3). On the other hand, the model naturally does not capture the collusive behaviour present in our data.

The model results confirm the intuition mentioned earlier. The fact that, as long as the market price is 24 or higher, the most successful choice is the lowest price, leads to a downward trend of the market price. If, however, a single firm has set the market price of 23 or lower, it has made a loss and therefore it will not be imitated. Thus, prices will fall below 24 only if – by experimentation – more than one firm has lowered the price in the same round. With a sufficiently high probability of imitation, however, this happens only occasionally, such that the price of 24 is very stable in the intermediate term.

A New Design

The two explanations for the phenomenon of the price of 24 have naturally occurred ex-post – we did not expect this to happen and therefore had no explanation for why it should happen. We have two competing explanations – focality (static) versus imitation (dynamic) – and, following the tradition of experimental hypothesis testing, we need a new experiment that is suitable to separate the two explanations. So we devised three new treatments, two of which we describe here in order to illustrate the power, but also the dangers of hypothesis testing with the use of artificial treatments.

One conjecture, the focal point hypothesis, refers to the fact that 24 is the number that, when unilaterally undercut, leads to absolute losses. The imitation hypothesis also makes use of this fact, but in a different way: A unilateral undercutter would make a payoff that is relatively lower than the competitors' – a negative payoff instead of zero. We can separate these two hypotheses by shifting all payoffs up or down by a constant, including the payoff obtained by firms not setting the lowest price. The price level that is focal according to the static explanation, is shifted upward or downward depending on whether the constant that is added is positive or

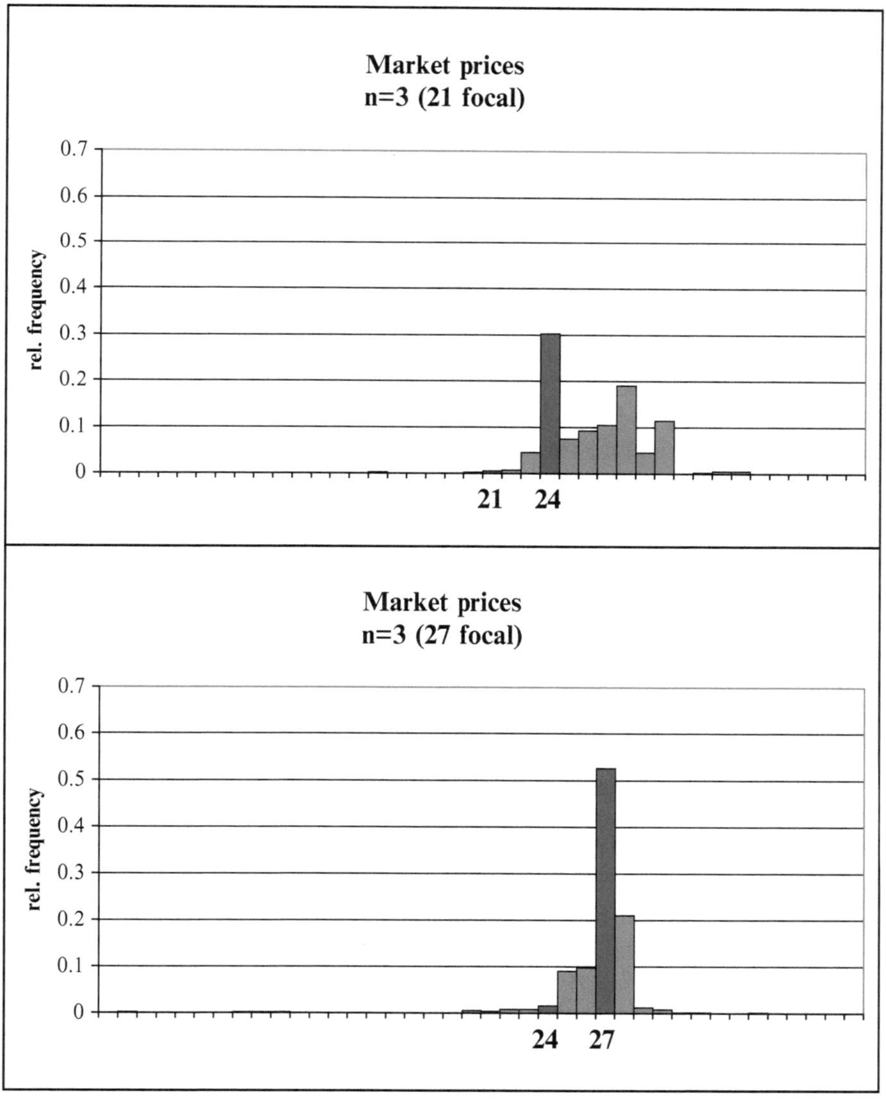

Fig. 6.6

negative, hence the number where unilateral undercutting leads to a negative absolute payoff. The imitation dynamic is invariant to this kind of manipulations, since relative payoff comparisons matter and those are unchanged since all payoffs are shifted.

We conducted two new treatments with such changes in payoffs for triopolies. We did this in ways in that the focal payoff would be 21 or 27 after this manipulation. So if we observe a predominant price of 24, we can interpret this as strong evidence against the focal point hypothesis and for imitation (or, of course, some third explanation we have not fathomed). If we observe these focal numbers to dominate, then we would have to discard our imitation explanation and accept the importance of the focal point. Figure 6.6 shows the results of the new treatments.

Concerning the separation of the two hypotheses, the evidence is not mixed but outright self-contradictory. While the treatment with focal point 27 provides strong evidence for the focal point hypothesis and an equally powerful refutation of the imitation hypothesis, the data from the treatment with focal point 21 allows the exact opposite conclusion: The number 21 plays no role, thus the focal point theory is rejected, but 24 remains the most frequently chosen number. So this is a clear case where the "modern" way of designing experiments to separate hypotheses with highly artificial, customised designs has found its limits. In this sense, our results can be seen as a warning not to put too much trust in the power of hypothesis testing. That we have run both a treatment with upward-shifted and one with downward-shifted numbers is owed to a mere coincidence that no direction of shift seemed natural and it was relatively convenient and cheap to run both treatments. Probably each of the two treatments alone would have been acceptable as a rigorous test of the two hypotheses. Each treatment would have clearly refuted one theory and strongly supported the other. But each conclusion would have misled us – in fact the real picture is more complex.

Exploratory Experiments or Hypothesis Test: Which Is the Way Forward?

Eventually we still can only speculate why the two hypothesis tests have yielded strong, but opposite results. We conjecture that the upward shift of payoffs – towards a focal point of 27 – has made the focal point a very attractive proposition, with high payoffs for all firms. Players would quickly spot his point, find that their aspiration levels are satisfied (aspiration adaptation being another one of Selten's early breakthroughs), and thus stay there. In the other treatment with a downward shift the focal point of 21 involves low, unattractive payoffs. Though players may recognise this point as focal, they are reluctant to choose it and try higher numbers. An imitation dynamic then leads them to coordinate on the price of 24. This explanation is consistent with our observation, but it is – at least according to the philosophy of strict hypothesis testing – still speculative, since we do not have subjected it to a test and frankly would not know how to devise one or what alternative explanation to test it against.

The results from our hypothesis tests show that seemingly clever and sophisticated separation methods are not always as reliable as one would wish. One problem might be that designs like ours are entirely artificial. Shifting all payoffs by a constant is not something that would occur naturally in real markets. This manipulation is solely applied to test two behavioural hypotheses against one another. This approach is likely to yield clear and strong results, but the danger is that the results may be very specific to the experimental setting and not necessarily transfer to real markets with their much richer environment.

The approach taken by Reinhard Selten is completely different. The design becomes very natural, the data very interesting. Analysis becomes an exercise in exploration. It becomes more difficult to clearly identify explanations for the observed behavioural phenomena. But we can expect them to be more robust against idiosyncratic effects as those we have observed in our own experiment, since they stay closer to the real-world environment in question. Our own current view is that one can learn about oligopolistic markets both from complex experiments in the Selten-tradition and from simple experiments like the ones we have recently conducted. Understanding imperfect market competition is not easy and if experimentalists can modestly contribute to it in two different ways this can only help. At the end of the day the two approaches are not mutually exclusive. There are now many researchers doing experimental work, it would be a pity if we all did the same.

Probably due to the strong influence of game-theory on experimental economics and industrial economics, there are now relatively few complex market experiments, some exceptions being Keser (1993) and Nagel and Vriend (1999), but perhaps the future will bring a new wave of more complex experiments, possibly in connection to a surge of interest in the study of complexity. In our view, this would be a good thing.

References

Abbink K, Brandts J (2005) Price competition under uncertainty: a laboratory analysis. Econ Inq 43:636–648

Abbink K, Brandts J (2008) 24. Pricing in Bertrand competition with increasing marginal costs. Games Econ Behav 63:1–31

Abbink K, Brandts J (2009) Collusion in growing and shrinking markets: empirical evidence from experimental duopolies. In: Hinloopen J, Normann H-T (eds) Experiments for antitrust policies. Cambridge University Press, Cambridge, UK

Abbink K, Brandts J, Herrmann B, Orzen H (2010) Inter-group conflict and intra-group punishment in an experimental contest game. Am Econ Rev 100:420–447

Armstrong M, Huck S (2010) Behavioral economics as applied to firms: a primer. working paper, University College London, Jan 2010

Bosman R, Hennig-Schmidt H, Van Winden F (2006) Exploring group decision making in a power-to-take experiment. Exp Econ 9:35–51

Brandts J, Pezanis-Christou P, Schram A (2008) Competition with forward contracts: a laboratory analysis motivated by electricity market design. Econ J 118:192–214
Brown-Kruse J, Rassenti S, Reynolds S, Smith V (1994) Bertrand-Edgeworth competition in experimental markets. Econometrica 62:343–371
Chamberlin EH (1948) An experimental imperfect market. J Polit Econ 56:95–108
Cooper D, Kagel J (2005) Are two heads better than one? Team versus individual play in signaling games. Am Econ Rev 95:477–509
Dastidar KG (1995) On the existence of pure strategy Bertrand equilibrium. Econ Theory 5:19–32
Davis D, Holt C (1994) Market power and mergers in laboratory markets with posted prices. Rand J Econ 25:467–487
Dufwenberg M, Gneezy U (2000) Price competition and market concentration: an experimental study. Int J Game Theory 18:7–22
Feri F, Irlenbusch B, Sutter M (forthcoming) Efficiency gains from team-based coordination – Large scale experimental evidence. Am Econ Rev (in press)
Gilet J, Schram A, Sonnemans J (forthcoming) Cartel formation and pricing: the effect of managerial decision making rules. Int J Indust Organ (in press)
Hansen RG (1988) Auctions with endogenous quantity. Rand J Econ 19:44–58
Hoggatt AC (1959) An experimental business game. Behav Sci 4:192–203
Holt C (1985) An experimental test of the consistent-conjectures hypothesis. Am Econ Rev 75:314–325
Huck S, Normann H-T, Oechssler J (2004) Two are few and four are many: number effects in experimental oligopolies. J Econ Behav Organ 53:435–446
Kahneman D, Tversky A (1979) Prospect theory. Econometrica 47:263–291
Keser C (1993) Some results of experimental duopoly markets with demand inertia. J Ind Econ 41:133–151
Nagel R, Vriend NJ (1999) An experimental study of adaptive behavior in an Oligopolistix market game. J Evol Econ 9:27–65
Plott C, Smith V (1978) An experimental examination of two exchange institutions. Rev Econ Stud 45:133–153
Ricciardi FM (1957) Top management decision simulation. The AMA Approach, New York
Sauermann H, Selten R (1959) Ein oligopolexperiment. Z Gesamte Staatswiss 115:427–471
Selten R (1967) Ein Oligopolexperiment mit Preisvariation und Investition. In: Sauermann H (ed) Beiträge zur experimentellen Wirtschaftsforschung. J. C. B. Mohr (Paul Siebeck), Tübingen, pp 103–135
Spulber DF (1995) Bertrand competition when rivals' costs are unknown. J Ind Econ 43:1–11

Chapter 7
Deviations from Equilibrium in an Experiment on Signaling Games: First Results [1]

Dieter Balkenborg and Saraswati Talloo

Introduction

In this paper we provide a summary of results concerning two series of experiments we ran based on a modified signalling game, which was presented graphically to subjects on a screen. The game for the initial experiment was selected by Reinhard Selten in coordination with the first named author. It has the interesting property that the strategically stable outcome (Kohlberg and Mertens 1986) does not coincide with the outcome of the Harsanyi-Selten solution (Harsanyi and Selten 1988). However, it is a complex game insofar as standard refinement concepts like the intuitive criterion, or the never-a-weak-best-response criterion, do not help to refine among the equilibria. The second motive for the design was to analyse, how the change in the reward at a particular terminal node would affect behaviour.

For the experiments we ran it turned out that the strategically stable equilibrium is never a good description of the data. While behaviour in some of the sessions converged to the Harsanyi-Selten outcome, there were systematic deviations from the equilibrium behaviour. Casual observations and discussions with participants suggested that a "collective reputation" effect might be at work within the random matching framework in which our basic games were played. [2] By this we mean that the subjects in the role of one player would abstain from a certain action which is in their short run interest, but would harm their opponent, in order to allow for

[1]We would like to thank Reinhard Selten for the design of the game used in the initial set of experiments; the staff working at the Bonn Laboratory of Experimental Economics and the Finance and Economics Experimental Laboratory in Exeter (FEELE), for their support while conducting the experiments; and Karim Sadrieh for the organisation of the "Scientific Excursions with Reinhard Selten" which motivated us to conduct the new experiments.

[2]The term is due to Reinhard Selten.

D. Balkenborg (✉) and S. Talloo
Department of Economics, University of Exeter Business School, Streatham Court, Rennes Drive, Exeter EX4 4PU, UK
e-mail: d.g.balkenborg@exeter.ac.uk

A. Ockenfels and A. Sadrieh (eds.), *The Selten School of Behavioral Economics*,
DOI 10.1007/978-3-642-13983-3_7, © Springer-Verlag Berlin Heidelberg 2010

coordination on a mutually beneficial outcome. Moreover, in the experiments we had given subjects information not only on the results of their own plays, but of all plays of the games that were simultaneously conducted. A reduction of this information should make it harder to build up a collective reputation. We hence conducted a new series of experiments where we made the mutually beneficial outcome even more attractive and hence gave a stronger incentive to build a collective reputation, while we varied the information on past outcomes given to subjects. We conjectured that more information would result in more coordination on the mutually beneficial outcome.

We did not find evidence for this hypothesis, but we did find systematic violations from equilibrium behaviour, very similar to those in the initial series of experiments. The purpose of the paper is to describe these. Due to time and space constraints the presentation of these observations must remain somewhat impressionistic; a more thorough quantitative and econometric analysis is planned for the future.

Previous experimental work on signalling games concentrated on the predictive power of refinement concepts (Brandts and Holt 1992, 1993; Banks et al. 1994). For an excellent survey see Camerer 2003. The analysis in these papers concentrated on pure strategy equilibria, but in our case the strategically stable equilibrium is mixed. More recent experiments study how changing a game or deciding in teams affects behaviour in signalling games (Cooper and Kagel 2003, 2005).

In section "Model and Experimental Design", we describe in more detail the extensive form games we are using and their normative solutions, as well as the experimental design. In section "Data Analysis and Results", we describe our findings. We conclude with a brief discussion in section "Discussion and Conclusions".

Model and Experimental Design

The experiments are based on the two signalling games shown in Figs. 7.1 and 7.2. Both signalling games have the following structure: Players always have to choose between a strategically safe and a strategically risky option. The game is one of incomplete information in which Player 1 can be of two possible types, 1a and 1b, which have equal probability. Player 1 chooses first, followed by Player 2. Player 1 can either end the game (the strategically safe option) or give the move to Player 2 (the strategically risky option). Player 2 can then choose between a strategically safe option which gives type independent payoffs and a strategically risky option, which gives type dependent payoffs. Player 1 only wants to use the strategically risky option, when Player 2 does so as well. For Player 2 it depends. The strategically risky option is better only when she faces type 1a. She is better off taking the strategically safe option against type 1b. The two games S and T have in common, that type 1a has a higher gain from both players taking the strategically risky option than type 1b. In the first experiment we varied the payoff "x" of type 1a at the terminal node following the play where both players chose the strategically risky strategy. x could be 4, 5 or 6.

Fig. 7.1 Basic game experiment S (where x = 4, 5, or 6)

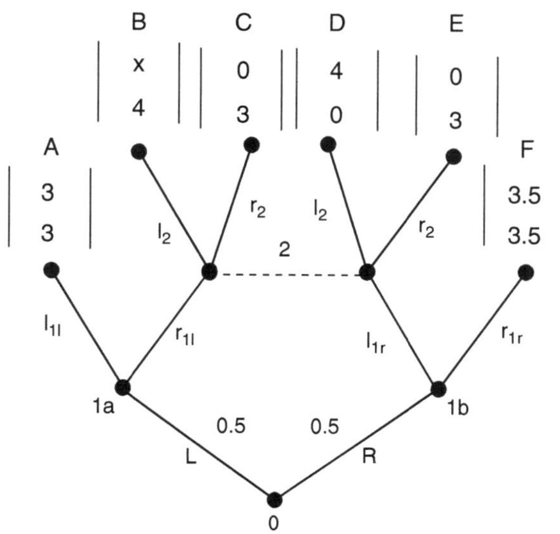

Fig. 7.2 Basic game experiment T

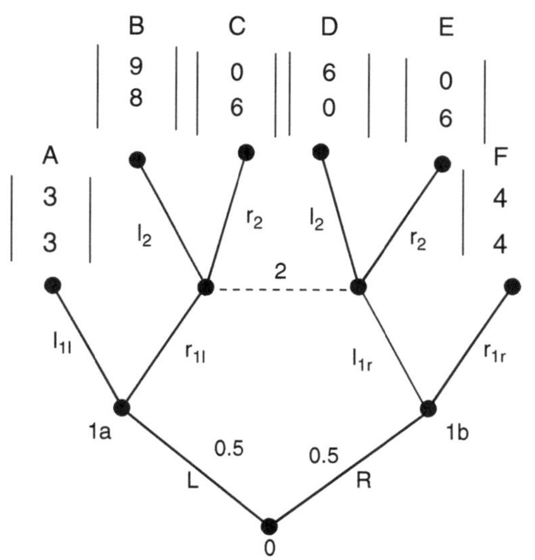

Normative Analysis

We work exclusively with behaviour strategies. Both games have two Nash equilibrium components. The first component consists of Nash equilibria where both types of Player 1 take the strategically safe option and Player 2 chooses the strategically safe option with a sufficiently high probability, namely at least with probability (x−3)/x in Game S (with x = 4, 5, 6) and at least probability 2/3 in Game T. This component contains, in particular, the Nash equilibrium where all players and types take their strategically safe option with certainty. The latter is uniformly stable and can be shown to be the equilibrium selected by the theory of Harsanyi and Selten (1988).

The second component consists of a single equilibrium where Player 1a takes the strategically risky option with certainty, while type Ib and Player 2 randomise. Namely, in Game S type 1b chooses the strategically safe option with probability 2/3 and Player 2 chooses the strategically safe option with probability 1/8. In Game T type 1b chooses the strategically safe option with probability 2/3 and Player 2 chooses the strategically safe option with probability 1/3.[3] Conditional on her information set being reached, Player 2 believes to face type 1b with probability 1/4. This equilibrium component can be shown to be the only strategically stable component of Nash equilibria in the sense of Kohlberg and Mertens (1986).

The purpose of the first set of experiments (Game S) was to test the two equilibrium refinements against each other. We expected the Harsanyi-Selten solution to arise for the parameter value x = 4, but did not rule out that terminal node B would be reached more often if x was increased.

In the new version (Game T), we made it more attractive to choose the strategically risky choice, but we made it more attractive for type 1a than for type 1b. We hence expected that Player 2's information set would be reached substantially more often in the new experiment.

The Extended Games

For most part of the experiments we used the extended models, S′ and T′ (See Figs. 7.3 and 7.4), which modified the basic games, S and T, as described above.

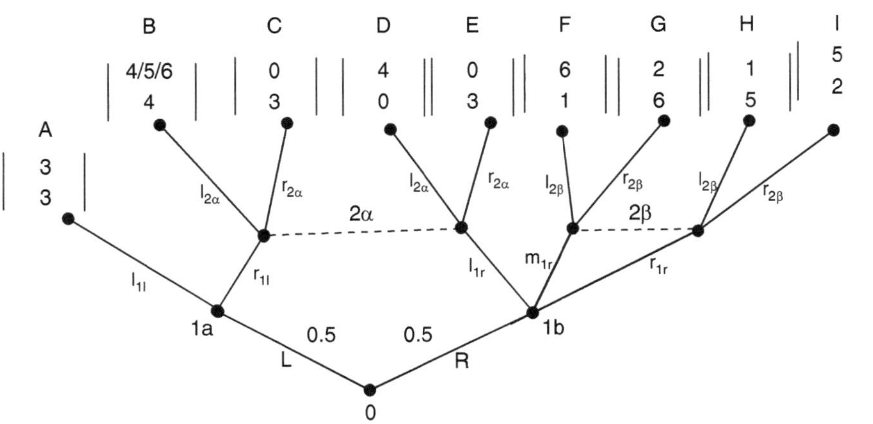

Fig. 7.3 The game S′

[3]Detailed proofs are available on request.

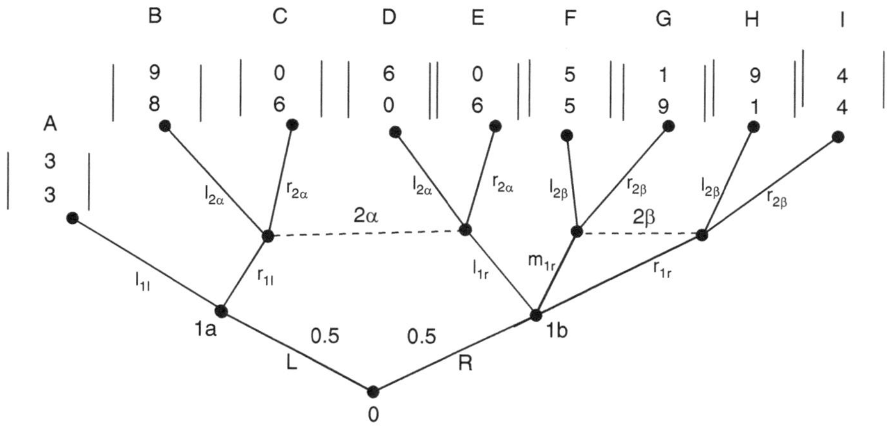

Fig. 7.4 The game T′

Table 7.1 Probabilities in the Nash equilibria

	r_{11}	m_{1r}	r_{1r}	$r_{2\alpha}$	$r_{2\beta}$
1st component Game S′	0	3/8	5/8	$\geq(x-3)/x$	5/8
2nd component Game S′	1	3/12	5/12	1/8	5/8
1st component Game T′	0	0	1	$\geq 2/3$	1
2nd component Game T′	1	0	2/3	1/3	1

In essence type 1b's strategically safe option was replaced with a 2×2 game with a unique equilibrium, having the same expected payoffs as the strategically safe option in the basic game.[4] In the game S′ for the first set of experiments, we used a game with unique mixed strategy equilibrium, where both players have to choose their right strategy with probability 5/8. In the game T′ for the second experiment, we used a prisoner's dilemma game, where the right strategy of both players was the dominant strategy. Since in all Nash equilibria of the basic game S and T the strategically safe choice of type 1b is always reached with positive probability, the Nash equilibria of the extended games are obtained by replacing the strategically safe strategy of type 1b with this equilibrium strategy in the 2×2 game and amend player 2's behaviour strategy with her choice at the new information set. See Table 7.1.

Experimental Design

The games above were used in two series of experiments, one conducted at the Bonn Laboratory of Experimental Economics,[5] and one at the Finance and Economics

[4]The 2×2 game was added following the strategically safe choice of type 1b, but then moves were coalesced.

[5]These experiments were designed and conducted under the supervision of Reinhard Selten.

Experimental Laboratory in Exeter (FEELE). The extensive games were shown graphically to the subjects on the computer screen. The subjects decided by highlighting the appropriate choice in the extensive form on the screen. Throughout games were repeated in a uniform random matching environment with 6 subjects in the role of player 1 and 6 subjects in the role of player 2. Subjects remained in the same role as long as the game was not changed.

In the first set of experiments conducted in Bonn, the game S′ was used in 9 sessions. For each value $x = 4, 5, 6$, three sessions were conducted. After the initial random allocation of roles, subjects played in the main part of the experiment the game S′, in strictly sequential order, 50 times. There was a short break after period 25. In the final part of the experiment consisting of 5 rounds, called the Tournament, subjects had to submit strategies for the extensive game. Each strategy of a player was then evaluated against all the strategies of all the players in the opposite role, and would receive the average payoff. Thus, we used the strategy method (Selten 1967), where subjects first learn to play the game sequentially and in the final part submit complete strategies.

In the second set of experiments conducted in Exeter, subjects played the simpler game T in the first 25 periods, and then switched to the more complex game T′ which was played sequentially for the next 25 rounds and the final 5 Tournament rounds. Due to a restriction of the computer software, roles had to be reallocated after Round 25. We stayed as close as possible to the design of the first set of experiments. However, we wanted to see whether it affects the results if subjects were only given the results of the play of their own game, or also the outcomes of the other 5 parallel plays occurring in each period (or respectively, of the other 30 parallel plays in case of the Tournament rounds). In the first set of experiments we had always given information on all plays to the subjects. In the second set of experiments we gave this full information only in five of the ten sessions conducted.[6]

The experimental sessions in Bonn lasted about 3½ hours and about 2½ hours in Exeter. Average payment per subject was about £12 in Exeter.

Data Analysis and Results

Summary Results

- *Player 1 Behaviour at Information Set 1a*

We evaluate how often the strategically safe option was taken by Player 1 at information set 1a. For the old set of experiments, a Mann-Whitney U-test shows that the strategically safe option is taken significantly more often in Rounds 1–50 in

[6]However, we did not see any indication that this difference of information mattered.

the sessions where Player 1's payoff at terminal node B was 4 or 5, as compared to when it was 6. Moreover, in each session of the game where the payoff was 6, Player 1 takes the safe option less often at information set 1a in Rounds 26–50, compared to Rounds 1–25.

As found in many experiments where players have an outside option there is a substantial fraction of subjects who take it (see e.g. Cooper et al. 1990). We see here for Rounds 1–50 that the strategically safe option is taken in at least 15% of the plays. We only find one session (with payoff 6) where the strategically safe option is practically not taken in Rounds 26–50 and Rounds 51–55 of the experiment. In only one session (with payoff 4) the strategically safe option is almost always taken (in 94% of plays), for all the others it ranges between 15–67%. The results are qualitatively the same for the tournament periods 51–55. Thus, behaviour is overall not consistent with either of the two Nash equilibrium components of the game where the strategically safe option is taken either with probability 0 or with probability 1 (Table 7.2).

Table 7.2 How often the strategically safe option is taken at Information Set 1a[7]

	Old experiments		New experiments	
	Avg	Std dev	Avg	Std dev
Rounds 1–25	0.515437	(0.26552)	0.208411	(0.1274)
Rounds 26–50	0.43788	(0.28489)	0.265718	(0.2042)
Rounds 51–55	0.517491	(0.33409)	0.19606	(0.15302)

Comparing the old and new experiments, the strategically safe option is foregone significantly more often in the new experiments, as we expected. This is significant by a Mann-Whitney U test conducted separately for Rounds 1–50 and tournament periods 51–55.[8] In the set of new experiments the percentage with which the strategically safe option is taken, is below 50% in each session, and separately for Rounds 1–25, 26–50 and tournament periods 51–55, with just one exception for period 26–50 (always significant by a sign test).

Arguably in the new set of experiments, Player 1 did not take the strategically risky option often enough at his information set 1a. Given the observed frequencies with which Player 2 chose her strategically risky option at information set 2α, he would have made a gain in each session. More precisely, (9*B%) – 3 is positive for each session, where for a given session and Rounds 1–50 or Rounds 51–55, B% is (number of times B is reached)/(number of times B or C are reached). In contrast, this "gain" varies considerably for the sessions in the old experiment.

[7]We calculated the average and standard deviation of the percentage of times Player 1 chose left at information set 1a, relative to the number of times this information set was reached for each session. The following tables are calculated in a similar manner.

[8]The test is highly significant for Rounds 1–25, but not for Rounds 26–50. Thus, the original stronger incentive for Player 1 to take the strategically risky option gets somewhat dampened by experience.

However, players are not simply irrational. The number of times the strategically risky option is taken is highly correlated with the gain to be made. This is significant at a 5% level of significance, using Spearman's Rank Correlation Test for both the new and old experiment sessions.[9]

- *Player 1 Behaviour at information set 1b*

In both the old and the new sets of experiments, Player 1 chose the left option at information set 1b significantly less often than the 33% predicted by the strategically stable (Kohlberg-Mertens) Nash equilibrium. This holds by a sign test for each session, in each part of the experiment (Rounds 1–25, 26–50 and 50–55) separately. In the new set of experiments game T′ was used in Rounds 26–55. Here, Player 1 chose right at information set 1b significantly more often than both left and middle, in each part of the experiment (Rounds 26–50 and 50–55) separately (Table 7.3).

Table 7.3 How often the left option is taken at Information Set 1b

	Old experiments		New experiments	
	Avg	Std dev	Avg	Std dev
Rounds 1–25	0.055979	(0.04221)	0.220375	(0.09273)
Rounds 26–50	0.049949	(0.03563)	0.103184	(0.06487)
Rounds 51–55	0.056884	(0.05144)	0.098769	(0.08681)

In the old set of experiments, left was never chosen in more than 12% of the cases for both Rounds 1–25 and 26–50, and 16% for the tournament periods 51–55. For the new experiments, the corresponding percentages of choosing left are 40, 22 and 27% for Rounds 1–25, 26–50, 51–55, respectively. We calculated various proxies for the gains Player 1 could have made at information set 1b by going left rather than right. These gains were sometimes positive and sometimes negative, varying greatly from session to session. We never found any significant correlation between the percentages of times Player 1b chose left and the gains. Subjects simply seemed to be reluctant to take the left option, which would be consistent with the aim to build up a collective reputation.

- *Observed frequency for Information Set 2α*

Since Player 1 rarely chooses his left option at information set 1b, the relative frequency with which the right node in information set 2α is reached is significantly below 25%. A sign test shows this for Rounds 1–25, 26–50 and 51–55 in the sessions of the old experiment and Rounds 26–50 and 51–55 in the sessions of the new experiment. This is consistent with a collective reputation effect and it would hence be the best for Player 2, in all these cases, to select the strategically

[9]Except for Rounds 51–55 in the new set of experiments which just misses the 5% level of significance.

risky option. Even for Rounds 1–25 in the sessions of the new experiment, the percentages are close to 25% or below (Table 7.4).

Table 7.4 How often the right node is reached at Information Set 2

	Old experiments		New experiments	
	Avg	Std dev	Avg	Std dev
Rounds 1–25	0.108521	(0.06229)	0.213985	(0.0844)
Rounds 26–50	0.082285	(0.06245)	0.126246	(0.08414)
Rounds 51–55	0.118831	(0.0875)	0.094254	(0.08086)

- *Player 2 Behaviour at Information Set 2α*

In the new set of experiments, Player 2 chooses the strategically safe option significantly more often than 33 1/3%, which is the maximal probability in a Nash equilibrium where information set 2α is never reached. This is significant for Rounds 1–25 and 51–55 by a Sign test, for Rounds 26–50 it still holds if we use a Wilcoxon Rank Test, based on the percentage of times left is taken minus 1/3. On average, the strategically risky option is taken in 60% of the cases, well below the 66 2/3% required by the Kohlberg-Mertens strategically stable Nash equilibrium. Given these averages, it makes sense for Player 1 to choose right at both information sets 1a and 1b, which is roughly consistent with actual behaviour. However, we are working here with very crude averages. In some of these sessions the percentage of Player 2 choosing left is well above 66 2/3% and so Player 1 would have an incentive to choose left at information set 1b (Table 7.5).

Table 7.5 How often the strategically risky option is taken at Information Set 2

	Old experiments		New experiments	
	Avg	Std dev	Avg	Std dev
Rounds 1–25	0.519377	(0.29827)	0.601518	(0.09404)
Rounds 26–50	0.653417	(0.1657)	0.577839	(0.20669)
Rounds 51–55	0.631131	(0.20074)	0.580699	(0.11262)

In the old experiments, the information set 2α is reached substantially less often. One may count the first session as one where the Harsanyi-Selten solution is played, because information set 2α is only reached 5 times in Rounds 1–25 and 26–50, and never in the final part. Disregarding this session, it is significant by a sign test that the strategically risky option is chosen in at least 33 1/3% of the cases. However, the percentages are significantly below the 87.5% required by strategic stability (there is only one exception with 93% in Rounds 1–25, and one with 89.5% in Rounds 51–55).

We wanted to see whether Player 2 at information set 2α responded to her experience and did not choose just randomly. For the new set of experiments we hence checked whether a player changed her behaviour more often after a "failure" than after a "success".[10] There are two ways in which Player 2 could make a failure, she may choose right and the left node of information set 2α is reached, resulting in a payoff of 6 instead of 8; or she may go left and the right node of information set 2α is reached, resulting in a payoff of 0 instead of 6.

We counted for each subject how often they switched after a failure or a success in Rounds 1–25 and 26–50, if information set 2α was reached. We took the difference of the 2 two percentages. We disregarded individuals for whom the difference was zero, or for whom the information set was never reached. In Rounds 1–25, we found overall 36 individuals who switched more often after failure, 14 who switched more often after success, and 6 who switched equally often. For Rounds 26–50, the corresponding numbers were 34, 8 and 7. Sign tests based on these numbers would indicate that most subjects switch more often after failure than success.

Looking at individual sessions, in Rounds 1–25, we found for eight sessions that the difference was positive for a majority of the subjects. In the other two sessions the number of subjects with a positive difference was equal to the number of subjects with a negative difference. A sign test then indicated that most subjects switch more often after failure than after success. However, with a corresponding analysis for Rounds 26–50 we just miss a significant result.

- *Player Behaviour in the embedded 2 × 2 game*

In the new experiments, we embedded a Prisoner's Dilemma type of 2 × 2 game into the signalling game in Rounds 26–50 and 51–55. As was to be expected, both players choose their strictly dominated action much less often than the undominated action. However, the percentages with which the dominated action is chosen are not negligible and can be as high as 23% in Rounds 26–50 and 32% in Rounds 51–55 in individual sessions.[11] Averaged over all sessions, Player 1 chooses the dominated action more often than Player 2, but a Wilcoxon Rank test does not yield significant results. For Rounds 26–50, the percentages of choices of dominated actions of Player 1 and Player 2 are positively correlated (correlation coefficient = 0.34), but negatively correlated in the tournament periods 51–55 (correlation coefficient = −0.25). It is interesting to observe that there is no significant difference between the number of times Player 1 chose left and the number of times he chose the dominated action middle at information set 1b (Table 7.6).

Table 7.6 How the right option is chosen in the 2 × 2 game in the New Experiment

	Player 1		Player 2	
	Avg	Std dev	Avg	Std dev
Rounds 26–50	0.899071	(0.08269)	0.937936	(0.0406)
Rounds 51–55	0.866204	(0.11283)	0.921962	(0.06393)

[10]We disregarded the sessions of the old experiments since information set 2α was not reached often enough.

[11]The fractions are calculated relative to the number of times Nature chose right and Player 1 did not choose left.

Table 7.7 How the right option is chosen in the 2 × 2 game in the Old Experiment

	Player 1		Player 2	
	Avg	Std dev	Avg	Std dev
Rounds 1–25	0.530085	(0.09553)	0.713404	(0.08242)
Rounds 26–50	0.578689	(0.06828)	0.663489	(0.07171)
Rounds 51–55	0.525137	(0.13771)	0.708392	(0.10936)

In the set of old experiments, we embedded a 2 × 2 game with a unique mixed-strategy Nash equilibria into the signalling game S, and used the resulting game S′ in all the rounds. The percentages of strategy choices in the nine sessions are roughly comparable with the mixed-strategy equilibrium, but as in many experiments with such 2 × 2 games (see for instance, Selten and Chmura 2008 and the literature they cite), one has strong own-payoff effects. For the main part of the experiment, periods 1–50, the percentages with which right is chosen are significantly below the equilibrium values for Player 1 and above for Player 2 (by a sign test). This finding is consistent with the predictions made by the alternative solution concepts for such 2 × 2 games in Selten and Chmura 2008 (Table 7.7).[12]

Discussion and Conclusions

We conducted an experiment on games in extensive form which had a signalling game with a 2 × 2 game embedded as its basis. The initial set of experiments was conducted with the aim of testing which of two competing theories of equilibrium selection or refinement theories better described behaviour. What we found is that there are systematic deviations from both types of theories. The second set of experiments was conducted in order to test for the appearance of a collective reputation. No statistically significant evidence was found. However, a number of interesting observations about the behaviour in these games could be made.

- At information set 1a, there tends to be a significant percentage of players who take the strategically safe option even if it would pay well to forego it. This is as observed in other games with outside options and could simply be explained by risk avoidance or low aspiration levels. Otherwise behaviour is fairly consistent with payoff maximization, actions yielding higher average payoffs are taken more often.
- Behaviour in the embedded 2 × 2 games is similar to the experimental results found when similar 2 × 2 games are played in isolation. In the Prisoner's Dilemma type game, the undominated actions were primarily chosen. However,

[12]In the impulse balance equilibrium, right is chosen with probability ½ by Player 1 and with probability 2/3 by Player 2. In the action sampling equilibrium, right is chosen with probability 0.56 by Player 1 and with probability 0.66 by Player 2.

the percentages of times in which the dominated action was chosen were not negligible. For the game with the unique mixed-strategy Nash equilibrium observed frequencies tended to be in a 10% range around the equilibrium. Deviations from the Nash equilibrium tended to be in the direction of behavioural concepts which adjust for the own-payoff effect as, for instance, impulse balance equilibrium or action sampling equilibrium (Selten and Chmura 2008).

- Throughout all sessions of the experiment, the action left at information set 1b is rarely chosen. This holds quite independently of the payoffs and how players behave at the other information sets. In the new set of experiments, it is taken roughly as often as the dominated action middle is chosen. Part of the explanation is presumably that the potential gains from taking this action are not very high. In fact, in many sessions it wouldn't pay given the behaviour of Player 2, but even when it would pay, the action left is not taken. Perhaps the subjects in the role of Player 1 are trying to build up collectively the reputation not to take this action, in order not to destroy a cooperation which leads to outcome B, when information set 1a is reached. Alternatively, the potential threat of a payoff 0 may deter Player 1 from taking the action. Both middle and right always yield non-negative profits. Right guarantees the payoff of 4 in the new set of experiments, while middle secures a payoff 2 in the old experiments.
- The percentage of times with which the right decision node is reached when play reaches information set 2α is systematically below ¼. Thus it would maximise Player 2's payoff if she chose her strategically risky choice.
- At information set 2α, left is typically chosen in at least 1/3 of the cases. Often this percentage is much higher although rarely above 2/3. We found some evidence for learning at this information set, but it isn't strong. Perhaps there is a substantial fraction of subjects who do not understand the strategic situation very well and choose both actions equally often, for instance, by always taking the highlighted choice randomly selected by the computer.[13] Such subjects would bias observed frequencies towards 50–50, and the behaviour of the other players may not fully compensate for this "irrationality".

An equilibrium concept that would lead to some of the behaviour as just described, is the notion of a cursed equilibrium (Eyster and Rabin 2005). This concept is a modification of the Nash equilibrium where players do not correctly Bayesian update based on the information they receive. Specifically for our games S and T, Player 2 would not correctly update the probability with which she is facing type 1a or 1b at information set 2α.

In the fully cursed equilibrium, Player 2 would believe at information set 2α that she is facing both types of Player 1 with equal probability. In a partially cursed equilibrium, her belief would be a weighted average of the former belief and the belief inferred by correctly Bayesian updating. If equal weight is put on both beliefs

[13] Our programme initially highlights each choice at the relevant information set with equal probability. Subject can then change which choice is highlighted with the left and right cursor keys. Once the desired choice is highlighted, subjects decide on it by pressing the Enter key.

there exists a partially cursed equilibrium where Player 1 always chooses right at both his information sets. The correct belief would then assign probability 1 to Player 2 facing type 1a. In the weighted average Player 2 would assign probability $\frac{3}{4}$ on facing type 1a, and thus she would be indifferent between her two options. It would hence be optimal for her to choose her strategically safe option with a probability between 1/3 and 2/3, thus making the choices of Player 1 rational. Thus we can construct a partially cursed equilibrium which fits better with most of our data than any of the Nash equilibria, but only for a highly specific and artificial weighting of beliefs.

A different approach we would like to explore in the future would be uncertainty aversion (See for instance, Eichberger, Kelsey 1996). Recall that it is optimal for Player 1 to choose right in Games S and T at both his information sets, if Player 2 chooses right with a probability between (x−3)/x for game S (x = 4, 5, 6) or 1/3 for game T and 2/3. The paper by Kelsey and Eichberger may explain why Player 2 may prefer to randomise in this way, even though she is not indifferent. Moreover, players without uncertainty aversion would choose left, and those with a high degree of uncertainty aversion would choose right, given the behaviour of Player 1, which could potentially explain the wide diversity of individual behaviour. Also, highly uncertain Player 1s would always choose the strategically safe option. How this and other models of bounded rational behaviour explain our findings, must be left for future research.

Appendix

Old Experiment, Terminal Nodes Reached in Rounds 1–25[a]

	A	B	C	D	E	F	G	H	I
1 (x = 4)	75	0	4	0	1	11	22	10	27
2 (x = 4)	15	59	5	8	0	12	20	6	25
3 (x = 4)	30	44	8	4	1	7	19	9	28
4 (x = 5)	66	3	10	0	2	9	24	17	19
5 (x = 5)	46	13	14	1	1	10	22	11	32
6 (x = 5)	53	9	20	3	3	10	19	11	22
7 (x = 6)	21	37	21	1	0	4	16	22	28
8 (x = 6)	38	26	18	6	2	4	34	2	20
9 (x = 6)	24	37	21	0	2	9	27	9	21

[a]"x" denotes the payoff of Player 1 at terminal node B

Old Experiment, Terminal Nodes Reached in Rounds 26–50

	A	B	C	D	E	F	G	H	I
1 (x = 4)	72	2	3	0	0	13	26	10	24
2 (x = 4)	51	13	13	4	2	2	21	17	27
3 (x = 4)	27	32	14	3	1	13	15	19	26
4 (x = 5)	32	38	7	1	0	13	16	13	30
5 (x = 5)	28	34	14	3	0	8	24	7	32

(continued)

6 (x = 5)	50	10	13	1	3	15	17	10	31
7 (x = 6)	3	54	20	1	1	11	24	18	18
8 (x = 6)	27	33	13	5	0	9	14	19	30
9 (x = 6)	7	55	11	5	4	8	21	11	28

Old Experiment, Terminal Nodes Reached in Rounds 51–55

	A	B	C	D	E	F	G	H	I
1 (x = 4)	91	0	0	0	0	10	26	15	38
2 (x = 4)	72	16	3	4	2	10	51	6	16
3 (x = 4)	69	12	14	3	1	22	25	20	14
4 (x = 5)	15	68	8	0	0	16	26	17	30
5 (x = 5)	50	31	11	1	0	5	20	12	50
6 (x = 5)	62	13	20	2	5	13	20	13	32
7 (x = 6)	0	72	19	3	1	3	24	12	46
8 (x = 6)	58	17	20	1	8	8	32	10	26
9 (x = 6)	16	44	35	7	6	13	25	10	24

New Experiment, Terminal Nodes Reached in Rounds 1–25[a]

	A	B	C	D	E	F	G	H	I
1	17	45	15	7	5	61	0	0	0
2	4	53	27	7	2	57	0	0	0
3	10	32	28	15	7	58	0	0	0
4	26	16	24	7	8	69	0	0	0
5	4	43	25	6	5	67	0	0	0
6*	28	31	23	5	5	58	0	0	0
7*	15	37	17	17	5	59	0	0	0
8*	14	40	26	3	8	59	0	0	0
9*	9	48	25	15	8	45	0	0	0
10*	30	23	26	17	11	43	0	0	0
$\sum$	157	368	236	99	64	576	0	0	0

[a]The * refers to experimental sessions where information on all simultaneous plays was not given to the subjects

New Experiment, Terminal Nodes Reached in Rounds 26–50

	A	B	C	D	E	F	G	H	I
1	2	49	14	5	1	2	16	6	55
2	1	61	12	13	3	0	2	2	56
3	24	29	34	2	0	2	3	5	51
4	33	21	26	0	1	0	1	4	64
5	17	38	23	6	6	0	14	2	44
6*	35	24	17	8	2	1	9	7	47
7*	36	7	33	4	5	0	2	1	62
8*	10	50	19	2	2	0	1	3	63
9*	5	48	16	5	0	1	7	5	63
10*	43	9	23	5	7	0	7	1	55
$\sum$	206	336	217	50	27	6	62	36	560

New Experiment, Terminal Nodes Reached in Rounds 51–55

	A	B	C	D	E	F	G	H	I
1	7	59	23	11	3	2	17	1	57
2	0	64	32	10	6	0	2	3	63
3	19	55	20	0	0	1	9	19	57
4	31	22	27	0	0	0	5	6	89
5	16	38	35	7	2	0	26	2	54
6*	39	35	30	2	5	0	15	3	51
7*	25	24	27	0	0	0	0	7	97
8*	7	45	41	4	3	0	2	11	67
9*	0	59	25	7	4	1	7	2	75
10*	32	28	36	13	9	0	15	6	41
$\sum$	176	429	296	54	32	4	98	60	651

References

Banks J, Camerer C, Porter D (1994) An experimental analysis of Nash refinements in signalling games. Games Econ Behav 6:1–31

Brandts J, Holt CA (1992) An experimental test of equilibrium dominance in signalling games. Am Econ Rev 82(5):1350–1365

Brandts J, Holt C (1993) Adjustment patterns and equilibrium selection in experimental signalling games. Int J Game Theory 22(3):279–302

Camerer CF (2003) Behavioral game theory: experiments in strategic interaction. Princeton University Press, Princeton, NJ, pp 408–473

Cooper DJ, Kagel JH (2003) The impact of meaningful context on strategic play in signalling games. J Econ Behav Organ 50:311–337

Cooper DJ, Kagel JH (2005) Are two head better than one? Team versus individual play in signalling games. Am Econ Rev 95:477–509

Cooper R, DeJong D, Forsythe B, Ross T (1990) Selection criteria and coordination games: some experimental results. Am Econ Rev 80:218–233

Eichberger J, Kelsey D (1996) Uncertainty aversion and preference for randomisation. J Econ Theory 71:31–43

Eyster E, Rabin M (2005) Cursed equilibrium. Econometrica 73(5):1623–1672

Harsanyi JC, Selten R (1988) A general theory of equilibrium selection in games. MIT Press, Cambridge, MA

Kohlberg E, Mertens J-F (1986) On the strategic stability of equilibria. Econometrica 54:1003–1039

Selten R (1967) Die Strategiemethode zur Erforschung des eingeschränkt rationalen Verhaltens im Rahmen eines Oligopolexperiments. In: Sauermann H (ed) Beiträge zur experimentellen Wirtschaftsforschung. J. C. B. Mohr (Paul Siebeck), Tübingen, pp 136–168

Selten R, Chmura T (2008) The American Economic Review 98(3): pp 938–966

Chapter 8
Strategy Choice and Network Effects

Siegfried K. Berninghaus, Claudia Keser, and Bodo Vogt

Social networks play an important role in life. We interact with our friends, neighbors and business partners, who in turn also interact with people who are not part of the community we directly interact with. How people form and maintain networks and how networks impact their behaviors raises behavioral questions that have been addressed by sociologists, economists, physicists, computer scientists and anthropologists.

In economics, the assumption of local interaction plays an important role in theoretical models of evolutionary learning. To analyze how conventions evolve in a population, models have been presented that apply the theory of evolutionary games (e.g., Kandori et al. 1993; Young 1993). In this framework members of a large population interact on the basis of pairwise random matching in simple coordination games. They are assumed to be boundedly rational in that they play best reply but deviate from this rule with a small probability. The assumption of pairwise random matching might be an appropriate assumption in a biological context, but in a socioeconomic context it seems less relevant. Anderlini and Ianni (1993), Ellison (1993), Blume (1993) and Berninghaus and Schwalbe (1996) consider local interaction structures, where each member of a population is assumed to be matched with a selected group of population members, called his reference group or neighborhood. These reference groups are supposed to be overlapping such that no group of the population can split off from the rest of the

S.K. Berninghaus
Karlsruher Institut für Technologie (KIT), Institut für Wirtschaftstheorie und Statistik, Lehrstuhl für Wirtschaftstheorie (VWL III), Postfach 69 80, D-76128 Karlsruhe, Germany

C. Keser (✉)
Claudia Keser Georg-August-Universität Göttingen, Platz der Göttinger Sieben 3, 37073 Göttingen Magdeburg, Germany
e-mail: claudia.keser@uni-goettingen.de

B. Vogt
Fakultät für Wirtschaftswissenschaft, Lehrstuhl für Empirische Wirtschaftsforschung, Otto-von-Guericke-Universität Magdeburg, Postfach 4120, 39016 Magdeburg, Germany

A. Ockenfels and A. Sadrieh (eds.), *The Selten School of Behavioral Economics*,
DOI 10.1007/978-3-642-13983-3_8,

population. As a consequence, each player interacts directly with a selected group only but indirectly with the whole population.

In this paper we survey our own work toward the understanding of the role of local interaction structures for coordination. We proceed in three steps to address the following questions. In the first step (Section "Coordination on Exogenous Networks") we examine the impact of exogenously given network structures on equilibrium selection in simple coordination games (Keser et al. (1998), Berninghaus et al. (2002)). In the second step (Section "Strategic Network Formation") we investigate strategic network formation: what kind of networks will be formed if connecting with other players implies costs (Berninghaus et al. 2006, 2007). In the third step (Section "Strategy Choice on Endogenous Networks") we combine both aspects, investigating equilibrium selection on endogenous networks (Berninghaus and Vogt 2004, 2006).

Coordination on Exogenous Networks

Keser et al. (1998) and Berninghaus et al. (2002) examine in laboratory experiments various aspects of equilibrium selection related neighborhood structures. These aspects include closed neighborhoods (where all players directly interact with each other) versus open neighborhoods (where players directly interact with their neighbors only and where neighborhoods are overlapping), neighborhood size, and neighborhood structure.

More specifically, the baseline game used in these experiments is a symmetric two-player coordination game, with two strict equilibria in pure strategies. One of the equilibria is payoff-dominant while the other is risk-dominant. Based on this game, we construct so-called neighborhood games in which each player plays the baseline game with a single strategy with several other players. An interaction structure determines for a population of players with which other players each player interacts. We consider closed neighborhoods and open neighborhoods with local interaction either around a circle or on a lattice.

The Two-Player Coordination Game

The baseline game G is a symmetric normal-form game with two players $i = 1, 2$. Each player i chooses a strategy $\sigma_i \in \{X, Y\}$. The payoff function $H_i(\sigma_i, \sigma_j)$ is illustrated in Table 8.1 with $d > a$, $a > c$, $d > b$, $a - c > d - b$. This game has two strict Nash-equilibria in pure strategies, (X, X) and (Y, Y). The

Table 8.1 The two-player coordination

	X	Y
X	a, a	b, c
Y	c, b	d, d

(*Y*, *Y*)-equilibrium is payoff-dominant as both players get a higher payoff than in the (*X*, *X*)-equilibrium. Furthermore, there exists an equilibrium in mixed strategies in which each player chooses *X* with probability $p* = \frac{(d-b)}{(a-c+d-b)}$. The (*X*, *X*)-equilibrium satisfies Harsanyi and Selten's (1988) criterion of risk dominance, as (a − c) > (d − b). This implies that $p^* < 1/2$.

Neighborhood Games

From the baseline two-player coordination game *G*, we derive "neighborhood" games in which each player out of a population of N ($N \geq 2$) players plays the game *G* with a single strategy with n_i ($1 \leq n_i \leq N - 1$) other players (his *neighbors*). Let I be the population of players, with $|I| = N < \infty$. Furthermore, let $\{N_i\}_{i \in I}$ be a given interaction or neighborhood structure with $|N_i| = n_i$, $N_i \subset I$, $j \in N_i \Rightarrow i \in N_j$, and $i \in N_i$. The interaction structure determines the structure of strategic interdependence *Neighborhood games.* A player *i*'s payoff function $\pi_i(\sigma_i, \{\sigma_j\}_{j \in N_i})$ in the neighborhood game is a function of $\{H_i(\sigma_i, \sigma_j)\}_{j \in N_i}$, the set of payoffs that player *j* realizes in his n_i plays of the game *G*.

We consider the payoff function with

$$\pi_i(\sigma_1, \ldots, \sigma_n) = \left(\sum_{j \in N_i} H_i(\sigma_i, \sigma_j) \right) / n_i.$$

To analyze Nash equilibria in a neighborhood game we require that each player satisfies the local best reply assumption: each player chooses the strategy which maximizes his payoff $\pi_i(\cdot)$, given the distribution of choices in his neighborhood.

Result 1. In neighborhood games based on game *G*, whatever *N*, the size of the population of players, and whatever the interaction structure, $\{N_i\}_{i \in I}$, two equilibria in pure strategies exist: in the (all *X*)-equilibrium all players of the population choose strategy *X*, while in the (all *Y*)-equilibrium all players of the population choose strategy *Y*. These equilibria need not be the unique ones; other equilibria might exist.

Closed Neighborhoods: In a closed neighborhood each player interacts with each of the other players in the population. Thus, $n = N - 1$. In our experiment we consider populations of size $N = 3$.

Open Neighborhoods: (a) Local Interaction along a Circle. *N* players are allocated around a circle. Each player interacts with his *n* (local) neighbors on the circle, *m* right and ($n - m$) left neighbors. In our experiment we consider neighborhood sizes of $n = 2$ and $n = 4$, with $m = n/2$. For example, in the case of two neighbors ($n = 2$), a player *i*, with $1 < i < N$, has player $i - 1$ as his left neighbor and player $i + 1$ as his right neighbor.

The left neighbor of player 1 is player *N*, while player 2 is his right neighbor. Similarly, the right neighbor of player *N* is player 1, while player $N - 1$ is his left neighbor.

(b) Local Interaction on a Lattice. To play a four-neighbor game ($n = 4$), 16 players ($N = 16$) are allocated on a lattice. Each player interacts with his four (local) neighbors on the lattice, his left, right, top, and bottom neighbor. If the players are allocated as illustrated in Table. 8.2, player 6, for example, has player 5 as his left neighbor, player 7 as his right neighbor, player 2 as his top neighbor, and player 10 as his bottom neighbor. Player 1 has player 4 as his left neighbor, player 2 as his right neighbor, player 13 as his top neighbor, and player 5 as his bottom neighbor.

Consider the four different neighborhood games described in Table 8.3. AVG2GROUP is a game with closed neighborhoods while AVG2CIRCLE is a game with local interaction around a circle. AVG4CIRCLE and AVG4LATTICE are average games with four neighbors. Both are games with local interaction, along a circle in AVG4CIRCLE and on a lattice in AVG4LATTICE.

Result 2. (a) In all games but AVG4LATTICE the (all X)-equilibrium and the (all Y)-equilibrium are the only equilibria in pure strategies.

(b) In game AVG4LATTICE additional equilibria in pure strategies exist in which 8 or 12 players choose strategy Y while the remaining players choose strategy X. Table 8.4 presents an example of such an equilibrium configuration.

Table 8.2 Allocation of 16 players on a lattice

	13	14	15	16	
4	1	2	3	4	1
8	5	6	7	8	5
12	9	10	11	12	9
16	13	14	15	16	13
	1	2	3	4	

Table 8.3 Four different neighborhood games

Game	n	Interaction structure	N
AVG2GROUP	2	Closed	3
AVG2CIRCLE	2	Closed	8
AVG4CIRCLE	4	Circle	16
AVG4LATTICE	4	Lattice	16

Table 8.4 An equilibrium configuration in game AVG4LATTICE

	X	X	Y	Y	
Y	X	X	Y	Y	X
Y	X	X	Y	Y	X
Y	X	X	Y	Y	X
Y	X	X	Y	Y	X
	X	X	Y	Y	

Experimental Design

In an experimental session, the respective game is played 20 times by the same population of players and with the same neighborhood structure. In each of the 20 rounds, each player chooses between strategy *X* and strategy *Y*. A player *i*'s payoff depends on his own choice and on the choice of his neighbors. A neighbor's payoff depends on the choices of the neighbor's own neighbors - of whom one is player *i*.

After each repetition, each player is informed about the distribution of his neighbors' decisions in the round just finished. He is not informed about the individual decisions of his neighbors. Nor is he informed, in case of open neighborhoods, about his neighbors' neighbors' decisions. A player's payoff is determined by the sum of his payoffs in all 20 rounds. Players have complete information about the game in the sense that they know each player's payoff function and that the game ends after 20 rounds. In the case of open neighborhoods, they know that their neighbors also interact with other neighbors. They are, however, not explicitly told that they are allocated around a circle or a lattice, and they are not informed about the size *N* of the circle or the lattice. Note that these informational assumptions on the players' side are not arbitrary but taken from the theoretical models on local interaction that we mentioned in the introduction. In most of these models, players are supposed to adapt their strategies from one round to the next by myopic best response to the distribution of their neighbors' strategies in the previous round. The experiments were run at the University of Karlsruhe.

Results

The experimental results show that

1. Local interaction matters
2. Neighborhood size does not matter
3. Interaction structure matters
4. Individual behavior can be characterized by local best reply combined with a lock-in effect.

The percentage of *X*-decisions over the 20 rounds in AVG2GROUP (closed neighborhood) to AVG2CIRCLE (open neighborhood) shows that risk dominance becomes significantly more important with local interaction than without.[1] This can easily be seen in Fig. 8.1 in the appendix, which shows the percentage of *X*-decisions over the 20 rounds in both games.

[1]This result is even stronger in similar experiments based on the minimum-pay-off function rather than the average-pay-off function in Keser et al. (1998).

The comparison of AVG2CIRCLE and AVG4CIRCLE does not show a significant difference in the percentage of X-decisions over all 20 rounds (see also Fig. 8.2). Thus, we have no evidence that the neighborhood size matters when we have local interaction around the circle.

Comparing the AVG4CIRCLE and AVG4LATTICE we observe actually in each round in AVG4LATTICE a higher percentage of X-decisions than in AVG4-CIRCLE (see Fig. 8.3). In the first round, this difference is statistically not significant. However, we observe a significant difference over all 20 rounds. We conclude, that everything else being equal, it matters whether players are allocated around a circle or on a lattice. This is particularly surprising, given that subjects received exactly the same instructions that contained information about the number of neighbors but not about the specific type of interaction structure.

Based on a logistic regression explaining the probability for choosing strategy X we can show that the number of neighbors who chose X in the previous round (local best reply), the players' own choice of X in the previous round (lock-in effect), and the average frequency of observed changes in the distribution of the neighbors' choices over all previous rounds have a significant positive impact. This statistical model allows us to explain the higher percentage of X-decisions on a lattice than around a circle. This is due to the observation of more individual decision changes on the lattice than around the circle. Thus, applying our statistical decision model, we should observe a higher number of X-decisions on the lattice than around a circle.

Strategic Network Formation

In Berninghaus et al. (2006) and Berninghaus et al. (2007) we focus on the network formation as a strategic game. In doing this we abstract from the strategic aspects of playing a 2×2 game when being matched with neighbors in the network, but make the simplifying assumption that agents only exchange fixed payoffs when being matched with their neighbors, instead. An example of this scenario is a pure *information exchange network,* where the only strategic aspect is to be connected with the appropriate partners in order to exchange information. In some sense this section serves to prepare the model in Section "Strategy Choice on Endogenous Networks", where we simultaneously analyze strategic network formation and strategy choice in a 2×2 base game.

A Simple Model of Strategic Network Formation

The network game is characterized by the set of players $I = \{1,..., n\}$, the strategy sets G_i, and payoff functions Π_i. An individual strategy g_i of player i is a vector of ones and zeros $g_i \in Gi = \{0, 1\}^n$ with the following interpretation. If $g_{ij} = 1$

player i establishes a link with player j, otherwise, we have $g_{ij} = 0$. By convention, a player cannot link with himself, that is $g_{ii} = 0$ for all i. A player has to pay for each link that he establishes. Note, that a bilateral connection between two players in our model is supposed to be already established if at least *one* player wants to open it, i.e. if $g_{ij} + g_{ji} \geq 1$ holds. If a bilateral connection is established both players benefit from the exchange of information even if only one player has to pay for the connection. In that respect our model belongs to the class of two-way flow models (see Bala and Goyal 2000).

Each strategy configuration $g = (g_1, \ldots, g_n)$ generates a directed graph[2] denoted by G_g, where the vertices represent players and a directed edge from i to j, i.e. $g_j = 1$, indicates that player i holds a link with j. According to the essential assumption of the model by Bala and Goyal we suppose that player j need not *agree* when player i wants to open a link with him. This distinguishes our model from any other type of network formation games in which selection of neighbors is not modeled as a non-cooperative game as in Jackson and Wolinsky (1996).

In a network generated by the strategy configuration g each player i may have a number of agents with whom he is connected. We call these agents *neighbors* of i. It is essential for our model to distinguish three types of neighbors. *Actively reached neighbors* are all players that i holds links with:

$$N_i^a(g_i) := \{j \in I | g_{ij} = 1\}.$$

Slightly abusing language, we call players in $N_i^a(g_i)$ *active neighbors*. Links of i with players in $N_i^a(g_i)$ are called i's *active links* or simply i's links. We call *passive neighbors* of i all players who hold a link with i:

$$N_i^p(G_g) := \{j \in I | g_{ji} = 1\}.$$

A link of j to i is called a *passive link* of i. We call *indirect neighbors* of i all active or passive neighbors of all active neighbors of i, i.e.

$$N_i^{ind}(G_g) := \{l \in I | \exists j \neq i \neq l : g_{ij} = 1 \text{ and } \max\{g_{jl}, g_{lj}\} = 1\}.$$

Thus, the set of *all neighbors* of player i is given by

$$N_i(G_g) := N_i^a(g_i) \cup N_i^p(G_g) \cup N_i^{ind}(G_g).$$

---

[2]Bala and Goyal (2000) introduce a network in the two-way flow model as a non-directed graph. This makes sense when an edge between two players denotes the two-way flow. In our model we emphasize the aspect of opening connections. We are interested in finding out which players initiate links with other players. For example, when both players ***i*** and ***j*** open a link with each other they are connected in the corresponding directed graph via two adversely oriented edges.

Let $n_i(G_g)$ denote the number of elements in $N_i(G_g)$ and $n_i^a(g_i)$ the number of elements in $N_i^a(g_i)$. Note that both active and passive neighbors are direct neighbors (in contrast to indirect neighbors). That is, they are connected in the associated directed graph by an oriented edge. Since direct neighbors in our model are treated differently depending on how the links with them are financed it makes sense to subdivide the class of direct neighbors into active and passive ones.

Costs for opening a link are supposed to be the same for each player and are denoted by k (>0). The benefit or return that player i can extract from being connected (either actively, passively or indirectly) with player j is the same for all players and supposed to be equal to a (>0). Given strategy configuration $g = (g_1, \ldots, g_n)$, player i's payoff or net return is given by

$$\Pi_i(g) := an_i(G_g) - kn_i^a(g_i). \tag{8.1}$$

We see from this definition that player i may benefit from a connection to j although he does not have to pay for it[3] which is an implication of our two-way flow model. We also find that player i's payoff is equal to zero if i is an isolated player in the network, that is, if $N_i(G_g)$ is an empty set.[4]

Our framework contrasts with the neighborhood concept used by, for example, Falk and Kosfeld (2003). They assume according to the theoretical model by Bala and Goyal (2000) that the set of neighbors of player i consists of all players connected with i by a path in the non-directed graph associated with G_g. First, we limit access of i to those players j to whom i has an active link (active neighbors) plus to all direct neighbors of j. Only from these players player i gets a return. Second, i has access to the return of those players l who have an active link to him (passive neighbors) but not with the neighbors of l. That is, if i pays for a link with a direct neighbor j, he also has access to the return of j's direct neighbors. However, if i does not pay for a link with a direct neighbor l, he does not have access to the return of l's neighbors. A player can only benefit once from every other player in the network, i.e. if he is, for example, actively and indirectly connected with the same player he will not get his return twice.

We believe that our neighborhood concept is more realistic for real-world networks than the assumption that players have access to the information of arbitrarily distant players (in the network) provided they can somehow be connected with each other.[5]

[3]For example, if $g_{ij} = 0$, but $g_{ji} = 1$ holds.

[4]This assumption contrasts with the framework of Bala and Goyal in which a player receives a payoff of $\boldsymbol{a}$ even if he is isolated in the network.

[5]In labor markets, for example, social networks play a significant role in getting a job. The empirical investigation of Granovetter (1995) concentrates on jobs found via social contacts. It reveals that the main part of the people, who got their job via social contacts, had heard about it from their friends, relatives or acquaintances (39.1% of the cases), or from acquaintances of those (45.3 %). That is, from a network point of view the person who is looking for a job has access to the information of his "neighbors" who are at the most two "steps" away.

For network games $\Gamma = \{G_1, \ldots, G_n; \Pi_1(\cdot), \ldots, \Pi_n(\cdot); I\}$ a *Nash network* is defined to be a vector of individual link proposals $g^* = (g_1^*, \ldots, g_n^*)$ so that no single player i has an incentive to open additional links different from g_i^* or to sever links prescribed by g_i^*.

Definition 1. *A strategy configuration g^* in Γ is a Nash equilibrium if*

$$\forall i : \Pi_i(g_{-i}^*, g_i^*) \geq \Pi_i(g_{-i}^*, g_i) \text{ for all } g_i \in G_i, \tag{8.2}$$

where $g_{-i}^* = (g_1^*, \ldots, g_{i-1}^*, g_{i+1}^*, \ldots, g_n^*)$. G_{g^*} is the Nash network generated by g^*.

Moreover, if in (8.2) the strict inequality holds for all i and g_i, the strategy configuration g^* is called a strict Nash equilibrium.

Because of the large number of Nash networks in this case, strict Nash networks are a reasonable refinement. Depending on the particular parameter values of a and k strict Nash networks can be characterized as *periphery-sponsored stars*[6] and/or the *empty network,* i.e. a network without any link. In the following we give sufficient conditions for a periphery-sponsored star and the empty network being a strict Nash network in our framework.

Result 3. If $n > 3$ trict Nash architectures[7] are:

(a) For $k \leq a$ the periphery-sponsored star (ps-star)
(b) For $a < k < (n - 1)$ a the ps-star and the empty network
(c) For $(n - 1)$ $a \leq k$ the empty network

Proof. See Berninghaus et al. (2006).

Can we observe these equilibrium network structures in the lab or are they purely theoretical constructs? And how much time will members of a population spend in equilibrium networks? To answer these questions we ran some experiments.

Network Formation Experiments

Experimental Design

The computerized experiment was performed at the University of Karlsruhe. It was organized in 14 sessions with two groups of six subjects each. The experiment was conducted in continuous time.

---

[6]A periphery-sponsored star is a graph where all $n-1$ players in $I-\{i\}$ have an active connection with i, no other connections exist.

[7]Two networks have the same architecture if one network can be obtained from the other by permuting the labels of agents. For example for $n = 6$ players the ps-star architecture has 6 configurations, i.e. the ps-stars are 6 of $(2^5)^6 = 1.073.741.824$ possible networks or 1 of 1.540.944 possible architectures.

Each network game lasted 30 min. In the beginning of the game all subjects had to decide for every possible link whether to open it or not. Thereafter, subjects could change their strategies, i.e. either open or sever links at any time. Information was updated by the computer five times per second. Particularly, the current payoff flow was computed every fifth of a second and accumulated payoff was "integrated" up to the given moment. The information presented on the subjects' computer screens throughout the game included the player's current payoff flow and his current accumulated payoffs. A subject's own and the other subjects' active links were illustrated on the screen by arrows. Moreover, the subjects to whom a player was connected had a different color on the computer screen than the remaining ones. The remaining time was indicated on each screen during the entire experiment.

The return per connected player was set equal to $a = 3$ ExCU (Experimental Currency Units) per minute and the costs per link differ with respect to four *treatments*. In *treatment I* we set $k = 2$ ExCU per minute which still guarantees a positive net return for each individual connection. In *treatment II* we assume $k = 7$, in *treatment III* we assume $k = 13$, and in *treatment IV* we set $k = 16$. In treatments II–IV opening a new link with another player without any active or passive links is no longer profitable. Nevertheless, in treatments II and III ps-stars and empty networks both are strict Nash (see Result 3), while in treatment IV only the empty network is a strict Nash network.

Eight groups participated in each of the treatments I–III. Only four groups participated in treatment IV. We restricted the number of participating groups in this treatment since the experimental results of these four groups were unambiguous.[8]

Experimental Results

Almost all groups (except treatment IV) reached several strict Nash networks, i.e. ps-stars, during the course of the experiment and, moreover, stayed in these networks for a considerable amount of time. There were significant differences in the results between the treatments.

In *treatment I*, the net return from being connected with another player is strictly positive irrespective of this player being connected with other players. Therefore, the empty network is not a Nash network in the network base game. The ps-star is the only strict Nash network and, furthermore, it is efficient. Seven (out of eight) groups reach the ps-star. Moreover, six groups leave the strict Nash network in order to form other ps-stars with a different center player.

[8]It is easy to see that the strategic problem for a single player in treatment IV is rather simple. The empty network is the only strict Nash network, no other network has the Nash property.

We observe that most groups leave the ps-star after some minutes but return to it again after some time or switch to another ps-star with another center player. A graphical illustration of the results for each group can be found in the appendix. Each chart exhibits for the course of 30 min who of the six players was the center player for all minutes where an equilibrium was reached. Let us consider group 3, for example (Fig. 8.4). Subjects in this group leave the same ps-star (with center player 5) eight times during the 30 min, but do not switch to another ps-star. Group 7, on the other hand, "visits" *all* six ps-star networks within 30 min.[9] The time spent in *non-strict* Nash networks is negligible.

In *treatment II,* the net return of a single active link is strictly negative (Fig. 8.5). There exist only two Nash networks, which are both strict: the ps-star and the empty network. Note, the ps-star is efficient, while the empty network is not. Half of all groups reach all ps-stars in this environment. The minimum number of different ps-stars reached by a group was equal to three (group 6) while half of all groups reach all different ps-stars within the 30-min period. Particular attention should be given to group 2 that switches three times from one ps-star to the next by interchanging the respective center player.

In *treatment III* (Fig. 8.6), the net return of a single active link like in treatment II is strictly negative. Compared with treatment II, it is more than twice as negative. Therefore, an individual player has to be careful when opening new links with members of his group. With each link, he should rather catch as many indirect neighbors as possible. There is no non-strict Nash network. There are only two strict Nash networks, the ps-star and the empty network. The former is also efficient while the latter is not. Compared with treatment II, we observe that on average less ps-stars are reached during the experiment but the number of different ps-stars reached during the experiment is still larger than in treatment I. The same conclusion holds for the average time spent in ps-stars. Redundant links are punished more severely in treatment III than in treatment II which may favor subjects who rather stay in empty networks than open non-profitable links.

Like in treatments II and III the net return of a single active link in *treatment IV* is strictly negative. However, in contrast to these treatments the ps-star is no longer strictly Nash. The only strict Nash network in treatment IV is the empty network. However, the ps-star is still efficient. Because of the assumed extremely high link costs we do not expect ps-stars to be formed at all during the course of the experiment. Indeed, the experimental results show that the groups spend most of the time in the empty network whereas no ps-star was reached at all.

Summarizing Our Main Findings: Groups exchange the center players in a ps-star. They leave a strict Nash network in order to form a new one with a different center player. Finding a new center player is not an easy task. Therefore, some groups do not succeed in forming new ps-stars but return to a center player on whom all have agreed before. This is observed very often in treatment I. In treatments II and III, we

[9] In a population of $n = 6$ players we find 6 essentially different ps-stars that are obtained by interchanging the center players.

observe an additional phenomenon. Groups exchange the center player several times in the same order. One group in treatment II and three groups in treatment III show exactly this behavior while many of the remaining groups also visit all different ps-stars several times (but not in the same order).

A possible explanation for our results may go as follows: In our design, center players are subsidized by the periphery players. Even if the instantaneous payoff difference between center and periphery player in a ps-star is not that large, accumulated payoff differences after 30 min show a considerable difference. In order to establish payoff equalization in the long run, groups have to exchange the center player. This generates payoff distributions in the long run, which show only a low degree of inequity. Therefore, inequity aversion might be an important motive that drives our experimental results.

Strategy Choice on Endogenous Networks

Above we analyzed how strategy configurations in a coordination game evolve when members of a population are connected via a fixed and exogenously given network (local interaction structure). Berninghaus and Vogt (2004, 2006) drop the assumption of a fixed and exogenously given network structure but assume that population members can autonomously choose their neighbors and simultaneously select an action in a 2×2 game which is then played against all direct neighbors in the network.

We consider a population of players who create their interaction structure via costly links and who choose an action in a 2×2 normal form game played with each of their interaction partners. Let $I = \{1,\ldots, n\}$ denote the set of n agents who are engaged in playing the same 2×2 game with each of their neighbors in the network of players (which describes the interaction structure). To avoid trivialities, we assume $n \geq 3$. If two players i and j are linked, they play the symmetric 2×2 normal form game $BG = \{\{i, j\}, \Sigma, H(\cdot)\}$ with strategy set $\Sigma = \{X,Y\}$ and payoff function $H(\cdot)$: $\Sigma \times \Sigma \rightarrow \mathbb{R}$ characterized by the payoff table in Table 8.5. The number of players choosing X and Y are denoted by nX and nY, respectively, with $nX + nY = n$.

In Berninghaus and Vogt (2006) we analyzed strategy choice in all types of symmetric 2×2 games as follows. Given four ordered potential payoff values

$$a > b > c > d > 0, \tag{8.3}$$

there are $4 \times 3 \times 2 = 24$ payoff tables. By eliminating symmetric cases, we can reduce the number of possible payoff tables to 12. Six payoff tables have exactly

Table 8.5 Payoff-table of *BG*

	X	Y
X	$H(X, X), H(X, X)$	$H(X, Y), H(Y, X)$
Y	$H(Y, X)$, H(X, Y)	$H(Y, Y), H(Y, Y)$

Table 8.6 Payoff-table of coordination and Hawk-Dove game as *BG*

	X	Y		X	Y
X	b, b	c, d	X	d, d	a, c
Y	d, c	a, a	Y	c, a	b, b

one pure Nash equilibrium (in dominant actions), while the remaining payoff tables have exactly two Nash equilibria in pure actions. In Berninghaus and Vogt (2006) simultaneous partner and strategy choice was analyzed for all 24 base games. Here we focus on pure coordination and Hawk-Dove games which are characterized by the payoff tables in Table 8.6.

In order to discuss the results of the interesting case of coordination games in which risk-dominant equilibria do not coincide with payoff dominant equilibria, we assume in addition to Requirement (8.4)

$$b - d > a - c, \tag{8.4}$$

which implies that in the coordination game the risk dominant equilibrium (X, X) does not coincide with the payoff dominant equilibrium (Y, Y). Concerning the Hawk-Dove game, we call Y the "dove action" and X "hawk action". Typically, dove players are better off when matched with other doves rather than hawks, although playing dove against a dove player does not constitute an equilibrium in *BG*.

We assume that networks are determined by individual decisions similar to section "A Simple Model of Strategic Network Formation", therefore, our presentation will be rather short on the strategic network formation aspect (for more details, see Berninghaus et al. 2010).

A strategy of player i in the *network game* in normal form is a vector of ones and zeros, $gi \in \{0, 1\}^n$. If $g_{ij} = 1$, player i activates a link to player j, otherwise $g_{ij} = 0$. A link between i and j allows both players to play the 2×2 game *BG*. A player cannot play the game with herself ($g_{ii} = 0$ for all i). Note that two players play the game if *at least one* of them has a link to the other. Each strategy configuration $g = (g_1, \ldots, g_n)$ in the network game generates a directed graph, denoted by Gg, whose vertices represent players and a directed edge between i and j (i.e. $g_{ij} = 1$) indicates that i has a link to j. The *set of i's neighbors* in a network G_g is again denoted by[10] $N_i(G_g)$, where

$$N_i(G_g) := N_i^a(g_i) \cup N_i^p(G_g).$$

It is defined as the set of all players to whom i has a link (i.e. $g_{ij} = 1$) and the set of all players who have a link to i (i.e. $g_{ji} = 1$): $N_i(G_g) = \{j \mid \max\{g_{ij}, g_{ji}\} = 1\}$.

[10]Note that we distinguish in this model only active and passive neighbors. In contrast to the model of network formation in the previous section we assume that players do not play a base game with *indirect neighbors*.

Like in the previous section we assume that costs $k > 0$ per link are constant and identical for all players and are payed by the player who activates a connection.

Combining the two components, that is, choosing an action in the base game *BG* and selecting a partner in the network game, we model the strategic situation of a player as a non-cooperative game in normal form in which an individual decision is composed of the choice of neighbors via links $g_i \in \{0, 1\}n$ in the network game and actions $\sigma_i \in \{X, Y\}$ in the 2×2 game. This *strategic-networking game* is denoted by $G = (S_1,\ldots, S_n; P_1(\cdot),\ldots, P_n(\cdot); I)$ with strategy set $S_i = \{X, Y\} \times \{0, 1\}^n$ and payoff functions P_i: $S_1 \times \ldots \times S_n \rightarrow \mathbb{R}$. Each strategy configuration $s = (s_1,\ldots, s_n) = ((\sigma_1, g_1),\ldots, (\sigma_n, g_n))$ induces a vector of each player's action in the *BG* games and a network represented by a directed graph G_g. Player i's payoff function is

$$P_i(s_i,s_{-i}) := H(X,X) \sum_{j \in N_i(G_g)} 1_{\{\sigma_j = X\}} + H(X,Y) \sum_{j \in N_i(G_g)} 1_{\{\sigma_j = Y\}} - kn_i^a(g_i) \quad \text{if} \quad \sigma_i = X,$$

$$P_i(s_i,s_{-i}) := H(Y,X) \sum_{j \in N_i(G_g)} 1_{\{\sigma_j = X\}} + H(Y,Y) \sum_{j \in N_i(G_g)} 1_{\{\sigma_j = Y\}} - kn_i^a(g_i) \quad \text{if} \quad \sigma_i = Y.$$

The value of the indicator function $1_{\{\sigma_j = x\}}$ equals one if $\sigma_j = x, x \in \{X, Y\}$, and zero, otherwise; s_{-i} is the vector of strategies $s_j \in S_j$ of players $j \in I \backslash \{i\}$; $n_i^a(g_i)$ denotes the cardinality of $N_i^a(g_i)$. A player's total payoff is determined by her action in the 2×2 game and the actions of the players with whom she is linked minus her total link costs. According to our payoff definition, player i may benefit from being linked to j even though she does not activate the link (if $g_{ij} = 0$ but $g_{ji} = 1$). In what follows, the term "strategy" refers to a strategy in the strategic-networking game G.

We are interested in the Nash-equilibria of the game G which are defined as follows

Definition 2. *The strategy configuration* $s^* = ((\sigma_1^*, g_1^*),\ldots, (\sigma_n^*, g_n^*))$ *in G is a Nash-equilibrium of G if* $\forall i : P_i(s_{-i}^*, s_i^*) \geq P_i(s_{-i}^*, s_i)$ *for* $s_i \in S_i$.

In a Nash-equilibrium no player has an incentive neither to change her neighbors nor to change her action $\sigma_i^* \in \{X, Y\}$ unilaterally.

How can the Nash-equilibria of G be characterized? We refer to Berninghaus and Vogt (2006) for more detailed results for arbitrary base games *BG*. Below we only present an outline of the Nash-equilibria for coordination games and Hawk-Dove games as base games (see also Berninghaus and Vogt 2004). The Nash-equilibria of G depend on the connection costs k. An equilibrium s^* in G is characterized by the resulting network structure G_g^* and the vector of actions σ^* in the base game.

In Table 8.7. we present the equilibrium networks under varying connection costs k.

Remarks. (a) Cases 1 and 2 are extreme benchmarks. Connection costs are either too low (case 2) to have a significant effect on network building or too large (case 1) to admit interesting network building.

Table 8.7 Survey of equilibria network structures in G

case		Hawk/Dove	Pure coordination
1	$k > a$	Empty network, n_X not determined	Empty network, n_X not determined
2	$k < d$	Complete network with $0 < n_X < n$	Complete network with n_X either $= 0$ or $= n$
3	$d < k < c$	X-players connected with all Y-players, Y-players connected with each other, $0 < n_X < n$	Complete network with n_X either $= 0$ or $= n$
4	$c < k < b$	X-players actively connected with all Y-players, Y-players connected with each other, $n_X > 0$ and empty network with $n_X = n$	Complete network with n_X either $= 0$ or $= n$ and a network composed of two disconnected complete networks of X- or Y-players
5	$b < k < a$	Bipartite graph (of X- and Y-players) with active X-links and empty network with $n_X = n$	Complete network with $n_X = 0$ and empty network with $n_X = n$

(b) There is no big variation in Nash networks for pure coordination games. We obtain either complete or empty networks. Only in case 4 when connection cost are large enough to exceed the coordination failure payoffs but smaller than the coordination payoffs two disconnected complete networks may arise.

(c) In Hawk-Dove games we observe more variation in network building. If connection costs are large enough ($b > k > c$) all hawk players have active connections with the dove players but not vice versa, while all dove players are connected with each other. Such a strategy configuration is in sharp contrast with playing the Hawk-Dove game bilaterally without connections. Therefore, network building adds a completely new aspect to strategic thinking.

In networks with pure coordination games all players choose either X or all choose Y except for case 4 where disconnected networks may exist in which all players either choose X or Y. Strategy choice in networks with Hawk-Dove games is more complicated. In some cases nX is strictly different from 0 and n. We determine the number of X players in the network by checking for a representative X-player whether it is profitable to deviate unilaterally by choosing Y. Depending on the prevailing equilibrium network structure this finally results in restrictions which nX has to satisfy. These restrictions are summarized in Table 8.8.

The game G is a purely static one, i.e., it is a one-shot game. We also ran experiments in which we repeated this game in "continuous time" with a *Hawk-Dove* game as a base game.[11] Preliminary experimental results showed that Nash configurations s^* are not a good predictor for configurations in the dynamic process of co-evolution of actions (in Hawk-Dove games) and networks. But our experimental data indicate that there may be different equilibrium concepts that may serve as better predictors in the co-evolution problem.[12]

[11]Concerning continuous time experiments we refer to the results in the section "Strategic network formation."

[12]A detailed report on these experiments can be found in Berninghaus et al. (2010).

Table 8.8 Conditions on equilibrium action choices

case		Without bilateral linking	
		Hawk/Dove	Pure coordination
a	$k < a$		
b	$k < d$	$\frac{n(a-b)-(a-b)}{a-b+c-d} < n_X^* < \frac{n(a-b)+(c-d)}{a-b+c-d}$	
c	$d < k < c$	$\frac{(n-1)(a-b)+n_j^X(k-d)}{a-b+c-d} < n_X^* < \frac{n(a-b)+(c-k)}{a-b+c-k}$	
d	$c < k < b$	$n_X^* > \frac{(n-1)(a-b)}{a-b+c-d} > 0$	$\frac{n(a-k)+(b-d)+n_i^X(k-d)}{a+b-k-d} < n_X^* < \frac{n(a-c)-(a-c)+n_j^Y(c-k)}{a+b-k-c}$
e	$b < k < a$	$n_X^* > \frac{(n-1)(a-k)}{a+c-k} > 0$	

Appendix

Experimental Results of Coordination on Networks: Graphical Illustration

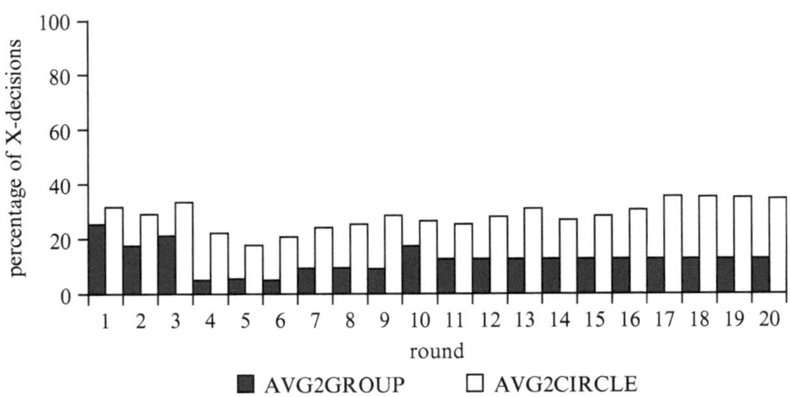

Fig. 8.1 Percentage of X-decisions in AVG2GROUP and AVG2CIRCLE

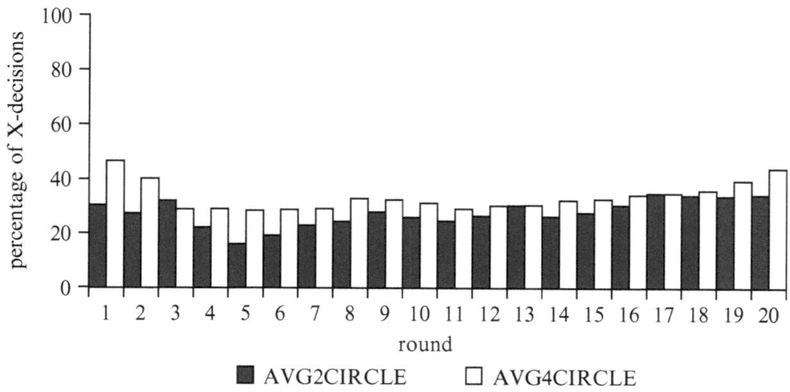

Fig. 8.2 Percentage of X-decisions in AVG2CIRCLE and AVG4CIRCLE

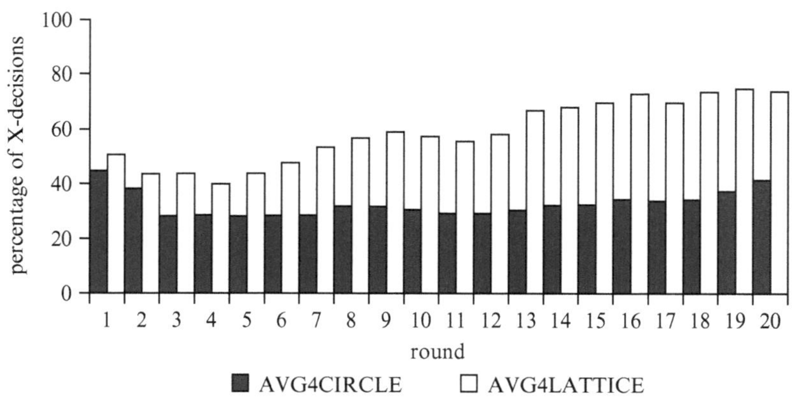

Fig. 8.3 Percentage of X-decisions in AVG4CIRCLE and AVG4LATTICE

Experimental Results of the Network Formation Game: Graphical Illustrations

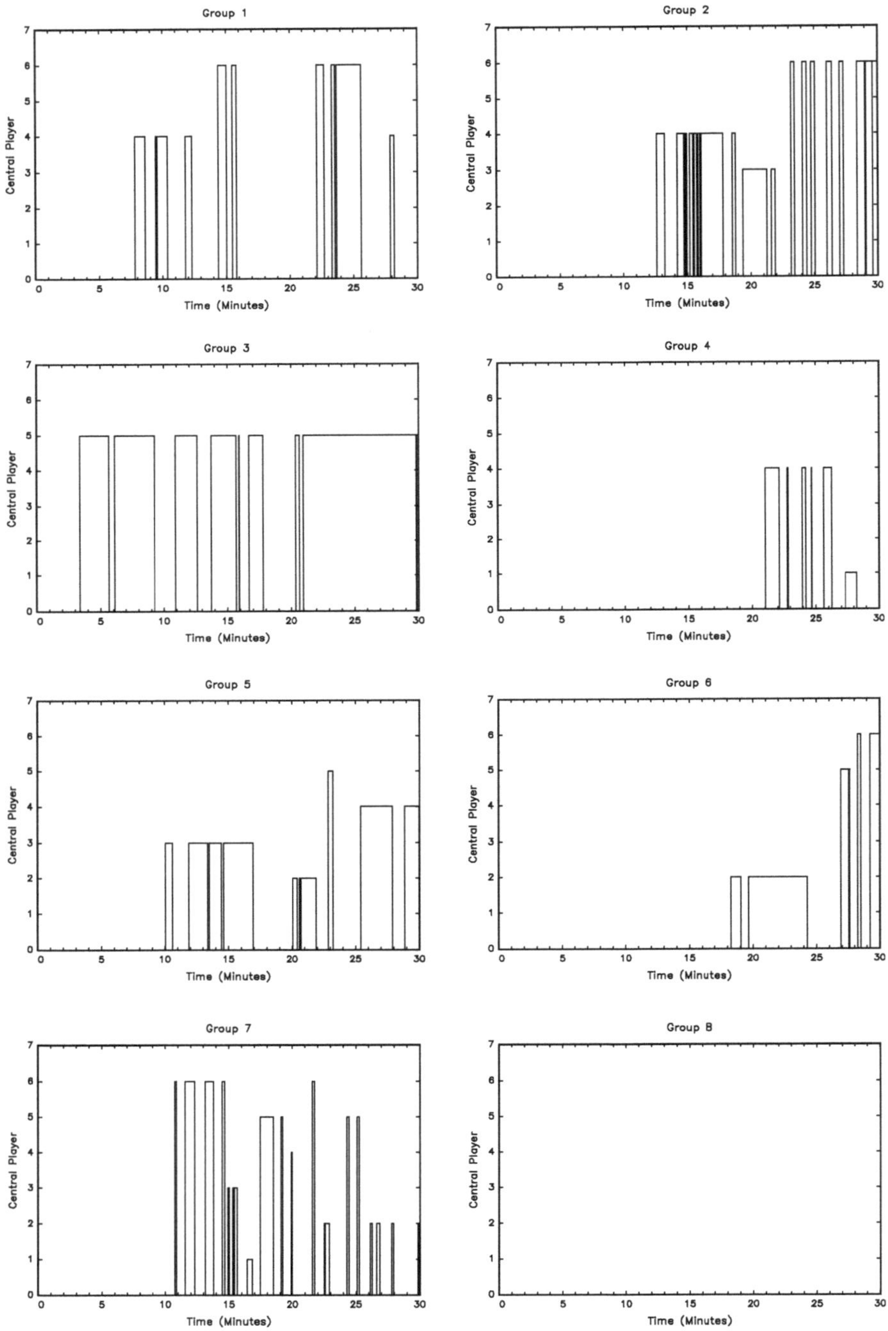

Fig. 8.4 Observed ps-stars in groups 1–8 (treatment I)

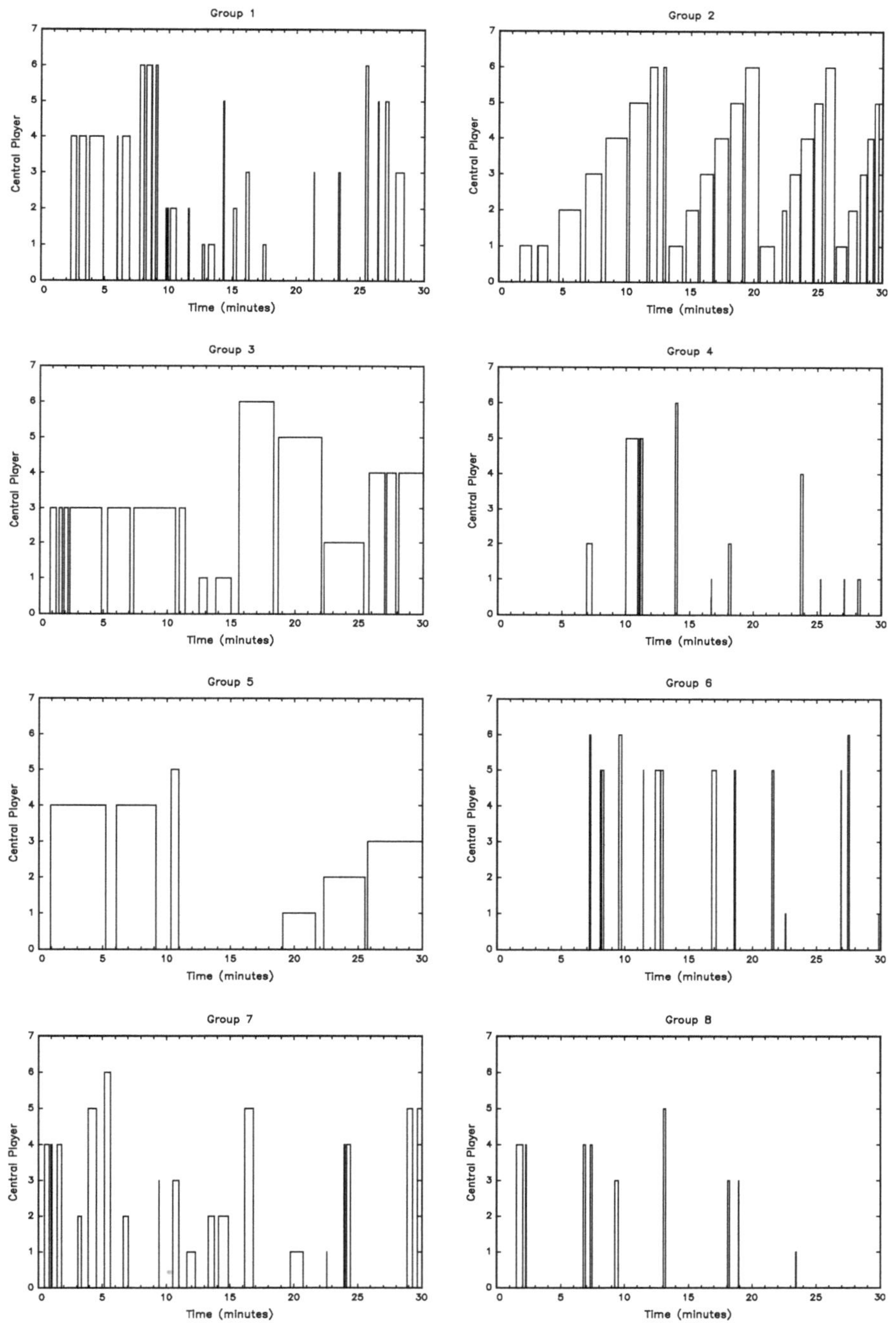

Fig. 8.5 Observed ps-stars in groups 1–8 (treatment II)

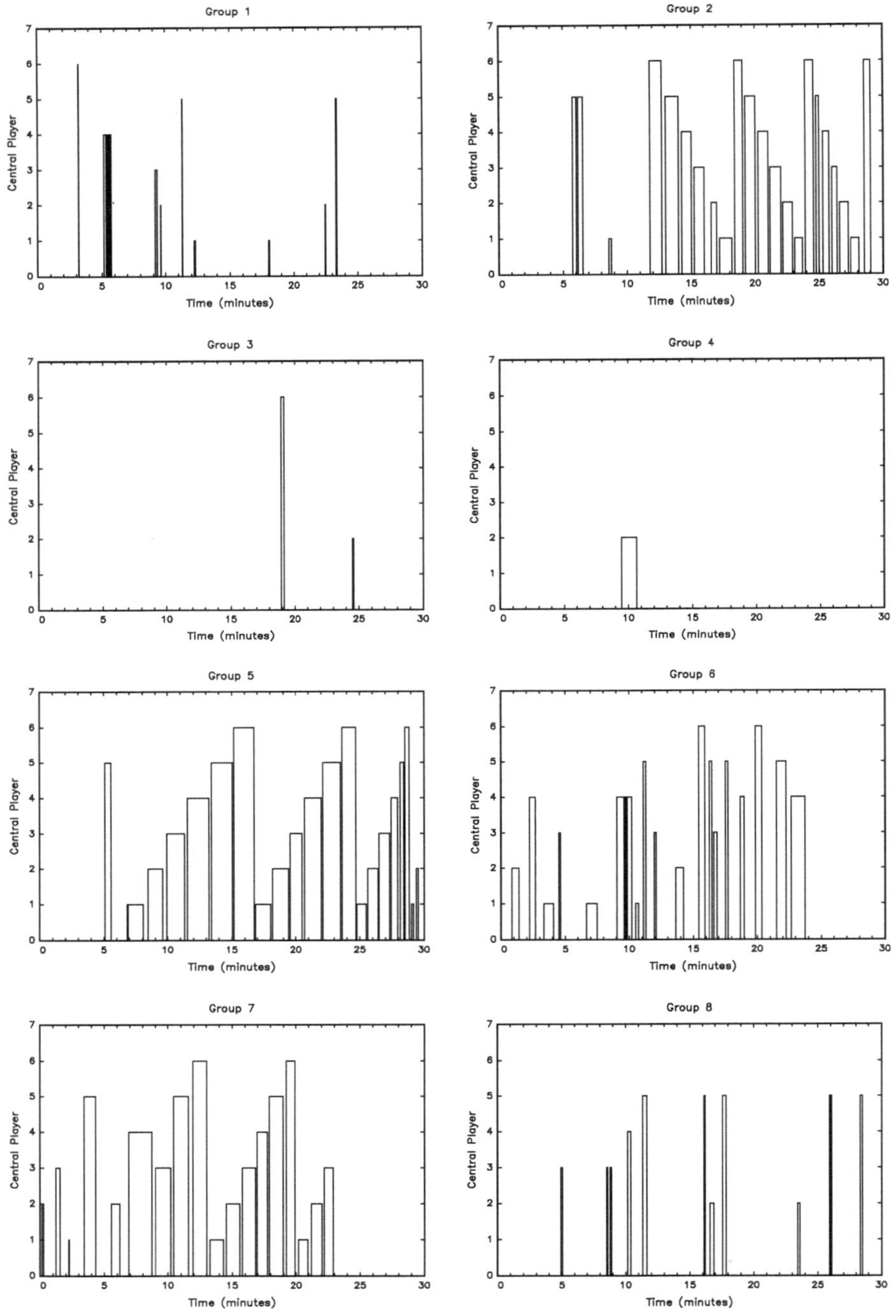

Fig. 8.6 Observed ps-stars in groups 1–8 (treatment III)

References

Anderlini L, Ianni A (1993) Path dependence and learning from neighbors. Games Econ Behav 13:141–177

Bala V, Goyal S (2000) A non-cooperative model of network formation. Econometrica 68:1181–1230

Berninghaus SK, Schwalbe U (1996) Conventions, local interaction, and automata networks. J Evol Econ 6:297–312

Berninghaus SK, Vogt B (2004) Network formation and coordination games. In: Huck S (ed) Advances in understanding strategic behavior: game theory, experiments, and bounded rationality. Palgrave, Basingstoke, UK, pp 55–72

Berninghaus SK, Vogt B (2006) Network formation in symmetric 2 × 2 games. Homo Oeconomicus 23:421–466

Berninghaus SK, Ehrhart K-M, Keser C (2002) Conventions and local interaction structures: experimental evidence. Games Econ Behav 39:177–205

Berninghaus SK, Ehrhart K-M, Ott M (2006) A network experiment in continuous time: the influence of link costs. Exp Econ 9:237–251

Berninghaus SK, Ehrhart K-M, Ott M, Vogt B (2007) Evolution of networks – an experimental analysis. J Evol Econ 17:317–347

Berninghaus SK, Ehrhart K-M, Ott M (2010) Cooperation and forward-looking behavior in Hawk-Dove games in endogenous networks: experimental evidence. Mimeo, Karlsruhe Institute of Technology, Faculty of Economics

Blume LE (1993) The statistical mechanics of strategic interaction. Games Econ Behav 5:387–424

Ellison G (1993) Learning, local interaction, and coordination. Econometrica 61:1047–1071

Falk A, Kosfeld M (2003) It's all about connections: evidence on network formation. IEW Working paper No. 146, University of Zurich

Granovetter M (1995) Getting a job: a study of contacts and careers, 2nd edn. University of Chicago Press, Chicago

Harsanyi JC, Selten R (1988) A General Theory of Equilibrium Selection in Games. MIT Press Books

Jackson MO, Wolinsky A (1996) A strategic model of social and economic networks. J Econ Theory 71:44–74

Kandori M, Mailath GJ, Rob R (1993) Learning, Mutation, and Long Run Equilibria in Games. Econometrica 61(1):29–56

Keser C, Ehrhart K-M, Berninghaus SK (1998) Coordination and local interaction: experimental evidence. Econ Lett 58:269–275

Young HP (1993) An Evolutionary Model of Bargaining. Journal of Economic Theory 59:145–168

Chapter 9
Walking with Reinhard Selten and the Guessing Game: From the Origin to the Brain

Giorgio Coricelli and Rosemarie Nagel

Fig. 9.1 Dinos thinking aloud about the guessing game. With the permission from Ryan North

G. Coricelli
Institut des Sciences Cognitives, CNRS, 67 Boulevard Pinel, 69675 Bron, France

R. Nagel (✉)
Department of Economics and ICREA, Universitat Pompeu Fabra, Ramón Trias Fargas, 25-27, 08005 Barcelona, Spain
e-mail: rosemarie.nagel@upf.edu

A. Ockenfels and A. Sadrieh (eds.), *The Selten School of Behavioral Economics*, DOI 10.1007/978-3-642-13983-3_9, © Springer-Verlag Berlin Heidelberg 2010

What could be the relationship between Reinhard Selten and the dinos' story? An easy answer is that he was Nagel's "Doktorvater" (PhD advisor) and the guessing game was one of the topics in her thesis. A more challenging connection is to Reinhard Selten's constant fear to be trapped and blinded by the mathematical reasoning of game theory as a description of human behavior like the dinos seem to be trapped in YEP, WOO. Doesn't the dinos' story how to play the game sound appealing? Certainly it does for many game theorists. When they play it a large proportion chooses 0 while the winning number among them is between 10 and 15! Other theorists, on the other hand, found the game rather boring since not playing the theory didn't seem a surprise and at the same time a clear structure of behavior was not at hand.

"Who here can name a game that sounds like it's way more fun than it actually is?" not much different from "don't waste your time", "there are better games, like the centipede game". So, dinos how would you like the centipede game?

The laughter of the dinos... this is what Reinhard Selten likes most, if people think something in his work or behavior is strange or unusual and at the end he is right. One hot day he went through the desert as usual with his umbrella. People passed by and were laughing at him. Hours later it was pouring with rain and the same people passed by and now Reinhard (American convention) was laughing, telling the story to us.

In the 1980s the dinos' interpretation of a boring game was similarly perceived by professors and students. It was then, when professors unfamiliar with experimental economics used it as a demonstration experiment in graduate game theory classes to explain concepts like rationalizability. Since the behavior was too dispersed to show any clear pattern, nobody got animated to use it as a basis for an experimental study. Furthermore, students were not taught in experimental economics and did not perceive the idea that experimental economics could become a serious field in economics.

That was very different in Selten's environment from the late 1950s. He entered economics and psychology as an undergraduate student with a major in mathematics, starting with experiments early with Heinz Sauermann. When he came to Bonn in the 1980s he changed the focus just on theory within the department to include also experiments, definitely for his graduate students. Everybody but one of his graduate students had to do experiments under his supervision. It was the opposite of what everybody was interested in for most of the profession. This is how Reinhard Selten can be best described: to be different, to be the opposite and to teach us to be the opposite. Recently he got a referee report from a respected journal: "this paper should not be published in this journal, moreover, this paper should never be published." This is what Reinhard Selten wants to produce, something a typical economist has never heard of but which at the same time is the heart of economic decision making. In the fifties until the early 1980s it was about experiments on oligopoly theory (see Chap. 6) and cooperative game theory (see e.g. Selten and Kuon 1993) when very few researchers would study these fields.

When he was asked after being awarded the Noble prize whether he sensed already in the 1960s that his sub-game perfect equilibrium concept would be outstanding, he replied he thought that the duopoly game with demand inertia was the interesting contribution of the paper (see also experiment by Keser 1993 on the same game). Thus, the study of dynamic games should be an important research question. Maybe it would be a good idea to produce a book for his 90th birthday with a collection of such experiments, invented and analyzed from 2010 to 2020. In search of equilibria in his game he started with backward induction and then realized that the equilibrium had very interesting properties as compared to the other equilibria he later calculated in the same game.

Some of his wishes clearly got fulfilled. In one of our many "Spaziergaenge" in 1993 (walks which end in a coffee place eating cakes), I asked him whether he thought he could get the noble prize? He answered: "I can't get it before John Nash and I don't want to get it without John Harsanyi." We joked that I would remind him if parts of it did not become true. Still being young I thought what a pity it must be that he got the Noble prize so late in life. However, Selten replied that getting it late was better since that way he could quietly work without so many obligations he now has.

Today again Selten runs experiments in which game theory cannot make any prediction because the structure of the situation is much too complicated. Many decisions have to be made in complex environments with many actors (see Pope et al. 2006). Given the complexity in this paper, he invented the concept of incomplete equilibrium, which allows us to leave out parts of the tree in the analysis. In an individual decision making experiment he studies behavior in a situation in which a decision maker has to achieve multiple goals, or aspiration levels instead of optimization (Selten et al. 2009). He does not agree that working on small problems is enough to understand complex situations by just adding small decisions. However, it wouldn't be Reinhard Selten, if he didn't find a small problem of a complex situation interesting like the ultimatum game, the end game of a long negotiation. But he does not like to study ideas others have already exploited. In 1989 I worked with Reinhard Tietz also talking to Werner Güth. When mentioning to run ultimatum games with incomplete information Selten replied "the train has left the station", that is, there are already too many ultimatum games around, find something nobody has worked on. Werner Güth and Eric Van Damme needed to convince him that the experiment should be done.

What would Reinhard Selten have chosen in the guessing game? He would ask for paper and pencil and ERASER! And think and think, and calculate, and erase again his thoughts and start all over; you wouldn't get an answer from him for a long time. He didn't need to be paid for that! You didn't want to have him in a pilot in which you just wanted to see whether everything worked. His computer would not spit out an answer before he had thought it all through. "Decisions emerge and are not calculated" in others not in him. He considers himself as a slow thinker, another reason not to work in fields where others could easily overtake him. We didn't actually ask him what he would choose. It would not be zero, the typical theorists' answer or 67 what theorists often choose knowing that zero is not a good answer. When he heard about the game he immediately saw the beauty in it. That is also Reinhard Selten, very quick in judgment!

The interest in the guessing game is now slowly surfacing outside of economics as already mentioned through the comics[1] or a party game even in Hollywood (Robinson 2004), but most easily through our neuro-economics work (Coricelli and Nagel 2009), connecting with the philosophical and psychological "theory of mind", an expression related to common knowledge or best reply concepts.

Let us summarize why the guessing game actually has some beauty for economists.

First of all it is a great demonstration experiment to show how game theory and human behavior can interplay for the understanding of core concepts in economics and bounded rational behavior. "What do I think what the others think, what the others think about what the others think . . ." and how can I outguess those others to win. The guessing game can be used by game theory instructors to explain many concepts by finding the equilibrium with fixed point arguments or a process of (in)-finite iterated elimination of dominated strategies. This is in line with rational expectations used in macro economics.

Secondly, at first sight behavior seems quite erratic. Game theory applies an infinite circular reasoning (Selten 1998) which, however, each human typically breaks at his own path: level 1 (best reply against random behavior), level 2 (best reply against level 1) etc. There is an underlying reasoning structure with a starting point of a human subject different from the theoretical starting point of 100. Of course changing the starting point might even more suggest a larger degree of freedom how bounded rational types of players can behave. The idea of finitely many steps of anticipation is similar as in Selten's paper on anticipatory learning (Selten 1991).

Thirdly, it is a strategic game which can be isolated from any fairness or other regarding preferences. But how should one play this game? Well it depends! It depends on who is playing with you. This is maybe the most important aspect. There is no clear answer. Game theory does give us one answer: how to play when everybody is rational and knows that everybody is rational, and so on, and there is no fool among them who plays tricks; they have to be in accordance as we imagine the dinos would play. But when we play with others we would need to inquire who the others are. Well trained economists play differently than beginners (Bosch-Domènech et al. 2002), and beginners who have been trained, play differently when they know that all others are beginners (Bob Slonim 2005). Teams are smarter than individuals not in the beginning but over time (Kocher and Sutter 2005). With two-person games you can test your intelligence and don't get tricked by the other chosen number (Grosskopf and Nagel 2008, 2009).

When you don't know him he appears to refute game theory completely for the understanding of human behavior. As said above this might be related to his fear that he as a theorist might not understand or be blinded by how humans really

[1]Ryan, the comic writer, e-mailed us how he got to know about the game: "I wish I had a better story: a mathematician friend of mine mentioned it once, and then I read about it on Wikipedia. :)". How would the dinos have discussed the game if their "father" had met one of us experimenters instead of a mathematician?

behave. But we as his students knew that he always requested a relationship to game theory as a bench mark. Even in the most difficult problems he asked us to search for equilibria (see thesis by Bettina Kuon 1994, who searched for equilibria in bargaining games with incomplete information about outside options or thesis by Karim Sadrieh 1998 on game theoretic equilibria in double auctions). For the guessing game he asked to search for Bayesian equilibria which might describe actual behavior.

Reinhard Selten taught us game theory, price theory, international trade theory, and experimental economics. He taught the theory classes as if he did not know anything about experiments. I can only recall one experiment he did in price theory about a three-person Cournot game. He collected all answers and wrote some in triplets on the board. He continued to teach us the equilibrium, showing that most were not playing the Cournot-Nash equilibrium. But as far as I remember in general he did not give us the idea that the theory was in great conflict with actual behavior. Later I learnt that his actual point of view is that game theory is like rational theology which has nothing to do with explaining actual behavior.

His strongest view is on "people don't optimize." It took us 20 years to understand. One could talk about an illusion of optimization, futility[2] of optimization or an optimization fallacy (which actually exists in programming language). When you think you have optimized, you probably have not searched enough, when you are still searching, you are probably never satisfied and thus one gets into a vicious circle. Aspiration adaptation or satisfying might be a natural way out (Selten 1996; Selten et al. 2009).

In seminar classes, which are like reading classes, he offered experimental strategy courses in which the students participated in experiments in a spontaneous way as in the lab and then had to design a strategy (called "strategy method," Selten 1967, Selten et al. 1997, 2003). Here we learnt that knowing behavior of real people really mattered. Selten was very interested in students' learning capacity. While his assistants shared the burden of seven different practice classes of a theory class, Reinhard Selten came to each one for some years. He walked around and looked us over the shoulder to see how we solved the problems, without saying much. Maybe for him it was like an interesting experiment, with seven independent sessions of the same treatment and that year for year! When he is in his institute, he always has time for us. We then sometimes need to tell him "Excuse me, Herr Selten, I have to go," since we know that he stays until 7:30 pm. He does never seem to be in a rush and is very patient.

Can we as economists also make a recommendation in our complicated guessing game environment with many types but only one equilibrium point? In a competitive world the game is to outguess with any method. One of the subjects in the newspaper experiment (Selten and Nagel 1998) taught us how to outguess: since we invited participants to comment on their choices, he wrote that after figuring out all possible ways how to answer the experiment, he got so

---

[2]This term comes from Robin Hogarth (private communication)

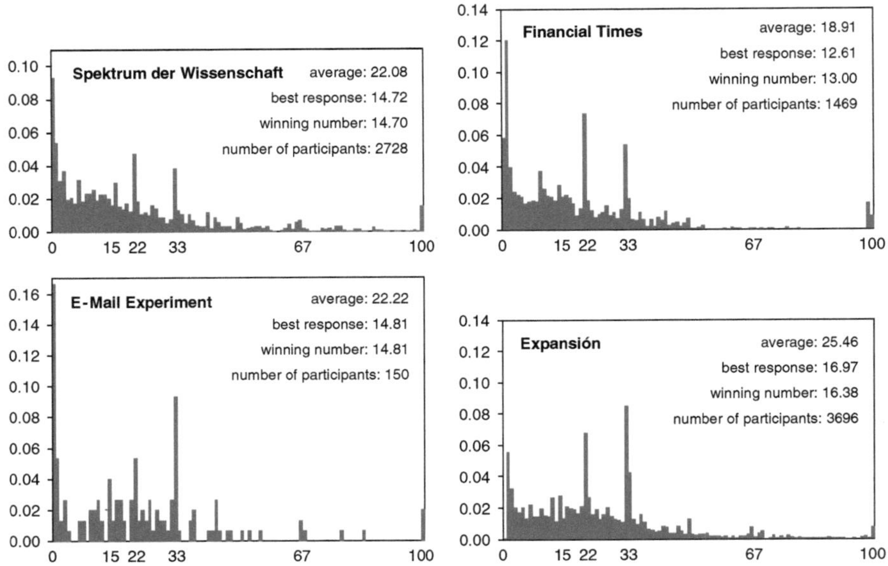

Fig. 9.2 Four different experiments with large numbers of participants with newspaper readers and one e-mail experiment by a participant (from Selten and Nagel, Spektrum der Wissenschaft (1998)

confused that he decided to do his own little experiment (see, e-mail game in Fig. 9.2). He proposed the experiment to a newsgroup which he thought was similar as the composition of the newspapers participants and then chose close to the answer he got from his own participants. He was off only by a few decimals (14.21 vs. 14.70)!

Selten was probably one of the first in economics who analyzed and structured comments and behavior in experimental economics to learn about the reasoning processes leading to a decision. He recommended us to follow social psychology classes and to carefully look at each single subject's choices. He used to record results on little cards collected in a box, gliding through it with his fingers. Economists were clearly not interested in comments or reasoning until recently. Nagel (1995) needed to cut the comment section describing subjects' thought processes. However, later in Bosch-Domènech et al. (2002) one referee supported the idea to carefully structure and classify comments of the 1,000 readers of the newspaper experiment.

The Origin of the Guessing Game

The origin of the experimental guessing game was unknown for about 30 years (see Bühren et al. 2009, and also blog by Roth on guessing game Sept. 2009, who makes some general comments about the origin of ideas). The first source for economists goes back to Herve Moulin's (1986) game theory book for social sciences. When he heard about the lab experiments he and Nagel started to search for the original source of the guessing game in the French version of Scientific American (Pour la Science) where he thought he had found the game. But our search was without success.

In 2009 Bühren and Frank run a lab experiment with 6,000 chess players. One of the participants happened to be Alain Ledoux (1981), former editor of Jeux & Stratégie who outed himself as the inventor of the guessing game which he named "psycho-statistique" in 1981. "Je suis sincèrement flatté que cette question ait pu intéressé un universitaire vingt-huit ans plus tard!" (I am sincerely flattered that this question could have been of interest to a university 28 years later.)

> "In 1981, the French magazine 'Jeux & Stratégie', a popular magazine devoted mainly to strategic board games, but also covering card games and mathematical games, arranged a big readers' competition consisting of mathematical puzzles but also problems from games such as chess, bridge and go. Ledoux (1981) reports on almost 15,000 participants, 4,078 of them being ex aequo, hence the winner had to be decided in a playoff. All first round winners received a letter with new puzzles, and to avoid another round with multiple winners, chief editor Alain Ledoux invented in the last question of this letter what is today known as the Beauty Contest (the name given to it by Ledoux, according to an email to us from July 9th, was "psycho-statistique", although this does not appear to have appeared in print). Readers were asked to state an integer between 1 and 1,000,000,000, the wining number being the one closest to two third of the average! The average turned out to be 134,822,738.26, two third of this being 89,881,825.51. This is 8.99% of the maximum number, markedly less than what is typically found in first rounds of Beauty Contest experiments (Bosch-Domènech et al. 2002). However, as explained above, the participants had been pre-selected, having solved a series of puzzles in the first round of the contest, and they knew that everyone else was pre-selected. Both facts should have resulted in a pretty high depth of reasoning." (Bühren and Frank 2009).

Keynes (1936) used a beauty contest metaphor choosing a face that the average would favor as an analog for clever investors. He does not provide an easy answer of what one should do. He just explains that there are many different types even "under professional investment". Definitely, a clever type should not ignore that there are others choosing "so that each competitor has to pick not those faces which he himself finds prettiest, nor even those that average opinion genuinely thinks the prettiest". Keynes states that there can be infinite types "and there are some, I believe, who practice the fourth, fifth and higher degrees." For anyone introducing types (of players) this is a big dilemma. However, the study on the guessing game and others games gives some optimism that most likely there are only finitely many types.

Keynes' game is not quite the same as the guessing game as it has many possible outcomes supported by game theory. Any face can be the equilibrium choice of all players. As in the stock market any point in time could be the final maximum or minimum of a cycle or start or crash of a bubble. Furthermore, the different thought processes actually cannot be distinguished through this metaphor, since a first guess is also the final guess, but Keynes' Gedankenexperiment is nevertheless understood: it connects nicely multiple equilibria with multiple types which is the deep problem of economic interaction between heterogeneous agents. Even rational players amongst themselves will get hopelessly lost.

The two mentioned differences between the Keynes contest and the guessing game were the main reasons why Reinhard Selten did not like to call it beauty contest: he always prefers to make a very clear distinction about everything. The name of the game could not be beauty contest and became the guessing game. Later we adopted the name from Keynes (Duffy and Nagel 1997) without forgetting to also use "the guessing game" as for this chapter.

Some Other Studies of the Guessing Game

According to Selten my own research about the guessing should have stopped right after the first experiment. That was typically his own approach, hit a new goal and then run as far away as possible, searching for new ideas, not being stuck in one which might or might not be successful. Furthermore, he recommended sending the paper to a journal with an easy acceptance process. One could never know whether some of his advices were just a test whether one had own but different ideas. However, it was also a guidance not to run with the crowd which he in his own past had seen to be misguided.

In a first wave of experiments on the beauty contest, typical treatments were tested as in other experimental settings: many repetitions, many players vs. few players, how an outlier can influence results, payoff changes, and interior vs. boundary solutions. The main topic was to formulize the reasoning process as mentioned above and model learning over time.

In a second wave of studies, new treatments have been introduced. Rather than observing only students, different subject pools have been invited to play the game; thousands of players participated in several newspaper experiments; more time than in the lab was given; games with equilibria in dominant strategies vs. many rounds of eliminations are compared; heterogeneity with respect to experience, different parameters, and team vs. individual behavior has been tested. The main topic in these new experiments was to introduce experimental designs with which reasoning could be improved. Nagel (1998) surveys these experiments and Camerer (2003) also includes a large class of dominance solvable games.

The Brain

Studying cognitive processes was not something strange to Seltens' students. Looking at his work and the many conversations with him we learnt that studying comments has been interesting. Therefore, the road to the brain was short. Already in the beginning when analyzing the comments for the guessing game I thought it would be better to look into the brain, however, I had no clue how. I came with a "cognitive mind" to Selten in 1986 and found very much support for that, unthinkable for a usual economist.

The most fascinating aspect with Reinhard Selten is, that any thought you might have from different scientific fields or even outside of any discipline, he easily can make comments. Nothing is strange for him and some of his favorites might be the very strange ones, for example his argument that cats can communicate with humans using different sounds, e.g. "rau" for "raus" (=outside=going outside). You can ask him for his cat sound-human translation dictionary. Whatever he does, he does it very systematically, with a love to invent new words and definitions, whether scientifically or non-scientifically. Definitely, he has thought a lot about religion, claiming if he started a religion, his first commandment would be "Sei dir nicht sicher" (don't be sure)... ah, the contrast to the dinos. And everybody would laugh about his thoughts which later might prove to be right... like the dinos laughed.

During my undergraduate studies I got into a phase of confusion. Selten wrote to me that I should come back and work, basically that he can help me. We never again talked about this "confusion," but instead he taught me the way to concentrate on small ideas in a very strict sense. This focus helped me to forget strange waves about which nevertheless much later we have been discussing. His way to see reality has a huge maybe infinite horizon encompassing almost everything also the impossible. However, as with almost everything he has a very strict separation between his scientific views and work and the other views. It is as if he closes the doors behind him when he goes from one to the other side (Garibaldi and Nagel 2010).

In 2004 my dream of looking into the brain became real with the study Coricelli and Nagel (2009) bringing the guessing game into the fMRI scanner. The design of this paper is actually one of two proposed designs of the guessing game, discussed with Selten in the beginning 1991. He asked me whether I wanted to study behavior over time or sessions with different parameters with the same person. I preferred to study behavior over time which was the starting point of the first pilots, encouraged by Guesnerie's experiment in a Ph.D.-class at the LSE. Also I had written a survey on learning models under Selten's supervision in my master thesis 1988, reading the paper on "Ĉu mi lernu Esperanton" (Should we learn Esperanto, Selten and Pool 1995) and the chain store paradox[3] (Selten 1978). Then, later in the study with ultimatum games and incomplete information, Mitzkewitz and I of course used learning direction theory (Selten and Stoeker 1987; Selten and Buchta 1998).

I wasn't so sure that levels of reasoning would be visible when repeating the game with different parameters (e.g. $p = 2/3$ or $p = 1/3$ etc.) without feedback between periods. However, this design proved to be very helpful in our fMRI study

[3]Many people know the chain store paradox. But do they know the second part in which he explains a program of decision making with imagination and similar words? When he submitted the paper to a well known journal the editor requested to cut out the reasoning part. When he refused, he was asked to cut it by some. In the end Selten send it uncut to where it is found today. In the working paper version he also explained his agony about the seemingly unreasonable result of the chain store. We don't know how this emotional story got eliminated in the final version.

(Coricelli and Nagel 2009). Here we also introduced a treatment in which we let subjects play against the computer who imitated random players and thus the average is clearly 50.

We categorized each player's behavior according to three categories: random behavior, low level (level 1), and high level of strategic reasoning (level 2 or higher). The high-level reasoning subjects clearly differentiated their behavior in the human compared to the computer condition. They behaved as level 1 in the computer condition but were classified as higher level of reasoning (level 2 or more) when interacting with human counterparts. The subjects classified as low level behaved similarly against the computer or the humans: at or close to level 1 in both conditions. Few subjects behaved in a random fashion.

It turns out that the brain activity between level 1 and level 2 can be clearly distinguished. When we analyzed separately high- and the low-level reasoning subjects, we found the activity in the medial prefrontal cortex to be stronger in subjects classified as high level. In the high reasoners, guessing a number in the human condition activated two main regions of the medial prefrontal cortex (mpfc), a more dorsal and a more ventral portion of the anterior mpfc (Fig. 9.2). The mpfc, is often related with "theory of mind" or mentalizing, what psychologists and philosophers define as the ability to think about others' thoughts and mental states in order to predict their intentions and actions.

High levels of reasoning in the guessing game implies thinking about how other players think about the others' (including yourself) thinking or behavior, and so on. In other words, high reasoners might assume that the same reasoning they are performing, – namely best replying to random players – is likely performed by others, thus inducing a process of iterative thinking. The prefrontal activity of the low-level reasoning subjects was found in the ventral anterior cingulate cortex (ACC, Fig. 9.3a), an area often attributed to self referential thinking in social

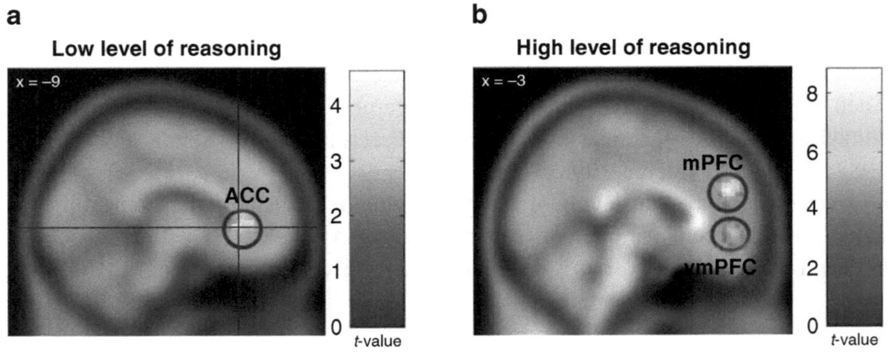

Fig. 9.3 Neuroanatomical dissociation between high and low level of reasoning. Choosing a number in the guessing game in the Human condition in contrast to the Computer condition was associated with relative enhanced activity in the rostral anterior cingulate cortex (*Left panel*, low level of reasoning subjects); and (*Right panel*, high level of reasoning subjects) activity in the dorsal portion of the medial prefrontal cortex (mPFC) and ventral mPFC

cognitive tasks. Thinking about the others as random players, thus, considering them as 'zero intelligent' agents, requires only a first person perspective of the interactive context. fMRI results show additional brain activities related to high- versus low-level reasoning in the right and left lateral orbitofrontal cortex and left and right dorsolateral prefrontal cortex, areas likely related to performance monitoring and cognitive control. This suggests that a complex cognitive process subserves the higher level of reasoning about others.

The pattern of brain activity in the right and left lateral orbitofrontal cortex and in the dorsolateral prefrontal cortex suggests a substantial jump in complexity when going from first to second level of reasoning. Actually I felt this jump of joy when I played the first time the guessing game in a master class at the LSE. I thought, given that I had participated already in many experiments of Selten's lab in Bonn, unlike the other students, I could outsmart them with level 2. The winning number was lower, and I didn't win. But in a later class I won in period 1 and 2. Most importantly I discovered only then the beauty of the beauty contest. Selten had taught us to jump far away from previous papers, and the distance between ultimatum games and guessing games is quite large. And he taught us that participating in experiments could be useful even for a researcher, which seems to be still unknown in most departments of economics.

The different brain activity between level 1 and 2 might be responsible for the observed limited step-level reasoning, either because subjects are not able to make this jump or because they believe that not everybody else is able to make this jump. This result provides a new interpretation that might be interesting for game theoretic modeling. This important difference has never been discussed in the experimental economics literature on strategic reasoning. Instead, the main difference has been thought to be between random behavior and higher level; mainly because level 1 contains already a best reply structure, a fundamental concept in economic theory. However data from Coricelli and Nagel (2009) suggest a discontinuity in the belief about other's behavior as naïve or random behavior (the underlying belief of level 1 players) vs. belief of best reply behavior (level 2 or higher).[4]

When first classifying behavior in the guessing game, I used the classification from Mitzkewitz and Nagel (1993) with focal types (choosing 50) and near focal types, anticipation types, equilibrium types, and random types, drawing stylist people in different colors for a talk in the European doctoral program (EDP) conference. All were quite astonished but Selten and Werner Hildenbrand. Selten advised me just to use the $50*p^n$, where the number n indicates the level type. My main concern was that this was rather ad hoc and very much guessing game specific. Another problem is that there could be 17 types (type 17 is pretty close to equilibrium). So far experiments

[4]However, a closer look on Keynes famous quote (see above) indicates that Keynes clearly distinguished level 1 (best reply to average behavior) from higher levels! Discussing this recently with Sergiu Hart, Sergiu rephrased it: "Level 1 is 1-player decision making: games against nature, or games against "non-thinking" players. Level 2 (and higher) is for multi-player games, against "thinking" players. Thus, level 1 is like statistics, whereas game theory starts entering in level 2".

show that there are not that many types. Now level of reasoning has also been introduced explaining behavior in games that do not have the property of a dominance structure as for matching pennies or entry games (Camerer et al. 2004). Given the brain data, level 1 could be a "near focal type," who is actually not an anticipation type, since he has no good model of the mind of the others.

The brain activity analysis also clearly distinguishes between the computer and the human condition in theory of mind areas, which indicates that facing humans requires a higher structure of complexity in thinking than facing a computer program.

We hope that with our analysis we can give a road map for more complex economic experiments in which the behavioral results cannot clearly identify different levels of reasoning. Then together with brain activities one might in the future structure a behavioral cloud in such a way that different level of reasoning types can also be distinguished.

What will Reinhard Selten think about what we have written about him? He will say, "Das kann man so nicht sagen, well, this is all not quite correct, you have to write it differently. Give me paper and pencil." But we would reply: "Herr Selten, we did not write what you think that we think what you think that you think... what you are, but just how we think how marvelous you are." Herr Selten would then respond: "Achso, well then you can write it like that ..." but we would respond, "Well you were right first, you are even more marvelous. Could we meet next Tuesday to write everything better?" And he would say, "no on Tuesday I go shopping," (one of these stationary choices – although he has not been seen recently with his umbrella) "but let me look in my calendar." and "Shall we first go on a Wanderung?"

References[5]

Selten R (1967) Die Strategiemethode zur Erforschung des eingeschränkt rationalen Verhaltens im Rahmen eines Oligopolexperiments. In: Sauermann H (ed) Beiträge zur experimentellen Wirtschaft-forschung. J. C. B. Mohr (Paul Siebeck), Tübingen, pp 136–168

Selten R (1978) The chain store paradox. Theory Decis 9(2):127–159

Selten R, Stoeker R (1987) End behavior in sequences of finite prisoner's dilemma supergames. J Econ Behav Organ 7(1):47–70

Selten R (1991) Anticipatory learning in two-person games. In: Selten R (ed) Game equilibrium models I. Springer, Berlin, pp 98–153

Selten R, Kuon B (1993) Demand commitment. Bargaining in three-person quota game experiments. Int J Game Theory 22:261–277

Selten R, Pool J (1995) Enkonduko en la Teorion de Lingvaj Ludoj -Ĉu mi lernu Esperanton. Akademia Libroservo, Institut für Kybernetik, Berlin-Paderborn

Selten R, Mitzkewitz M, Uhlich GR (1997) Duopoly strategies programmed by experienced players. Econometrica 65(3):517–555

[5]In order to put Reinhard Selten's literature first, we turn around the alphabetical order and invert one name (Bob Slonim)

Selten R (1996) Aspiration adaptation theory. Discussion Paper Series B 389, University of Bonn, Germany

Selten R (1998) Features of experimentally observed bounded rationality. Eur Econ Rev 42(2–5):413–436

Selten R, Buchta J (1998) Experimental sealed bid first price auction with directly observed bid functions. In: Budescu D, Erev I, Zwick R (eds) Games and human behavior: essays in honor of Amnon Rapoport. Erlbaum, Hillsdale, NJ

Selten R, Nagel R (1998) Das Zahlenwahlspiel. Spektrum Wiss 1:89–91

Selten R, Abbink K, Buchta J, Sadrieh A (2003) How to play 3 × 3-games – a strategy method experiment. Games Econ Behav 45:19–37

Selten R, Pittnauer S, Hohnisch M (2009) Experimental results on the process of goal formation and aspiration adaptation. Mimeo, University of Bonn

Sadrieh A (1998) The alternating double auction market. A game theoretic and experimental investigation, vol 466, Lecture notes in economics and mathematical systems. Springer, Berlin

Roth A (2009) A famous experiment in game theory. http://marketdesigner.blogspot.com/2010/02/famous-experiment-in-game-theory.html

Robinson S (2004) How real people think in strategic games. SIAM News 37(1): 1–3

Pope R, Selten R, Kube S, von Hagen J (2006) Experimental evidence on the benefits of eliminating exchange rate uncertainties and why expected utility theory causes economists to miss them. Labsi experimental economics laboratory, University of Siena 010, University of Siena

North R (2008) Dinos thinking about the 2/3 of the average game. http://www.qwantz.com/index.php?comic=1360

Nagel R (1995) Unraveling in guessing games: an experimental study. Am Econ Rev 85(5):1313–1326

Nagel R (1998) A survey on experimental 'beauty-contest games': bounded rationality and learning. In: Budescu D, Erev I, Zwick R (eds) Games and human behavior, essays in honor of Amnon Rapoport. Lawrence Erlbaum Associates, New Jersey, pp 105–142

Moulin H (1986) Game theory for social sciences. New York Press, New York, NY

Mitzkewitz M, Nagel R (1993) Experimental results on ultimatum games with incomplete information. Int J Game Theory 22:171–198

Ledoux A (1981) Concours résultats complets. Les victimes se sont plu à jouer le 14 d'atout. Jeux & Stratégie 2(10):10–11

Kuon B (1994) Two-person bargaining experiments with incomplete information, vol. 412, Lecture notes in economics and mathematical systems. Springer, Berlin

Kocher MG, Sutter M (2005) The decision maker matters: individual versus group behaviourin experimental beauty-contest games. Econ J 115(500):200–223

Keynes JM (1936) The general theory of interest, employment and money. Macmillan, London

Keser C (1993) Some results of experimental duopoly markets with demand inertia. J Ind Econ 41:133–151

Grosskopf B, Nagel R (2009) Unraveling in two person guessing games. Working Paper, Texas A&M

Grosskopf B, Nagel R (2008) The two-person beauty contest. Games Econ Behav 62(1):93–99

Garibaldi P, Nagel R (2010) The in(de)finite skyscraper: a metaphor-morphosis of the mind. Non-scientific mimeo

Duffy J, Nagel R (1997) On the robustness of behavior in experimental beauty-contest games. Econ J 107:1684–1700

Coricelli, G. and Nagel, R. (2009), Neural Correlates of Depth of Strategic Reasoning in Medial Prefrontal Cortex., Proceedings of the National Academy of Sciences USA 106 (23):9163–9168

Camerer CF, Hua HT, Chong J-K (2004) A cognitive hierarchy model of games. Q J Econ 119(3):861–898

Camerer CF (2003) Behavioral game theory. Russell Sage Foundation/Princeton University Press, New York NY/Princeton NJ
Bühren C, Frank B, Nagel R (2010) A historical note on the beauty contest. UPF-discussion paper
Bosch-Domènech A, García-Montalvo J, Nagel R, Satorra A (2002) One, two, three, infinity, ...: newspaper and lab beauty-contest experiments. Am Econ Rev 92(5):1687–1701
Bob S (2005) Competing against experienced and inexperienced players. Exp Econ 8(1):55–75

Part III
Pro-Social Behavior

Chapter 10
Understanding Negotiations: A Video Approach in Experimental Gaming

Heike Hennig-Schmidt, Ulrike Leopold-Wildburger, Axel Ostmann and Frans van Winden

Motivating Video Experiments

Reinhard Selten since the beginning of his scientific career has been concerned with developing descriptive theories that take account of the boundedly rational behavior of human subjects. The concept of bounded rationality was introduced by Herbert Simon in the 1950s (Simon 1955, 1957), and Selten was immediately convinced by his arguments (Selten 1995). Together with a group of other researchers in Frankfurt around the economist Heinz Sauermann who shared the view that behavior of economic agents is not adequately modeled by "homo oeconomicus" theories Reinhard Selten started to run economic experiments and to develop a corresponding methodology already in the 1950s (see Sauermann and Selten 1967). In Selten's own words: "We invented experimental economics". [1] In these early days, the group's conferences and publications (in particular the Contributions to Experimental Economics, 8 volumes, 1967–1978) propagated interdisciplinary and internationally linked research to establish experimental economics as an accepted field of economics (Selten and Tietz 1980). Founding the "Gesellschaft für experimentelle Wirtschaftsforschung (GEW, now GfeW)" – "Society for Experimental Economics – An Association of Austrian, German, and Swiss

[1]Personal communication with Heike Hennig-Schmidt

H. Hennig-Schmidt (✉)
*B*onn*E*con*L*ab, Laboratory for Experimental Economics, University of Bonn, Adenauerallee 24-42, D - 53113 Bonn, Germany
e-mail: hschmidt@uni-bonn.de
U. Leopold-Wildburger
Institut für Statistik und Operations Research, University of Graz, Universit 15, A-8010 Graz, Austria
A. Ostmann
Universität des Saarlandes, Building FZU (50.40), D-66123 Saarbrücken, Germany
F. van Winden
CREED, Faculty of Economics and Business, University of Amsterdam, Roetersstraat 11, Amsterdam 1018 WB, The Netherlands

A. Ockenfels and A. Sadrieh (eds.), *The Selten School of Behavioral Economics*, DOI 10.1007/978-3-642-13983-3_10, © Springer-Verlag Berlin Heidelberg 2010

Experimental Economists" – in 1977 had a further impact on spreading the new methods, in particular in Germany.[2]

Selten has always emphasized that the structure of boundedly rational economic behavior cannot be invented in the armchair; it must be explored experimentally (Selten 1993, 1995). This was his research program between 1972 and 1984 in Bielefeld, and since 1984 in Bonn where he founded *B*onn*E*con*L*ab, the Laboratory for Experimental Economics, the oldest such institution in Europe.[3] Selten's view on human behavior implied looking deeper into peoples' decision making processes in economic experiments. To this end, Selten on the one hand very early applied methods uncommon in economics – as were experiments in the 1950s and 1960s per se. For instance did he observe subjects in face-to-face negotiations (Selten and Schuster 1968). On the other hand, he developed new methods himself like the strategy method (Selten 1967) which requires experimental subjects to formulate strategies for a game involving a sequence of decisions. For an application see Selten et al. (1988, 1997), see also Keser (1992).

Selten used and fostered the interdisciplinary scientific dialogue on negotiation and decision making experiments. These fields were of great importance in psychology the paradigms being investigated, however, lacked a game-theoretically well-defined incentive structure. Moreover, decisions were not incentivized according to the interactively derived experimental outcomes. The studies of the 1970s and early 1980s revealed research questions Selten found crucial to understand the processes and outcomes of negotiations. Results from face-to-face negotiations could not be derived from analyzing the incentive structure and the available strategic options alone, in particular as the subjects did not behave like "fully rational automata". The concept of bounded rationality as promoted by Simon and others required new methods of collecting data to make cognitive, emotional and social processes observable. By publishing a survey on such methods in negotiation research Chertkoff and Esser (1976) made them available to a larger audience. This article induced Selten and his group to employ new techniques also in game theoretically motivated experiments. The methods used are for instance protocol analysis and thinking-aloud (Ericsson and Simon 1984/1993, see also Austin and Delaney 1998) in individual decision tasks, and group discussions to analyze motivations and cognitions in negotiation processes.

The GEW, being strongly shaped by Selten, and the "Sozialwissenschaftlicher Ausschuß des Vereins für Socialpolitik (VfS)" (Research Committee on Social Sciences of VfS) were institutional forums for methodological discussions and the

---

[2]Concepts of bounded rationality were not of equal importance to experimental economists in the US. See Roth (1995) who states: "Interestingly the aspiration hypothesis has attracted different amounts of attention on different sides of the ocean. American experimenters and theorists have subsequently come to regard aspirations as, at most, an intermediate variable, rather than as a primary explanatory variable. Our German counterparts have been more inclined to regard aspirations as a primary explanatory variable (p. 88, Fn. 28).

[3]Austin Hoggatt had built up the first computerized laboratory for experimental economics at the University of California in Berkeley several years before.

further development of experimental economics. It was here that Axel Ostmann presented his research on how individual agents behave compared to player groups (cf. Ostmann 1990, 1996). These studies resulted from a research project by Werner Tack and Axel Ostmann in Saarbrücken between 1983 and 1986 on cooperative three person games and involved direct face-to-face negotiations. Discussions were recorded on audio tape and were transcribed into text protocols.

The main advantage of groups as negotiators was the insight into cognitive processes that usually go on within the individual hided from being observed. This was the reason why Selten pursued this research approach also in other bargaining situations. He knew that Marlies Klemisch-Ahlert audiotaped bargaining in two person groups in Bielefeld (Klemisch-Ahlert 1996) and found recording the discussions a very promising approach. Yet, when transcribing the audio-tapes it became obvious that with more than two group members identification of speakers and actions became extremely difficult. Replacing audio taping by video taping was an obvious solution for this problem.

In 1990 and 1991, Selten encouraged the PhD and Diploma students at the Bonn Laboratory for Experimental Economics to engage in video experiments. Several of them did as the 'Sonderforschungsbereich 303 and Laboratory for Experimental Economics Data Documentation Series 1996–2002' prove, as does the large archive of meanwhile digitized video recordings, and quite a few papers and publications.[4]

All four authors of this essay have long-lasting personal and scientific connections with Reinhard Selten and have been greatly influenced by his personality and his work. Axel Ostmann got into closer contact with him in 1978, when Ostmann started his PhD studies at the University of Bielefeld. Ulrike Leopold-Wildburger's first connection also dates from 1978 when Reinhard Selten gave a guest lecture at the Wittgenstein Symposium in Kirchberg/Wechsel. Frans van Winden has met Reinhard Selten already in 1976 at the Econometric Society European Meeting in Helsinki. Heike Hennig-Schmidt got into touch with him in 1984 when he moved to the University of Bonn. All of us were or still are involved in video experiments.

It was in 1991 that Selten inspired Heike Hennig-Schmidt to run a video experiment on two person coalition bargaining with groups as players. She readily accepted this suggestion not imagining this endeavor to guide her research for the next 20 years. Axel Ostmann is the 'veteran' of video experiments in our group. He started video taping experiments in 1990. Hennig-Schmidt visited Ostmann in Saarbrücken to learn about the video technique, organizational procedures and evaluation methods. Frans van Winden entered the video enterprise after Hennig-Schmidt gave a lecture on video experiments at the ENDEAR Summer School on Experimental Economics in Bari in 1999. The joint research following from our discussions not only focuses on evaluating the text protocols, it also compares group decision making with individual choices. Some side remarks on the Bari Summer School: Students and teachers notably overlap with the authors of this

[4]In addition to papers and publications reported in this essay, see e.g. Artale (1996), Jacobsen and Sadrieh (1996), Rockenbach et al. (2007).

volume. Selten to whom this volume is dedicated also gave lectures to make students familiar with his research approach. A remarkable number of the Bari students are reknown experimentalists by now. Even though Ulrike Leopold-Wildburger has started joint negotiation research with Reinhard Selten much earlier and is still engaged in doing so[5] she is the 'newcomer' in the video research group. Joint research with Hennig-Schmidt began in 2003 'only'.

When reading this essay the reader will realize the great time lag between running video experiments and writing up/publishing the results. The reasons will become clear from our discussion on the pros and cons of video experiments below. It needs quite some patience, persistence and determination to do this kind of research. Yet, in our view the results are worth the effort.

Why did we make negotiation video experiments the topic of our essay? Bargaining has been in the focus of Selten's experimental research since the early 1960s (Selten 1993) not the least because of the theoretic challenges, from both the game theoretic and the descriptive point of view. After all, negotiating is one of the most frequent social interactions: in business life, international trade, national and international politics but also for individuals engaged in private purchase of goods and services. It is important for successful negotiations to understand the behavior of bargaining partners, their goals and motivations as well as socio-emotional factors influencing the decision process. A lack of understanding will make mutually satisfactory agreements extremely difficult. Joint appreciation is even more important if the cultural backgrounds of the negotiators differ.

Field data on motives underlying the bargaining process are not easy to obtain and if available, are difficult to interpret. Video experiments are a tool to gain the required empirical evidence under controlled conditions the game to be studied being well defined. The players are motivated by monetary incentives. Observing experimental negotiators when making their decisions can reveal their motives and goals (Selten 2000) and also provide insight into the dynamics of interpersonal relations shaping the group process.

Our essay surveys a body of studies the authors did by applying the video approach. We report on remarkable regularities across different types of games. We proceed as follows. Section "Video Techniques and Analysis of Verbal Data" gives an account of video techniques, video experiments' pros and cons and the analysis of verbal data. In section "Observing Groups as Players", we report on experiments where groups act as players in strategic games and cooperative settings. We use the in-group communication to learn about cognition and reasoning processes of individuals; but we also investigate choice data. Section "Observing Group Processes in Coalition Games" analyzes situations where individuals are in conflict situations. We analyze the impact of socio-emotional factors and interpersonal relations, i.e. the social field, on group processes. In section "Cross-Cultural and Inter-Cultural Settings", we compare behavior in negotiation experiments run in Germany and PR China. The last section concludes.

---

[5]Selten and Leopold-Wildburger (1982a, b), Becker et al. (2007).

Video Techniques and Analysis of Verbal Data

Video Technologies and Documentation of Video Data

According to our different research interests, we use two video technologies. In the experiments of section "Observing Groups as Players" and section "Cross-Cultural and Inter-Cultural Settings" single players are represented by groups of three to five persons having to take a consensus decision (Picture 10.1). They interact with another player group of three persons in a dyad. Intra-group discussions are videotaped; inter-group contacts are anonymous with the experimenter transmitting decisions. Player groups are paid according to their performance. Each member within the group receives the amount the group has been achieving. As the research interest focuses on the decision making process *within* a player group and not in the *strategic* argumentation between the groups an anonymous inter-group design was implemented. In one experiment, groups bargained face-to-face with each other these between-group discussions being videotaped in addition to the within group interaction.

All experiments of section "Observing Group Processes in Coalition Games" use a face-to-face videotaping technology. Six subjects in the role of managers directly bargain with each other on the disposition of a fixed sum of money for important investments. Subjects are paid according to their performance. Here, the research interest focuses on the direct interactions during the bargaining process and on the socio psychological field that is expected to arise automatically and might influence the outcome of the coalitional bargaining (Selten and Schuster 1970). Therefore, the whole face-to-face bargaining process was videotaped.

Video experiments provide records of the discussions on different media (tapes, CD, DVD). It is, however, inconvenient and inefficient to use the records directly. Rather, what is required is a type of documentation that meets the requirements of easy accessibility to the discussion content, translation into foreign language in cross-cultural and intercultural studies, comparability of different text segments, computer-aided text analysis, comfortable verbal data management,

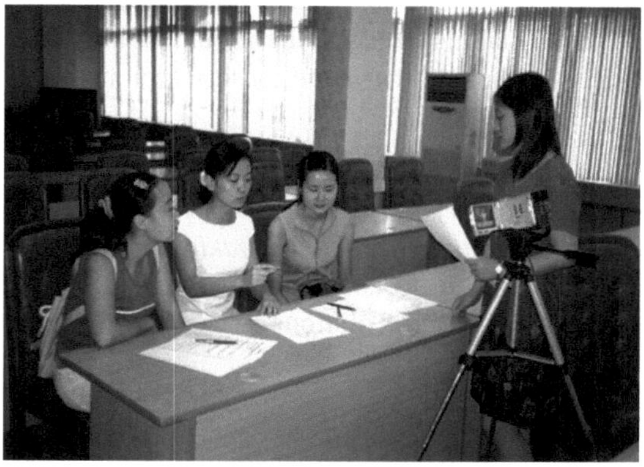

Picture 10.1 Instructing player groups

Picture 10.2 Transcribing the videos

simultaneous analysis by several persons, also from different countries, and easy reproduction in research papers. Transcripts meet these demands. Therefore, it is advisable to make full transcripts of the video-taped discussions by persons who are especially trained for this task (Picture 10.2). Different transcription systems exist emphasizing different features of interaction. We focused on the verbal statements by literally transcribing the discussions. We are interested in what subjects actually say to avoid own interpretations as far as possible e.g. of facial expressions or body movements; see also the paragraph on subjective interpretation of verbal data in section "Potential Weaknesses of Video Experiments". The text protocols do not replace the original media, though, and it is often helpful to work in parallel with both.[6]

Pros and Cons of Video Experiments

Observing participants during decision making in video experiments has its pros and cons. For one thing, the video method enables an authentic insight into the decision making process, the argumentation and the motives of the experimental subjects. On the other hand, observation might change behavior.

Advantages of Video Experiments

Observing authentic decision making processes. Video experiments enable a direct insight into the decision making process of the experimental subjects and their spontaneous argumentations. They allow an authentic access to the cognitive

[6] See Potter (1996) for several arguments of this paragraph.

processes underlying participants' decisions. Participants reveal their motives themselves and we need not infer from subjects' behavioral data what intentions and conceptions they might have had. This important information is often not available or not noticeable by other methods (Bakeman 2000) like paper-and-pencil and computer experiments usually documenting choices but not the process leading to these choices. Questionnaires[7] or the exchange of written messages[8] require less effort than video experiments as no transcripts have to be produced and less verbal data are generated. These data are usually poorer than observing behavior and recording spoken language and allows a reduced view on the decision process only. Moreover, in questionnaires, important influencing factors might be suppressed that are harder to conceal in discussions (see Trommsdorff and Wilpert 1994). Finally, asking for hypothetical behavior does not always allow inferences on actual behavior.

Advantage over audio taping. Audio taping is a great progress compared to note taking. Yet, video taping has decisive advantages. Speakers can be identified more easily and thus their characteristics as well as arguments can be attributed to the respective individuals (Orbell et al. 1988; Kerr et al. 2000). Non-verbal expressions, such as gestures and facial expressions are also recorded and give additional information. Finally, lip movements are observed and may facilitate understanding subjects' speech, in particular in heated debates, when several people talk at the same time.

Spontaneity and non-introspection. Experimental subjects in video experiments verbalize their arguments spontaneously when discussing the decision task. There is no retardation and rationalization effect as in retrospective reports (Gilhooly and Green 1996, Reis and Gable 2000). It has been shown that introspectively derived data can only incorrectly – if at all – reveal decision mechanisms because subjects are not able to adequately describe their own cognitive processes (Nisbett and Wilson 1977). Non-introspective verbal data derived from observation avoid this problem (Ericsson and Simon 1984, 1993; Anderson 1990).

Avoiding an experimenter demand effect. Subjects' attention is not drawn to the research question. Participants are not asked to make certain statements which might provide an experimenter demand effect by inducing them to align the required information with their choices. Moreover, spontaneity of statements permits to investigate which issues become a focus of participants' discussions.

Variety of arguments. In discussions, group members have to articulate their thoughts, motives and expectations in order to make their point. The interaction of several discussants most likely induces a large range of arguments as participants view the task from different perspectives (Kerr et al. 2000).

Complexity of cognitive processes. Complicated and complex cognitive processes often need time to develop. Discussions like in video experiments are a suitable means to make such processes observable as arguments and counterarguments are likely to advance complex thought processes. Verbal data derived from written

---

[7]See e.g. Burton et al. (2000), Bosman and van Winden (2002), Bosman et al. (2006).

[8]See e.g. Charness and Dufwenberg (2006), Cooper and Kagel (2005), Gächter and Riedl (2005).

statements are much poorer in this respect (see Walkowitz et al. 2010, Björn Frank, personal communication). We will touch the complexity of cognitive processes in section "Observing Groups as Players".

Cross-cultural and intercultural applicability. Video experiments are well-suited for cross-cultural and intercultural research[9] (see also Kerr et al. 2000; Smith 2000). Running video experiments with exactly the same experimental design in different countries allows to directly comparing discussions and arguments across cultures. These discussions not only refer to the same well-defined experimental context but also to identical situations during a decision making process. When running experiments in different cultures one has to pay attention to issues like language, stakes, and experimenter interactions (cf. Roth et al. 1991, Herrmann et al. 2008a, b) as they can affect cross-cultural comparability. We will come back to this issue in section "Cross-Cultural and Inter-Cultural Settings".

Use of transcripts for various research questions. Generating the transcripts of the video-taped discussions is a tedious and time-consuming procedure. When available, however, text protocols provide a rich data base for various research questions. Therefore, it is advisable to make full transcripts of the video-taped discussions. This enables evaluating the data from different points of view if the focus of analysis changes or if new research questions arise. This advantage will become obvious in the analyses of section "Observing Groups as Players" and section "Cross-Cultural and Inter-Cultural Settings".

Potential Weaknesses of Video Experiments

We acknowledge that video experiments have potential weaknesses as well.

Observation effect. The video technique may influence participants' behavior due to the impact of the observation medium. Our video records indicate that this effect need not be important. Participants seem to feel pretty comfortable and forget after several minutes that they are video taped (see also Potter 1996). They talk about private matters and sometimes they give even negative comments on their university teachers. In addition, the available evidence in the literature suggests that observation does not systematically affect behavior. Endres et al. (1999) and Bornstein et al. (2004) find no difference in the behavior of participants that were audiotaped and videotaped, respectively, and those who were not.

Subjective interpretation of verbal data. Analyzing verbal data involves elements of subjectivity and interpretation as the researcher evaluates the text from his/her perspective. To make the extent of subjectivity transparent it is important to give account of the full analyzing and interpretative process (Pidgeon 1996). To add

[9]In cross-cultural experiments, behavior in two or more cultures is compared with no direct interaction of subjects belonging to different cultures; inter-cultural experiments investigate behavior of subjects in different cultures interacting directly with each other. In intra-cultural experiments, participants from one country/culture interact with each other. See Walkowitz et al. (2009).

objectivity to this process, at least two persons different from the researcher should independently code the text according to the pre-specified and well documented categories (Smith 2000; Bartholomew et al. 2000) the coding being checked for inter-coder reliability (Merten 1995; Krippendorff 1980). Thus, validity criteria are applied to the qualitative data analysis (Smith 1996). We will come back to this issue when discussing the procedure of content analyzing the verbal data and the inter-rater reliability check in sections "Analysis of Verbal Data" and "Inter-rater Reliability".

Small number of observations. Data analysis of video experiments is usually based on fewer observations than mere choice data because of the financial resources required for running the experiments and the manpower necessary for evaluating the verbal material. Getting the data is expensive because in group experiments at least two times as many participants are required as in individual experiments. In the experiments reported in this paper, three to six times as many subjects were needed. A limitation of independent observations per experiment is advisable as the process of making the verbal data available for data analysis is extremely labor intensive. It comprises the extensive process of transcribing, the amount of time it takes to develop the skill for content analyzing the verbal data and the long-lasting coding process (Gill 1996). Video experiments create an extensive body of verbal data that has to be handled. For instance, the 16 sessions of the Alternating Offer experiment run in Germany and China (see section "Observing Groups as Players" and section "Cross-Cultural and Inter-Cultural Settings") provided about 1,100 pages of transcripts from 65 h of video tapes. Cumulative evidence on certain research topics based on different video experiments compensates for the relatively low numbers of observations in single experiments, though.

Conclusion on the Video Method

We do not deny potential weaknesses of the video technique. We have argued, however, that potential shortcomings can be overcome or can be taken into account by appropriate measures. We believe video experiments to be a valuable method as it provides data that are difficult or impossible to gain by other elicitation methods, in particular when individual decision mechanisms are not observable. Video experiments can valuably complement other experimental techniques.

Analysis of Verbal Data

Content analysis is a suitable method to derive the required verbal data base. Content analysis is a technique to extract the information a researcher needs to answer his/her research questions from a body of – in our case verbal – material by systematically and objectively identifying specified characteristics of the data (Smith 2000). This information is operationalized by coding the verbal material.

Coding denotes the process of assigning text segments to categories. The categories are to be carefully designed before coding is done to capture the relevant aspects of the research questions.

At least two native raters provided with a detailed coding manual independently assign segments of the video-taped discussions to these categories. Coding is to be made very restrictive in order to rule out raters' own interpretations as far as possible. Only when a category characteristic is explicitly mentioned during a discussion, this text segment is to be assigned to the category.

Often a hierarchical coding process is used (Bakeman 2000; Smith 2000) involving the design of one or more main categories. These are identified by independent raters or by auto-coding software for qualitative data analysis like e.g. ATLAS.ti that can be applied to Chinese transcripts as well. Then, sub-codes are created and the instances of the main categories are assigned to the subcategories again by independent raters.

Inter-rater Reliability

Data are reliable only if inter-rater agreement on category assignments is high. A generally accepted measure is Cohen's Kappa K which accounts for the agreement that would result if raters merely make random assignments (Siegel and Castellan 1988). K is the ratio of the proportion of times the raters agree, P(A), (corrected for chance agreement P(E)) to the maximum proportion of times they could have agreed (corrected for chance agreement), i.e. $K = [P(A) - P(E)]/[(1-P(E)]$. $K = 1.00$ indicates complete agreement among the raters whereas $K = 0$ means no agreement above chance. $K \geq 0.80$ indicates satisfactory inter-rater reliability (Merten 1995; Smith 2000). As for exploratory studies slightly lower values are acceptable (Krippendorff 1980) $K = 0.70$ can be taken as the reliability threshold. As the vast majority of the categories of our coding systems met this standard we conclude that on average our classification systems yield satisfactory data.

In order to base the analysis on as much data as possible, rater disagreement has to be resolved. Bartholomew et al. (2000) suggested the following procedure. If two raters disagree on a categorical assignment, a third (and if necessary a fourth) rater is added, and the classification agreed upon by at least two of them becomes the final rating. The remaining disagreements are excluded from the analysis. In the experiments reported below, the above procedure allowed to resolve nearly all disagreements.

Observing Groups as Players

In this part of our essay, we report on a number of different video experiments where single players are represented by groups of three to five persons having to take a consensus decision. In-group discussions are used to learn about motivations, cognition and reasoning processes of individuals because individual decision making processes are not observable in general. Thus, the video experiments of section "Observing Groups as Players" and section "Cross-Cultural and Inter-Cultural Settings" change the standard experimental procedure in that individuals are substituted by groups. The evidence of whether the group effect matters is inconclusive. Based on final outcomes, some studies find that groups behave more in accordance with standard game theory[10] while others suggest that groups behave less like gamesmen.[11] In some analyses no difference is found.[12]

Looking at within-group processes, Chertkoff and Esser (1976) argue that group representatives bargain harder than individuals. Even though this behavior did not affect the bargaining outcomes it might do so in other settings. Based on experiments with a large set of different three person cooperative games Ostmann (1990a, section 5) reported a clear superiority of groups with regard to the quality of statements and the quantity of arguments given in the negotiation. See also Kocher and Sutter (2005) who found groups to learn faster and to reason with more depth.

Eliciting individual motives like in the reported video experiments involves a tradeoff. On the one hand, directing individual subjects' attention to the research interest may influence their behavior. On the other hand, decisions might be affected as well if the attention impact is avoided by going from individual to group settings. Existing procedures that make individual motives and decision processes obvious may create a very unnatural environment like for instance the thinking-aloud method (Ericsson and Simon 1984/1993). An environment for spontaneous discussions in a group setting is more natural and more familiar to the subjects from outside the lab and might therefore have a smaller effect on subjects. When choosing the latter approach it is advisable to control for group behavior and observation by repeating the experiment with individuals who are not observed and ask them to state the motivations for their choices.

The experiments in this part of the essay are based on the one-shot simultaneous ultimatum game (Güth et al. 1982) in symmetric and asymmetric settings; on the one-shot, simultaneous tripled take game (Sadrieh and Hennig-Schmidt 1999); on the one-shot, sequential power-to-take game (Bosman and van Winden 2002); on a repeated alternating offer bargaining game with anonymous interaction (Selten 1981), on a cooperative version of Selten (1981) with face-to-face inter-group contacts and on a repeated prisoner dilemma game (Flood 1958).

In all but the cooperative games the following procedures were applied in these experiments. Player types are assigned randomly. Full anonymity between player

---

[10]See Bornstein and Yaniv (1998) and Robert and Carnevale (1997), focusing on an ultimatum game; Bornstein et al. (2004), using the centipede game; Bornstein et al. (2008), analyzing a Bertrand duopoly; and Cooper and Kagel (2005) studying a signalling game.

[11]For example, Cason and Mui (1998) find that groups are more generous in a dictator game, while Cox and Hayne (2006), analyzing common value auctions, and Kocher and Sutter (2007), focusing on gift exchange, report some mixed evidence.

[12]See Raab and Schipper (2009), examining Cournot competition, and Bosman et al. (2006) analyzing a power-to-take game.

groups was guaranteed by either inviting them to separate rooms or by applying another procedure; see the description of the power-to-take game. Groups interact with another player group of several persons in a dyad. The group decision is to be taken jointly and filled in on corresponding forms. All group members have to agree by signing the forms. No explicit decision rule is imposed. Instructions are read aloud by native experimenters. Intra-group discussions are videotaped; inter-group contacts are anonymous with the experimenter transmitting decisions. In the one-shot games, groups have 10–15 min to make their decision; in the repeated games there is no time constraint. Participants are fully informed on all features of the experimental design and the procedure. Decisions are made by paper and pencil. At the end of the experiment, participants are paid out in their groups and dismissed. When experiments were run in China exactly the same design and organizational procedure was used.

In the cooperative alternating offer bargaining game, role assignment is not random but groups earn their position by a quiz. Moreover, player groups in a dyad also bargain face-to-face with each other, the between-group discussions being video taped as well. If not reported otherwise, the experiments were conducted in *B*onn*E*con*L*ab, the Laboratory for Experimental Economics at the University of Bonn. The Chinese experiments were run at Sichuan University, Chengdu, China, and the experiments in Austria were conducted in the Max Jung Labor at the University of Graz.

Before reporting the results we briefly describe the games and the design of the video experiments and refer to the papers they are based on.

The Games

The Ultimatum Game (UG) (Hennig-Schmidt et al. 2008, Walkowitz et al. 2010)

In the standard UG, a proposer can decide on how to split a given amount of money (the pie) between herself and a responder. The responder can either accept or reject the proposal. In case of acceptance, both receive the amounts as allocated; in case of rejection both receive nothing. In our video experiment with groups as players the proposer A has to decide on the division of a pie of 20 tokens that she can allocate between herself and the responder B. A decides on the amount $x \in \{0,1, \ldots, 20\}$ to be sent to B. Simultaneously and independently, B states acceptance or rejection for any possible offer. In case of acceptance of x, A receives the payoff $20-x$, and B gets x. In case of rejection, both receive nothing (treatment T1). In two additional treatments, A is guaranteed a positive outside option of 8 in case of rejection whereas B's outside option is 0 or 2 respectively (T2 or T3). The sub-game perfect equilibrium in T1 is $x = 0$ if money is infinitely divisible. A will keep the whole endowment that B will accept. With a smallest money unit of 1 token a second sub-game perfect equilibrium exists, namely $x = 1$. With a similar argument, the second sub-game perfect equilibrium in T2 (T3) is $x = 1$ ($x = 3$).

UG was run at Bonn University in 2002 and at Sichuan University in 2001 and 2002. In Bonn, we ran 35 sessions comprising 200 students; in Chengdu, the number of sessions was 36 involving 208 students. In each session, only one proposer and one receiver group interacted. Each proposer (receiver) group provides one independent observation and we had 12 of them in each treatment except for T3 in Germany where only 11 of each participated. At the end of the experiment, the experimenters matched proposal and acceptance or rejection and informed the groups about the result.

The Tripled Take Game (TTG) (Sadrieh and Hennig-Schmidt 1999)

In TTG, player B is endowed with 12 tokens, and player A has an endowment of 0. B decides on an acceptance limit $y \in \{0, \ldots ,12\}$ for the amount x which A can take away from her. Simultaneously and independently, player A is deciding on the amount $x \in \{0, \ldots ,12\}$ he is taking away from B. For $x \leq y$ B receives the payoff $12-x$, and A receives $3x$. For $x > y$ both receive nothing. The subgame perfect equilibrium of this game is $y = 12$ and $x = 12$ if money is infinitely divisible, and $y = 11$ and $x = 11$ with a smallest money unit of 1 token. A will take away all the money but one token which B will accept since she is better off than when rejecting.

TTG was conducted in 1999. We run 12 sessions comprising 66 students. In each session, only one A and one B group interacted each of them providing one independent observation. At the end of the experiment, the experimenters matched acceptance limit and the amount taken away and informed the groups about the result.

The Power-to-Take Game (PTT) (Bosman et al. 2006)

One player group, the 'take authority', is paired to another player group, the 'responder'. Before the game is played, each participant in the experiment has to earn an income E_i by doing an individual real effort task and instructions regarding the game are given. The game has two stages. At the first stage, the randomly chosen take authority A decides on the take rate $t \in [0,1]$, which is the part of the responder B's prior-to-the-take income E_B (the sum of the group members' incomes) that will be transferred to the take authority after the second stage. At the second stage, the responder decides on the destruction rate $d \in [0,1]$, the part of E_B responders will destroy. The responder group can only destroy its own income E_B and not E_A, the total income from the real effort task earned by the take authority members. The take authority's total earning from the experiment is thus equal to the transfer $t(1-d)E_B + E_A$. Responders' total earnings equal $(1-t)(1-d)E_B$.[13]

[13]Group payoffs were equally divided among the group members, or proportional to E_i in the few cases where these individually earned incomes differed.

After having completed the individual effort task, participants were randomly divided into take authorities and responders. Then the instructions (see Bosman et al. 2006) were read and two responder groups and two take authority groups were randomly formed by letting take authorities draw a coded envelope from a box. The envelope contained a form on which the total income of the members of the matched responder group from the real effort task was stated. Each take authority group was then brought to a separate room to guarantee full anonymity between groups. After the decision time of 10 min take authorities had to fill their own total earnings from the effort task into the decision form as well as the take rate, and put the form back in the envelope which was brought to the matched responders. After having learned about the take rate chosen responders were put in their group and each group was brought to a separate room. Responder groups also had 10 min to decide on their destruction rate. The envelopes containing the forms were then returned to the take authority groups for their information. In between, members of both player types had to answer questionnaires on expectations and emotions.

The experiment was run in 1999. 70 student subjects participated in six experimental sessions, each session providing two independent observations.

The Alternating Offer Bargaining Game with random assignment of conflict payoffs (AOBr) (Hennig-Schmidt 1999; Hennig-Schmidt and Yan 2009)

AOBr is based on a non-cooperative model of characteristic function bargaining (Selten 1981) with anonymous interaction and no communication. Two players bargain on the distribution of a coalition value $v(AB)$, the pie, by alternating offers. No bargaining costs are involved; there is no time limit. The players have four decision options: (1) to make a proposal, (2) to shift the initiative to the opposite player, (3) to accept the proposal of the opposite player, and (4) to break off the negotiation. If players settle on an allocation of $v(AB)$ they receive the amounts agreed upon. If they do not reach an agreement they receive a conflict payoff $v(A)$, $v(B)$, respectively, with $v(A) > v(B)$, $v(A) + v(B) = v(AB)/2$. The experimental variables are $v(AB) = 320$, $v(A) = 128$, and $v(B) = 32$, $v(A)/v(B) = 4/1$. We also had a treatment with $v(A) = 96$, $v(B) = 64$, $v(A)/v(B) = 3/2$. Groups with the high (low) conflict payoff are called strong (weak) groups.

Participants playing together in one group were assigned to different rooms. Groups were assigned randomly to be strong or weak groups. After group members were provided with the introduction and the bargaining protocol (see Hennig-Schmidt 1999) the first proposal round started. Offers were transmitted by the experimenter. In case of acceptance or break off, the other group was informed and the subjects were paid.

German sessions were run in 1991 and 1992; Chinese sessions in 1999. In total, 127 students participated, 73 in Bonn and 54 in the Chengdu providing 20 sessions in Germany and nine sessions in China.

The Alternating Offer Bargaining Game with earned assignment of conflict payoffs and face-to-face bargaining (AOBe) (Hennig-Schmidt and Yi 2009, Hennig-Schmidt et al. 2009a, b)

AOBe is the cooperative version of AOBr. Two teams of five persons each bargain on the distribution of a coalition value $v(\text{AB}) = 300$ Euro and receive conflict payoffs $v(\text{A}) = 120$ Euro and $v(\text{B}) = 30$ Euro in case of no agreement. Teams earned their position as a high or low outside option player in a quiz preceding the experiment. Between-team bargaining was face-to-face. Between the negotiation phases, teams had internal within-group discussions to elaborate on their strategies.

The experiment was conducted during the first and second Sino-German Summer School on Experimental Economics in Chengdu and in Bonn in 2006 and 2007.[14] Sessions lasted for about 3 h. All discussions were videotaped. We had two German–German, two Chinese–Chinese and two Chinese–German bargaining sessions. We analyzed the transcripts of 12 teams; half of them were high and half of them were low outside option players.

Iterated PD (IPD) (Hennig-Schmidt and Leopold-Wildburger 2010)

In the IPD, four players (12 subjects) participate in a session each player interacting repeatedly with the same partner. After a certain number of rounds not known to the participants in advance they are rematched with a new player, one they had not interacted with before. There are three matchings per session. Payoffs are 3/3 (1/1) if both players cooperate (defect). In case that one player defects he receives 5 whereas the cooperator gets 0.

In each session, 12 subjects were jointly introduced by the experimenter. Then, subjects were randomly assigned to four groups of three people in a way that guaranteed group composition to be anonymous. The four groups played simultaneously, two groups interacting during a matching. Groups could take as much time as they wanted to decide on cooperation or defection. The experimenter informed the teams about decisions and the start of a new matching. 120 students in 40 teams participated in ten sessions, six sessions being run in Bonn in 2004 and four sessions in Graz in 2003.

Results

The Equity Principle

Selten in his papers on the equity principle in economic behavior (1978, 1987) strongly suggested its relevance for the resolution of economic distribution

---

[14]The summer schools were supported by the Sino-German Center for Research Promotion, Beijing.

conflicts. One major result of our video experiments is that equity is a major guiding principle of subjects' behavior in all allocation tasks. In that respect, the results of our symmetric and asymmetric bargaining experiments remarkably corroborate Selten's conjecture.

Allocations seem to attract subjects' attention that in some sense entail an equal division of the amount to be distributed. They are chosen because they correspond to social norms and are focal with regard to equity and justice. Selten (1978, 1987) has given a general definition of the equity principle: The amount in question is distributed to the parties involved in such a way that each party is treated equally according to a certain standard. Consider a group of n members 1,2, ... ,n and an amount r of money to be divided among these members. A division of r is a vector (r_1, ... ,r_n,) with $r_i \geq 0$ for $i = 1, \ldots, n$ and r_1,+···+r_n, $= r$. r_i is called the share of member i and w_i the weight of member i, with $i = 1, 2, \ldots, n$. The equity principle requires the following relation to hold:

$$\frac{r_1}{w_1} = \frac{r_2}{w_2} = \cdots = \frac{r_n}{w_n} \tag{10.1}$$

Table 10.1 shows equitable allocations due to different amounts being at stake for allocation – the *standard of distribution* in the terminology of Selten's (1978) – and the quota being applied to the standard of distribution to arrive at the final outcome – the *standard of comparison*. These allocations have been mentioned during group discussions and were in participants' focus because of their equality feature. Selten (1987) emphasizes that only in special cases the application of the equity principle gives rise to an equal division. In many situations, in particular in the asymmetric ones, there are good reasons for an uneven split of r.

Exactly that is what has been found in the video experiments. Apart from the symmetric UG, *more than one* equity norm has been identified. And except for the Proportional Split, all allocations are characterized by an egalitarian standard of comparison. Since the standards of distribution differ, however, rather unequal payoffs for both players result. ESSD is a compromise between different standards of distribution and is found by an 'iterated equity principle'.[15]

Selten (1978, 1987) stresses a feature of the equity principle that may cause problems in its practical application. When the standard of distribution and the standard of comparison are known or undisputed the final outcome is clear. No general rule exists, however, how both should be determined. Even though the number of reasonable equity standard might be quite small there is potential for essential conflicts. In AOBr, for instance, both players may agree on the coalition value as the standard of distribution. Yet, player A may want the standard of comparison to be proportional to conflict payoffs, whereas player B strives for an

[15]Reinhard Selten: "Lectures on Bounded Rationality", University of Bonn, winter term 1995/ 1996.

Table 10.1 Equity norms, standards of distribution and comparison in our experimental games

Game	Equity norm	Standard of distribution	Standard of comparison	Payoff Player A	Payoff Player B
Ultimatum (UG)					
Symmetric T1	– Equal split of endowment (ES)	– Endowment	– Egalitarian	10	10
Asymmetric T2	– ES	– Endowment	– Egalitarian	10	10
	– Split the difference (SD)	– Endowment minus sum of outside options	– Egalitarian	14	6
Asymmetric T3	– ES	– Endowment	– Egalitarian	10	10
	– SD	– Endowment minus sum of outside options	– Egalitarian	13	7
Tripled Take (TTG)	– Equality of payoffs	– Joint payoff	– Egalitarian	9	9
	– Equal split of endowment player B	– Endowment player B	– Egalitarian	18	6
Power-to-take (PTT)	– Equality of payoffs	– Joint earned income	– Egalitarian	20	20
	– Equal split of earned income player B	– Endowment player B	– Egalitarian	30	10
Alternating Offer (AOBr)[a]	– Equal split (ES)	– Coalition value	– Egalitarian	160	160
	– Split the difference (SD)	– Coalition value minus sum of conflict payoffs	– Egalitarian	208 (176)	112 (144)
	– Proportional Split (P)	– Coalition value	– Proportional to conflict payoff	256 (192)	64 (128)
	– Equal split of difference between SD and ES (ESSD)	– (SD–ES)	– Egalitarian	184 (168)	136 (152)
	– Equal split of difference between proposals of bargainers	– Difference between proposals of bargainers	– Egalitarian	Depending on proposals	Depending on proposals
	– Equal split of difference between aspiration levels of bargainers	– Difference between aspiration levels of bargainers	– Egalitarian	Depending on aspiration levels	Depending on aspiration levels
Alternating Offer (AOBe)	– Equal split (ES)	– Coalition value	– Egalitarian	150	150
	– Split the difference (SD)	– Coalition value minus sum of conflict payoffs	– Egalitarian	205	95
	– Proportional split (PS)	– Coalition value	– Proportional to conflict payoff	240	60

[a]Payoffs in parentheses are those of the second treatment of AOBr

egalitarian standard of comparison. Both players will arrive at rather divergent payoffs which may render an agreement impossible.

How does the equity principle materialize in A-players' final choices? Either the final outcomes will result if a player B accepts player A's decision or if player B does not destroy own income in PTT. Remarkably, the relative frequencies of final equitable choices are not higher than 33% in the anonymous games and 66% in the face-to-face AOBe. From choices alone one would not have been able to infer that participants discussed and were guided by several other equity norms (Table 10.1). Only the transcripts reveal that these norms influenced the bargaining process. In the following two sections we analyze how equity norms affect decisions by their impact on aspiration levels (section "Aspiration Levels and Equity") and by their fairness characteristic (section "Fairness and Equity").

Aspiration Levels and Equity

Selten attaches great explanatory power to the concept of aspiration levels, i.e. distributive goals negotiators want to achieve (Sauermann and Selten 1962, see also Selten 1998). Goal-directed behavior is more adequate than optimization to comply with the boundedly rationality humans are subjected to (Selten 1998, 2000).

Hennig-Schmidt and Yan (2009) analyze the transcripts of AOBr to check for evidence of goal-directed behavior.[16] A hierarchical coding was applied. First, two independent raters coded for aspiration levels, being the payoff a group member is willing to accept as final payoff of the negotiation for at least one round. Then, the raters coded the aspiration levels for equitable allocations (see Table 10.1). The category system also accounted for the phenomenon that round numbers attract attention and are often preferred to non-round numbers (Albers and Albers 1983; Selten 1987). Moreover, prominent allocations derived from the equity principle were found. This is in line with Selten (1987) emphasizing that even when results are based on equity considerations small deviations due to prominence can occur.

Behavior in AOBr was found to be consistent with the aspiration level approach.[17] Subjects build aspiration levels and adapt them during the bargaining process. Most initial aspiration levels are formulated quite at the beginning of the negotiation.

Selten also hypothesized that "considerations of equity may have an [...] influence on the formation of aspiration levels" (Selten 1978, 280). Hennig-Schmidt and Yan (2009) check his conjecture by analyzing the impact of equity on aspiration levels in AOBr.[18] Initial aspiration levels are found to be strongly influenced by modifications of the equity principle. Strong groups are motivated by Proportional Split, Split the Difference and Equal Split whereas weak groups are motivated mainly by the Equal Split. Strong groups start the negotiation by demanding more or less

[16]See also Hennig-Schmidt (1999).

[17]The same is true for AOBe, see Hennig-Schmidt and Yi (2009).

[18]See also Hennig-Schmidt (1999).

their initial aspiration level. Weak groups, however, try to conceal their true initial bargaining goals by demanding significantly more (offering significantly less) than they really strive for in the beginning of the bargaining process.

The motives for final aspiration levels are not as consistent as for initial bargaining goals. This is plausible as initial aspiration levels are formed in the beginning of the bargaining process and are not yet influenced by the opponent's behavior. Throughout the bargaining process, however, groups have to take their opponents' offers and demands into account and have to adjust their own aspiration levels if they do not want to risk a break off. Equity and Prominence are still of importance, even though the equity principle loses its importance during the bargaining process. Yet, it has the important influence on aspiration formation Selten predicted: It delineates the range of intended negotiation outcomes and serves as a reasonable, well-justifiable and well-justified baseline for goal formation.

Fairness and Equity

Fairness was found to be an important issue for the decision process in the video experiments. Moreover, a very close relation between equity and fairness was found, a connection Selten has repeatedly emphasized (see e.g. Selten 1978, 1987, 2000).

The experimental and theoretical literature on social preferences assumes that persons concerned with fairness behave differently from those who are not (e.g. Fehr and Schmidt 1999; Bolton and Ockenfels 2000; Rabin 1993; Dufwenberg and Kirchsteiger 2004; Falk and Fischbacher 2006). The distinction between fair-minded and 'selfish' people is usually based on choices or questionnaires. The verbal data of the video experiments allow a clear distinction between those groups who actually argue about fairness and those who don't. The fraction of fair-minded

Table 10.2 Number of player groups discussing fairness at least once

	# Fairness concerned players[a]	Total # players	Percentage
UG (Walkowitz et al. 2010)	50	69	72.46
TTG (Sadrieh and Hennig-Schmidt 1999)	11	24	45.83
PTT (Bosman et al. 2006)	14	70	20.00
AOBr (Hennig-Schmidt 1999)	30	40	75.00
AOBe (Hennig-Schmidt and Yi 2009; Hennig-Schmidt et al. 2009a)	6	6	100.00
Total # player groups/Mean percentage	111	209	53.11

[a]*Note:* In all experiments but PTT numbers refer to player groups. In PTT, numbers comprise individual group members

participants can easily be calculated and behavior of these players can be checked for differences. See Table 10.2 for the experiments included in this analysis.

A hierarchical coding was used to check for fairness concerns. On the first level, we analyzed whether fairness was discussed or not. The verbal data were coded by two independent raters with regard to subjects mentioning the words *(un)fair*, *(un) just* (in German "(un)gerecht, (un)fair") in all word combinations like *fairness, justice, unfairness, injustice*. In UG, the transcripts were auto-coded by the text analysis software ATLAS.ti. Subsequently, two independent raters coded the fairness discussions for equitable allocations as in Table 10.1 and other categories. These, however, turned out not to be of major importance and are neglected in the following analysis.

One of the main results from coding the transcripts is that with very few exceptions, fairness is attributed to equitable allocations. All allocations mentioned in Table 10.1 have been connected to fairness at least once in a session.[19] Not all players make fairness an issue, though (Table 10.2). On average, 53% of them do so at least once. The relevance of fairness for justifying specific equitable allocations like Equal Split, Split the Difference and Proportional Split is intensively discussed.

The content analysis of Walkowitz et al. (2010) provides a clear tendency in UG. The Equal Split – being the predominant fairness norm for those groups who make it an issue – seems to capture what participants associate with fairness in the symmetric *and* in the asymmetric settings. This is an astonishing finding as it has been maintained in the literature on self-serving biases in fairness perceptions that in asymmetric situations subjects tend to those fairness notions that give them a higher payoff (see e.g. Komorita and Kravitz 1979; McClintock et al. 1984; Babcock et al. 1995; Babcock and Loewenstein 1997). Equal-Split related fairness discussions are also found to have a pronounced effect on A-players' decisions. Those who connect the Equal Split to fairness increase their offers significantly. Fairness concerns per se do not have such an effect.[20] The Equal Split predominance also found in TTG corroborates the assumption in the experimental and theoretical literature that the notion of fairness is closely related with the Equal Split.

In PTT and the alternating offer experiments, the picture looks different. Other equitable but unequal allocations become important as reference points for fairness as well: In PTT, it is the equal division of player B's endowment which involves a take rate of 50%; in AOBr and AOBe, it is Split the Difference and Proportional Split (see Table 10.1). The equality feature not only loses relative importance but also a large discrepancy between the fairness notions of both player types is found. In many cases, the strategic asymmetry induces B-players to show aversion against

[19]Exceptions are the Equal Split of the difference between proposals/aspiration levels of bargainers.

[20]Hennig-Schmidt and Geng (2006) investigate the impact of fairness on behavior in PTT using the transcripts of Bosman et al. (2006). Here, fairness-oriented take authorities claim lower take rates than groups without such a concern. Their payoffs, nevertheless, are higher, as are the payoffs of the corresponding responder groups as the latter had no incentive to destroy.

inequality as Fehr and Schmidt (1999) have assumed, whereas A-players exhibit an aversion towards *equality*. The self-serving bias, which has been well established for individuals, also shows up in the group setting. It appears that these individual biases are not corrected in the group discussion.

Content analyzing the above video experiments reveal that different fairness norms exist even in one game, and that fairness norms differ across games. Fairness perceptions coincide with the different appearances of the equity principle. When explaining behavior, it seems problematic to base general models on one specific fairness concept only.

The IPD differs from the above games in that it does not involve an allocation task. Participants nevertheless might be motivated by fairness criteria and inequity aversion if they are concerned with equality of payoffs. For instance, they can achieve the same payoffs if both players always cooperate or always defect. The same holds for a regular alternating pattern of one player cooperating and the other one defecting. Different final payoffs result if the latter pattern occurs irregularly. The transcripts of Hennig-Schmidt and Leopold-Wildburger (2010) reveal that fairness does come up during the discussions but not very frequently (in around 35% of the player groups). In particular is fairness associated with the cooperative move or with some kind of alternation after a sucker situation. Fairness with regard to equal final payoffs is not explicitly made an issue.

The verbal data also allow studying whether players anticipate an end effect (Selten and Stöcker 1986) and how they are affected by past cooperation/defection. This analysis is beyond the scope of this essay and can be found in Hennig-Schmidt and Leopold-Wildburger (2010).

Mentalizing

The UG, TTG and PTT video experiments are played as non-cooperative one-shot games with anonymous interaction and without communication. In PTT, the first-moving take authority has the difficult task to predict how much the responder will destroy as a reaction to the take rate. In UG and TTG, both players choose simultaneously as we use the strategy method, and thus player B has to predict player A's decision as well. How do players cope with this situation? What are the cognitive processes involved?

It is plausible to expect that subjects intensively think about their counterpart, like in Selten's model of imaginary bargaining (Selten 2000)? Reasoning about others has been referred to as "theory of mind", "mentalizing", or perspective taking and comprises the ability of humans "to make attributions to the mental states (desires, beliefs, actions) of others" (Singer and Fehr 2005, 340). Studies in psychology (Hertwig and Ortmann 2001), neuroscience (Singer 2006) and neuroeconomics (Bhatt and Camerer 2005) found evidence that subjects mentalize.

Mentalizing is also observable in the discussions of the video experiments; see Geng (2004), Geng and Hennig-Schmidt (2007), Walkowitz et al. (2010) for analyses of the UG text protocols and Hennig-Schmidt and Geng (2005, 2006) for investigations of the PTT transcripts.[21] When mentalizing, subjects form a mental image of the counterparts' actions, goals, and beliefs. Cognitive mechanisms of different complexity are involved. On the one hand, participants merely form an opinion about their counterpart's actions and intentions. On the other hand, subjects explicitly put themselves into the counterpart's shoes and view the situation as if they were that person. To distinguish both mechanisms Walkowitz et al. (2010) call the first one *Own perspective*; the more complex cognitive scheme is termed *Counterpart's perspective*. In the latter, subjects actually put themselves into the perspective of the counterpart and proceed a cognitive step further than in *Own perspective* mode. Hennig-Schmidt and coauthors also found that participants – in both modes – reason on first and higher-order levels.[22] This is in line with findings from e.g. Nagel (1995, 1998) and Costa-Gomes and Crawford (2006).

The cognitive processes involved are complicated, in particular when mentalizing in higher-order levels. For instance, a person in *Own perspective*-Level 2 (*OP2*) contemplates about what the counterpart may think the person (group) herself will do or think. In *Counterpart's Perspective*-Level 2 (*CP2*) a subject contemplates about what she being in the counterpart's shoes would think that she herself (her group) would do or think. The following examples are translations from UG transcripts.[23]

- *OP2*: In my [*Proposer*] opinion they [*Responder*] also reflect upon how we [*Proposer*] decide.
- *CP2:* If I [*Responder*] were in the other [*Proposer*] group then I [*as Proposer*] would not accept that [allocation]. Because I [*as Proposer*] would say to myself perhaps the others [*Responder*] will say that they [*Proposer*] will get much more than we [*Responder*] get.

Even though rather sophisticated cognitive processes are found, group members seem to ignore the decision rule of their counterpart and typically view them as if they were single agents (Bosman et al. 2006). This is surprising, as the reaction of a responder group to a particular take rate in PTT may depend on the decision rule that responders use. If, for instance, in each responder group at most one member would want to destroy and the simple majority rule is used, take authorities could claim whatever they like. If responders use other rules, a lower take rate may be optimal for the take authorities.

The ability to understand other people is not restricted to the capacity to understand their actions, beliefs and intentions (Frith and Singer 2008). Successfully solving distributive tasks in particular in our one-shot non-cooperative games

[21]Hennig-Schmidt and Geng (2005) also use the transcripts of Bosman et al. (2006).

[22]Again content analysis and a hierarchical coding procedure is used. Two raters in a first step coded the transcripts for *Own* and *Counterpart's perspective*. In a second step they coded for *Levels of Reasoning*.

[23]Parentheses [..] are inserted to facilitate understanding the complicated thought process.

also requires empathetic feelings, i.e. understanding the counterpart's emotional states. Take PTT as an example. High take rates may induce angry responders to destroy own income in order to punish greedy take authorities. To avoid this kind of reaction, it is very important for take authorities to anticipate responders' emotions before deciding on a take rate. See van Winden (2001), Bosman and van Winden (2002), Bosman et al. (2005), Reuben and van Winden (2008, 2010) on how emotions affect responders' behavior in individual PTT experiments. The content analysis of the PTT and UG video experiments (Hennig-Schmidt and Geng 2005, 2006) reveals that A-players do take B-players' emotions into account.[24] They discuss that emotions may induce their counterparts to reject offers in UG or destroy own income in PTT. Imagining the counterpart's feelings, arguing in one or both mentalizing modes on higher levels of reasoning can help getting a more realistic picture of the situation. Hennig-Schmidt and Geng (2006), for instance, find a significant correlation between the relative frequency of A-players' arguments in *Counterpart perspective* mode and discussions of B-players' potential emotions.

Singer and Fehr (2005) regret the insufficient knowledge on how individuals build beliefs about others. They suggest future research on the relative importance of mentalizing and empathizing for the prediction of others' motives and actions. The research using verbal data provides new insights that might help to fill the gap. Moreover, it would be interesting to align the above findings with neuro-economic research. Do neural correlates exist when subjects think in one or the other of the two mentalizing modes like the ones Corricelli and Nagel (2009) found when analyzing depths of strategic reasoning? If so which regions of the brain are involved?

Observing Group Processes in Coalition Games

In this section we report on observing group processes in coalition games. Here, individuals are in conflict situations. We analyze the impact of cognitive and socio-emotional factors on the negotiation process.

Experimental Setup

The data are collected in two series of 13 sessions each. In each single session, six subjects were involved. No subject took part twice. After a 10-min warm-up, the

[24] The transcripts were content analyzed for emotion words like anger, rage, sadness, defiance, revenge, pain.

subjects were instructed to perform a group task in discussing and analyzing a bargaining game with six players (open-ended, with a mean duration of about 20 min). After this group task, each subject had to fill in a first interpersonal rating form (mean duration: 15 min). The instruction for the bargaining task followed. This bargaining task was based on a specific six-person game. After the random assignment of roles (players were given pseudonyms like "Hem" and "Dex") a questionnaire on aspirations and intentions had to be completed. Single letters designating players' names in the tables below are abbreviations of theoretical characteristics of these players. For example, can player S be seen as "strong". The characteristics of the players will be explained below.

Next, the subjects had to perform the bargaining task. The two treatments differed in presentation and rules for bargaining. The $2 \times 13 = 26$ sessions were run in the years 1990 and 1991 in the Video Laboratories of the Department of Psychology of the University at Saarbrücken. In all but one session, all subjects were students. In one session, managers of a local firm served as participants.

In the first treatment, bargaining took place within committees. Each committee could decide on proposals to distribute a fixed sum of money. The only way to reach a valid decision was for the same payoff vector to pass two subsequent committees. Bargaining took place in a sequence of sessions of different committees (each session lasted for 10 min). The starting committee was randomly drawn. Every committee was free to decide on a payoff vector and on the committee to decide next. A proposal became valid if the majority level was reached. If the proposal was one on payoffs and if it was identical to the decision of the previous committee, it became the final decision and the bargaining was closed. The committee structure is given in Table 10.3.

In the second treatment, bargaining took place in the whole group of six participants. The subjects were instructed that they individually are endowed with two resources that are needed to perform a joint project. They had to negotiate a contract by forming a coalition of contributors and by agreeing (within this coalition) on the distribution of profits. Table 10.4 shows the distribution of endowments and the minimum amounts needed to engage successfully in the project.

After bargaining, the subjects filled out the second interpersonal rating form.

Table 10.3 Distribution of votes (committee treatment)

Committee/player	S	P	B	W	Z	F	Majority level
Committee "Red"	1	2	–	–	–	1	3
Committee "Blue"	2	3	2	1	1	–	5
Committee "Yellow"	2	–	2	1	1	–	5

Table 10.4 Distribution of endowments (resources treatment)

Resource/player	S	P	B	W	Z	F	Amount needed
Resource 1	1	–	–	–	–	1	1
Resource 2	2	3	2	1	1	–	5

The Game

The game can be represented by listing its five minimal winning coalitions: SP, SBW, SBZ, PBF, PWZF. Every coalition equal to or larger than one of these five coalitions is winning and can distribute the payoff while all other coalition are losing and would get a payoff of zero.

It is evident that players W and Z are symmetric. A simple substitution analysis (for determining the so-called desirability relation of that simple game), shows that S and P are stronger than any of the other players, because on the one hand, a substitution for one of both by one of the others would leave the set of winning coalitions and, on the other hand, a substitution for another player by S or P can enter that set and will never leave it. If we denote "x is stronger than y determined by substitution" by $x > y$, knowing that the relationship is transitive, we can represent that strength by the following shorthand:

- S, P > B, F, and B > W, Z.
- W and Z symmetric.

For each of the following pairs, we cannot determine which is stronger: S and P, F and B, F and W, and F and Z.

If we relate the membership of P and F in the five minimal winning coalitions, we can see that F is dependent on P insofar as each of the minimal winning coalitions of which F is member requires the membership of P too, whereas P has an alternative coalition without F being a member. In the game given here, no other pair of players is related in such a way.

According to the relations of strength and dependence, a player is in one of five roles, which can be named in an intuitive way as follows: the strong, the protector, the bourgeois, the weak-symmetric, and the follower (s, p, b, w, and f). This game originates from experiments by Kravitz (1987).

Data Set and Results

The video lab was equipped with two cameras and a separated control room; here the cameras and the recording were operated. The cameras could be moved and the recorded pictures could be modified by zooming, splitting and composing; title marks and time counters were added. The records are used to produce transcripts, sequences of coded acts (with time, speaker, target and content) and other material to further understand the bargaining process. The videos are confronted with other data from (1) questionnaires on aspirations and intentions and (2) from two sets of ratings on the perceived socio-emotional orientations of the self and the partners, the so-called social-field data – coded according to SYMLOG rules, Bales and Cohen (1979).

In some experiments, conflict induction failed. Subjects did not enter the conflict and distributed the payoff equally among all partners without bargaining. In the remaining 17 conflict-based negotiations 13 minimal-winning and 4 over-sized coalition were formed. For purely self-interested subjects, over-sized coalitions are "irrational" – subjects would give up money. On the other hand did it become clear in our experiments that the over-sized coalitions in some cases not only were seen as socially more acceptable but also as strategically more stable, since a single person's extortion becomes unlikely.

Symmetry plays a major role in the generation of arguments. Symmetric players tend to be treated equally. The attempt to play off the subjects in symmetric roles against each other is quite dangerous. In such a case the usual answer is strong resistance and in most cases such moves end in the exclusion of the claimant.

A comparison of the first social field ratings before bargaining, the second social field ratings after bargaining and the video data not only showed that behavior and socio-emotional orientation of the subjects adapt to the role in the game. It also became apparent how this process worked. Subjects in weak roles tend to avoid conflict and show an increase of "withdrawel"-scores, whereas subjects in powerful roles show the opposite (Ostmann 1996).

A first analysis of the committee treatment (Ostmann 1992a) showed that membership in the final coalition was negatively related to high "fight"-scores, while subjects with extremely low "withdrawel"-scores ("initiators") nearly always were included in the final coalition. A second study compared this data with the aspirations reported in the questionnaires and revealed in the bargaining process (Ostmann 1992b). This analysis showed that in addition to the social field data extreme aspiration values are important for explaining membership.

The above results were also found in the resources treatment (Ostmann 1994a, b). Since the data from both treatments are quite similar the data were pooled for a further analysis (Ostmann 1996). Aspiration data were found to be the best predictor for conflict. Very low minimal aspiration values of the powerful subjects favor agreements whereas extreme aspiration values increase the probability of conflict.

Cross-Cultural and Inter-Cultural Settings

This part of our essay reports comparative studies across subject pools in Germany and China, and intra-cultural research in a Chinese subject pool. Why are studies concerned with different countries and cultures interesting and important? And why is China the focus of the studies reported?

It has been argued that stylized facts taken for granted in one's own culture may have no or minor significance in another. As Allison (1998, 3) put it "We may be required to alter our modes of understanding in order to interpret the signs of another culture correctly". He also emphasizes that we should be as much aware of the similarities that exist as of the differences. Globalization induces interaction between countries far apart in distance as well as in philosophical and cultural backgrounds. These differences might cause frictions, in particular in the negotiation context. If bargaining partners are ignorant about each others' behavior, goals and motivations they might fail to reach a mutually satisfactory agreement. Successful business relations require the reduction of uncertainties caused by misunderstanding. Cross-cultural and inter-cultural studies, including economic experiments, can help in this respect.

Methodological and theoretical issues might also call for comparative studies across countries/cultures. If researchers study behavior in their own environment only they might fall prey to biases like the false consensus effect (Ross et al. 1977). They might not be able to imagine that behavior in other societies/cultures differ from their own because of motives not apparent to them from their own experience. This bias and ignorance may also result in constructing decision models that neglect important explaining variables from other parts of the world (see also van de Vijver and Leung 1997). Research in cross-cultural psychology has shown that decision models based on a Western cultural research tradition seem not adequate for explaining Chinese decision making behavior (Leung and Bond 1984; Chiu 1990). A famous example is Hofstede's 5-dimension model designed to explain cultural differences (Hofstede 1980, 1991). The fifth dimension was only added after researchers concerned with the Confucian value system found these values not adequately represented in the original 4-dimension model (Chinese Culture Connection 1987).

In the studies surveyed below, Chinese subjects served as comparison for German participants because both countries are different in many respects. Germany is influenced by the Greek/Christian culture and philosophy and China by Confucianism/Taoism (Jullien 2004). Both societies diverge strongly according to criteria like trust, rule of law, GDP, democracy, power distance, individualism and uncertainty avoidance developed in order to characterize societies (Herrmann et al. 2008b, Table S1). These differences are likely to influence behavior and the underlying motives. Similar behavior and guidance by the same principles, however, would make a strong case for cross-cultural validity.

When running experiments in different cultures one must pay attention to issues like language, currency effects, stakes, and experimenter interactions as all of them can affect cross-cultural comparability (cf. Roth et al. 1991; Henrich et al. 2001, 2004; Herrmann et al. 2008a, b). These effects were taken into account by using the back translation procedure (Brislin 1970), and calculating participants' rewards to equal the hourly wage in a typical students' job in both countries. Sessions were run by native experimenters. Moreover, in each experiment one of the experimenters (Hennig-Schmidt) was present in the German *and* the corresponding Chinese sessions. Moreover, all Chinese and German experimenters were provided with an extensive protocol in English to guarantee that sessions were conducted as similar as possible.[25] For a more detailed analysis of all relevant factors important

[25]See Walkowitz et al. (2009) for an extensive description of the organizational and procedural steps in an inter-cultural experiment.

in cross- and inter-cultural research see the excellent papers by Herrmann et al. (2008a, b).

Aspiration Levels and Equity: Cross-Cultural Analysis

Hennig-Schmidt and Yan (2009) compare the German and Chinese sessions of AOBr[26] and find remarkable similarities across countries. Most importantly, also in the Chinese subject pool aspiration formation is strongly influenced by the equity principle with strong negotiators striving for unequal and weak bargainers pursuing equal allocations. As argued in Hennig-Schmidt (2002), two kinds of aversion seem to exist; weak players exhibit *inequality* aversion, whereas most strong players exhibit an aversion against *equality*. Using the language of Fehr and Schmidt (1999), strong bargainers may dislike egalitarian outcomes and may not suffer from inequity being to their material advantage. In the course of the bargaining process, equity loses significance for aspiration adaptation; instead, prominence gains importance. Aspiration levels in both cultures are guided by equity and prominence making a strong case for these principles' cross-cultural validity.

Fairness-related aspiration levels mainly correspond to the equity principle and appear more important in the German than in the Chinese treatment.

Hennig-Schmidt and Yan (2009) do not find evidence for differences in bargaining outcomes between the subject pools in both countries. Yet, strong discrepancies with regard to the level of resistance against downward aspiration adaptation exist. Germans tend to steadily reducing their goals whereas Chinese display long periods of stagnation or minimal concessions (Fig. 10.1). Part of these delay strategies implies false goals which are meant to make opponents perceive the false information as the true aspiration level. Interestingly, the delay strategies do not result in active negotiation break offs as shown by German participants. If groups of both countries had negotiated directly with each other without adapting their behavior the different bargaining practices most probably would have resulted in many sessions ending in conflict.

The importance of the equity principle in both the German and Chinese subject pool are corroborated by AOBe (Hennig-Schmidt and Yi 2009; Walkowitz et al. 2010). Equity plays an important role in all teams independent of their nationality with Split the Difference being central for aspiration formation. Moreover, all demands (offers) in the first negotiation round pertained to the equity principle, Proportional Split for strong and Equal Split for weak groups.

[26]We used the same coding system and procedure as in Germany, see section "Aspiration Levels and Equity".

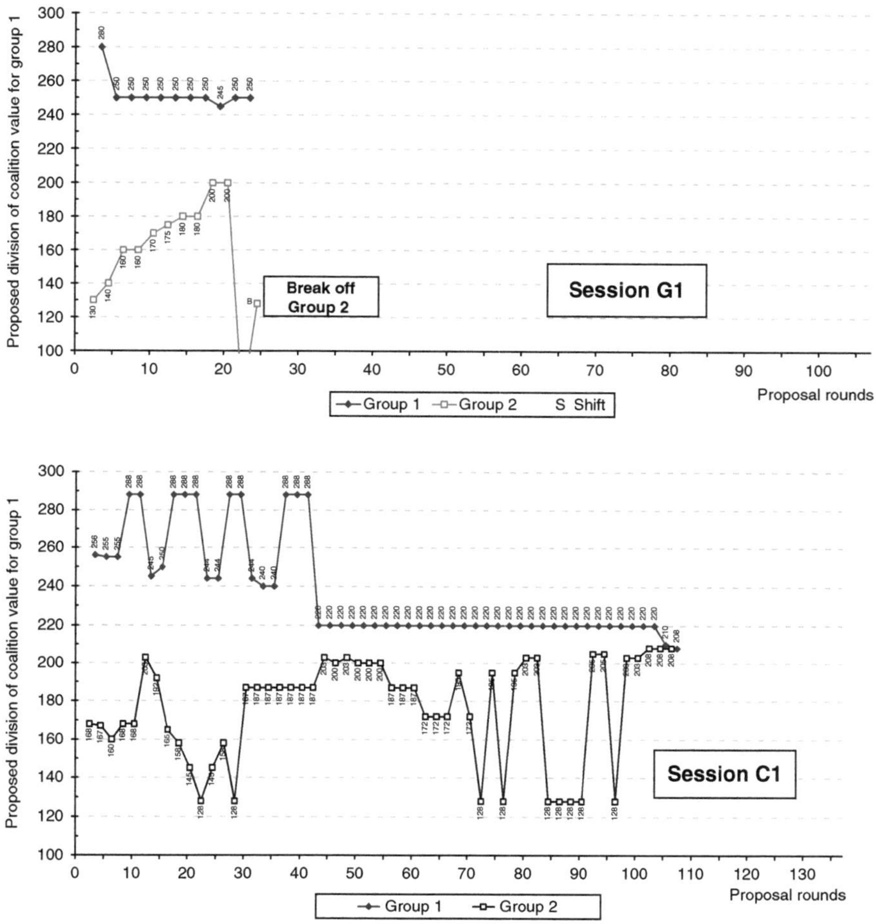

Fig. 10.1 Examples of negotiation processes in AOBr in Germany above and China below
Note: Group 1 is an A-player and group 2 a B-player.

Fairness and Culture

Walkowitz et al. (2010) also analyze the role fairness plays in the discussions of the Chinese subjects in UG and provide a comparison between the two subject pools. Justice and fairness in the sense of impartiality and lack of bias towards everybody are of central concern in Western thinking rooted in the Greek philosophy. It does not go without saying that fairness is of comparable importance in Chinese societies. Here, a different notion of fairness prevails because justice is conceived as the fulfillment of role expectations (Chiu and Hong 1997) and different justice standards apply for different social relations (Zhang and Yang 1998). This also holds for modern Chinese societies. Given the differences in fairness conceptions in the East and the West, the authors conjecture that fairness

Table 10.5 Fairness discussions in individual and video experiments

	Country	# Player groups		# Individuals	
		abs. (all)	rel. (%)	abs. (all)	rel. (%)
Video experiments, group discussions					
• Alternating offer bargaining game (AOBr) (Hennig-Schmidt 1999)	Germany	30 (40)	75.00		
• Tripled Take game (TTG) (Sadrieh and (Hennig-Schmidt 1999)	Germany	12 (24)	50.00		
• Power-to-take game (PTT) (Bosman et al. 2006)[a]	Germany			14 (70)	20.00
• Alternating offer bargaining game (AOBr) (Hennig-Schmidt and Yan 2009)	Germany	[b]	[b]		
	China	*17 (18)*	*94.44*		
• Alternating offer bargaining game (AOBe) (Hennig-Schmidt and Yi 2009; Hennig-Schmidt et al. 2009a)	Germany	6 (6)	100.00		
	China	*6 (6)*	*100.00*		
• Ultimatum game (Walkowitz et al. 2010, *UGTeam*)	Germany	50 (69)	72.46		
	China	*44 (72)*	*61.11*		
Individual experiments, answers to open questions					
• Ultimatum games					
–Kohnz and Hennig-Schmidt (2005)	Germany			10 (80)	12.50
–Hennig-Schmidt et al. (2009b)	Germany			63 (218)	28.90
–Hennig-Schmidt et al. (2010)	Germany			99 (192)	52.56
	China			*26 (160)*	*16.25*
–Walkowitz et al. (2010), UGInd	Germany			64 (180)	35.56
	China			*39 (179)*	*21.79*
• Power-to-take game (Bosman et al. 2009)	*China*			*7 (76)*	*9.21*

[a]In PTT, the numbers refer to individual group members
[b]Subsample of Hennig-Schmidt (1999)

and related arguments are likely to play an important role in Germany and a minor role in China. Similarities across countries would make a strong case for its cross-cultural validity.

The analysis reveals several cultural differences with regard to fairness perceptions.[27] Fairness does not appear irrelevant for the Chinese but seems more important for the German participants: (1) Fewer Chinese player groups discuss about fairness. (2) They start discussing it much later during the negotiation process. (3) Although the Equal Split is the pertinent fairness norm in both countries less Chinese groups discuss this fairness norm. (4) When mentalizing, fewer Chinese teams are concerned with their counterparts when discussing fairness. (5) More Chinese player groups argue on the notion of fairness as a general principle. These results suggest that culture might matter when fairness is concerned.

[27]The transcripts were first auto-coded by ATLAS.ti for Chinese subjects mentioning the words (不)公平 (bu) gong ping, (不)合理 (bu) he li, (不)平等 (bu) ping deng, as equivalents for (un)fair, (un)just (in German "gerecht, fair") in all word combinations. Then the same coding system and procedure as in Germany was applied, see section "Fairness and Equity".

Walkowitz et al. (2010) analyze the importance of fairness in the experiments mentioned in this survey plus in several additional ones (see Table 10.5). They also include experiments where participants decide individually and answer open questions on the influencing factors of their decisions; their written statements were coded for fairness as in the studies reported above. The survey comprises around 1.350 independent observations consisting of 1,800 participants in Germany and China. Fairness in spontaneous arguments was found to be of minor influence as only 35% of all players/player groups talk about fairness (27% in China, 40% in Germany). The findings from UG were corroborated in that highly significantly more German than Chinese participants talk about fairness.

Non-monotonic Strategies in Ultimatum Bargaining

The video transcripts also allow analyzing the underlying motives for an astonishing phenomenon in the Chinese UG treatments: More than half of the player groups – and individuals in later experiments as well – stated non-monotonic strategies, i.e. they rejected low *and* high offers (Hennig-Schmidt et al. 2008). Similar findings were reported by Henrich et al. (2001), and later by other authors (e.g. Bahry and Wilson 2006; Bellemare et al. 2008); yet the literature was far from giving a consistent explanation.

When offers are low, responders are likely to perceive disadvantageous allocations as unfair because being treated unfairly corresponds to low payoffs. The monetary and the motivational incentive are not at odds and rejecting a low offer becomes a likely action. When confronted with advantageous allocations of offers higher than 50% of the pie being treated unfairly corresponds to high payoffs. Now, the monetary and the motivational incentive are in conflict. If the motivational incentive prevails what are the reasons for such seemingly implausible behavior?

The content analysis of Hennig-Schmidt et al. (2008) provides important insights to this question.[28] Social concern was found to be the main motivation for refusing advantageous proposals. This is in agreement with models of inequity aversion. An inequity-averse responder suffers a loss in utility when he is worse off *and* when he is better off than the proposer (Bolton and Ockenfels 2000; Fehr and Schmidt 1999). Other motives turn out to be important as well. Among these are beliefs about proposer behavior, in particular non-expectancy of high offers, but also emotional, ethical, and moral reasons, special decision rules employed in some groups and aversion against unpleasant numbers. Reciprocity avoidance was advanced as a motive for high offer refutations by Henrich et al. (2001) and captures arguments in favor of status-seeking through gift-giving; see also Ostmann (2003) and Kohlert et al. (2005). Subjects may reject advantageous offers

[28]The coding system was designed to capture the motivations for rejecting advantageous offers.

because they experience the experimental setting like a familiar real-life situation where they would have to reciprocate a large gift. Hennig-Schmidt et al. (2008) expected to find comparable arguments during the group discussions because in the Chinese society, large gifts do create strong obligations to give in return at a later date. No such argument was found, yet subjects felt uncomfortable about advantageous offers because they may involve a trick or a ploy.

The robustness of the results in the group experiment was assessed by repeating the experiment with participants that decided individually and were not observed. These individuals show the same rejection behavior regarding high offers as groups do. Thus, motivations revealed during group discussions in UG seem to pertain to individuals as well.

As social concern was found to be the main motivation for non-monotonic strategies, Hennig-Schmidt et al. (2008) argue that models of social preference, in particular models of inequity aversion, capture an important behavioral aspect in ultimatum bargaining. On the other hand, advantageous offer rejections cannot be handled by these models due to restrictions on the parameter space. The empirical findings on high offer rejections speak to reconsider the assumptions on rejection behavior.

Conclusive Remarks

The video studies surveyed above give interesting and new insights into determinants of decision processes. Moreover, they reveal remarkable regularities – also in a cross-cultural German-Chinese context. Apparently, analyzing verbal data based on observing group behavior is a valid complement to traditional techniques of analyzing behavior in economic experiments.

Many of the findings reported in this essay are based on ideas Selten has formulated already at the beginning of his scientific career, i.e. more than 40 or 50 years ago. His personality might be one reason for these ideas being far ahead of others. "Since I am slow, I have to try to be early" he characterized himself in his paper "In search of a better understanding of economic behaviour" (Selten 1995, 134). Slowness seems only a minor part of the story, if at all. Certainly, Selten had and still has visions and path-breaking, innovative ideas. He never gets tired to promote them, to defend them and to even fight for them if necessary. Frans van Winden recalls that Reinhard Selten at the Amsterdam Workshops in Experimental Economics in the early 1990s, often strongly took stance against the more lenient statistical/methodological attitudes of experimentalists from the US – e. g. regarding the number of independent observations. And this issue occupies Selten to date.

The problems Selten advances and solves are complicated ones. To make the audience listen, short and clear sentences are required. Reducing the problem's complexity by simple linguistic constructions is his credo. Despite German being

Selten's mother tongue he shares this view with Mark Twain who wrote a wonderful story about the awful German language's complexity.[29]

Where do Selten's amazing ideas come from? In an interview he told, "Sometimes I am asked where I have my best ideas. As a scientist you are always occupied by your scientific investigations. You always reflect on your research and most often it is not clear when you had your ideas. All at once they are there! One idea I had .... in a dream".[30]

We wish Reinhard Selten many innovative and fruitful ideas, when contemplating on his own, when discussing with friends and students, on hiking tours and... in productive dreams.

Acknowledgements The work reported in this essay could not have been done without the support of numerous people and institutions.

Besides Reinhard Selten's great scientific support Heike Hennig-Schmidt is grateful to Dr. Gari Walkowitz, Dr. Hong Geng and Dipl. Übersetzer Chaoliang Yang for their invaluable research assistance over the past years at the University of Bonn. They were involved in designing and running the experiments in Germany and China, translating the transcripts and evaluating the verbal data. Without them the reported analyses would not have been possible. Special thanks go to Dr. Ronald Bosman for the joint work on PTT at the University of Amsterdam and now at the Dutch Central Bank. Axel Ostmann's thanks go to the Department of Psychology at the University of Saarbrücken for providing access to the video laboratories. He is grateful to Werner Tack for long lasting research collaboration and scientific support.

Financial support is gratefully acknowledged for Heike Hennig-Schmidt by Deutsche Forschungsgemeinschaft (Sonderforschungsbereich 303, HE 2790/2, 446 CHV-111/1/00, CHV 113/174/0-1), Sino-German Center for Research Promotion, Beijing (GZ379, GZ414) and University of Bonn; for Frans van Winden and Heike Hennig-Schmidt by EU-TMR Research Network ENDEAR (FMRX-0238); for Ulrike Leopold-Wildburger by Austrian Science Foundation, Grant No. 13919 and by Karl-Franzens-Universität, Graz; for Axel Ostmann by Deutsche Forschungsgemeinschaft (TA 56/3 and OS 94/2).

Last but not least, Heike Hennig-Schmidt thanks Simone Albus, Christoph Blumert, Haye Cao, Wei Deng, Nicole Feyen, Yue Fu, Christine Hottenrott, Valentine Hunecke, Malte Jakubowski, Stephan Kober, Jan Meise, Yi Na, Ping Ni, Rainer M. Rilke, Meike Ritzer, Holger Schmidt, Heidi Schrader, Yunhui Wan and Ziyin Yan for transcribing the videos and Julia Bernd, Wei Deng, Hao Fu, Christine Hottenrott, Valentine Hunecke, Liming Li, Jan Meise, Rainer M. Rilke, Ying Shen, and Ying Wang for their assistance in text analyzing the transcripts.

[29]"An average sentence, in a German newspaper, is a sublime and impressive curiosity; it occupies a quarter of a column; [...]; it is built mainly of compound words constructed by the writer on the spot, and not to be found in any dictionary – six or seven words compacted into one, without joint or seam – that is, without hyphens; it treats of fourteen or fifteen different subjects, each enclosed in a parenthesis of its own, with here and there extra parentheses, which re-enclose three or four of the minor parentheses, making pens with pens; finally, all the parentheses and re-parentheses are massed together between a couple of king-parentheses, one of which is placed in the first line of the majestic sentence and the other in the middle of the last line of it – after which comes the *verb*, and you find out for the first time what the man has been talking about; and after the verb – merely by way of ornament, as far as I can make out, – the writer shovels in "haben sind gewesen gehabt haben geworden sein," or words to that effect, and the monument is finished" (Mark Twain 1880/2010, 11).

[30]Universität Bonn (2009), own translation from German.

References

Albers W, Albers G (1983) On the prominence structure of the decimal system. In: Scholz RW (ed) Decision making under uncertainty. Elsevier Science, Amsterdam, pp 271–287

Allison RE (1998) An overview of the Chinese mind. In: Allison RE (ed) Understanding the Chinese mind. The Philosophical Roots. Oxford University Press, Hong Kong, pp 1–25

Anderson JR (1990) The adaptive character of thought. Lawrence Erlbaum Associates. Hilldale, New Jersey

Artale A (1996) Rings in auctions: an experimental approach, vol 447, Lecture notes in economics and mathematical systems. Springer, Berlin

Austin J, Delaney PF (1998) Protocol analysis as a tool for behavior analysis. Anal Verbal Behav 15:41–56

Babcock L, Loewenstein G (1997) Explaining bargaining impasse: the role of self-serving biases. J Econ Perspect 11:109–126

Babcock L, Loewenstein G, Issacharoff S, Camerer C (1995) Biased judgments of fairness in bargaining. Am Econ Rev 85:1337–1343

Bahry D, Wilson R (2006) Confusion or fairness in the field? Rejections in the ultimatum game under the strategy method. J Econ Behav Organ 60:37–54

Bakeman R (2000) Behavioral observation and coding. In: Reis HT, Judd CM (eds) Handbook of research methods in social and personality psychology. Cambridge University Press, Cambridge, UK, pp 138–159

Bales RF, Cohen SP (1979) SYMLOG: a system for the multiple level observation of groups. Free Press, New York

Bartholomew K, Henderson AJZ, Marcia JA (2000) Content analysis and narrative analysis. In: Reis HT, Judd CM (eds) Handbook of research methods in social and personality psychology. Cambridge University Press, Cambridge UK, pp 286–312

Becker O, Feit T, Hofer V, Leopold-Wildburger U, Selten R (2007) Educational effects in an experiment with the management game SINTO-Market. Cent Eur J Oper Res 15(4):301–308

Bellemare C, Kröger S, van Soest A (2008) Measuring inequity aversion in a heterogeneous population using experimental decisions and subjective probabilities. Econometrica 76:815–839

Bhatt M, Camerer CF (2005) Self-referential thinking and equilibrium as states of mind in games: fMRI evidence. Games Econ Behav 52:424–459

Bolton G, Ockenfels A (2000) A theory of equity, reciprocity and competition. Am Econ Rev 90:166–193

Bornstein G, Yaniv I (1998) Individual and group behavior in the ultimatum game: are groups more rational players? Exp Econ 1:101–108

Bornstein G, Kugler T, Ziegelmeyer A (2004) Individual and group decisions in the centipede game: are groups more rational players? J Exp Soc Psychol 40:599–605

Bornstein G, Budescu DV, Kugler T, Selten R (2008) Repeated price competition between individuals and between teams. J Econ Behav Organ 66:808–821

Bosman R, van Winden F (2002) Emotional hazard in a power-to-take game experiment. Econ J 112:146–169

Bosman R, Sutter M, van Winden F (2005) The impact of real effort and emotions in the power-to-take game. J Econ Psychol 26:407–429

Bosman R, Hennig-Schmidt H, van Winden F (2006) Exploring group decision making in a power-to-take experiment. Exp Econ 9(1):35–51

Bosman R, Hennig-Schmidt H, van Winden F (2009) A power-to-take experiment in China and the EU: Do culture and stake size matter? mimeo

Brislin RW (1970) Back-translation for cross-cultural research. J Cross Cult Psychol 1:185–216

Burton AC, Chilton SM, Jones MK (2000) The WTP/WTA discrepancy: a preliminary qualitative examination. Environ Values 9:481–491

Cason TN, Mui V-L (1998) Social influence in the sequential dictator game. Math Psychol 42:248–265
Charness G, Dufwenberg M (2006) Promises and partnership. Econometrica 74:1579–1601
Chertkoff JM, Esser JK (1976) A review of some experiments in explicit bargaining. J Exp Soc Psychol 12:464–486
Chinese Culture Connection (1987) Chinese values and the search for culture-free dimensions of culture. J Cross Cult Psychol 18:143–164
Chiu CY (1990) Distributive justice among Hong Kong Chinese college students. J Soc Psychol 130:649–656
Chiu CY, Hong YY (1997) Justice in Chinese societies, a Chinese perspective. In: Kao HSR, Sinha D (eds) Asian perspectives on psychology. Sage, Thousand Oaks, CA, pp 164–184
Cooper DJ, Kagel JH (2005) Are two heads better than one? Team versus individual play in signaling games. Am Econ Rev 95:477–509
Corricelli G, Nagel R (2009) Neural correlates of depth of strategic reasoning in medial prefrontal cortex. Proc Natl Acad Sci USA 106:9163–9168
Costa-Gomes MA, Crawford VP (2006) Cognition and behavior in two-person guessing games: an experimental study. Am Econ Rev 96:1737–1768
Cox JC, Hayne SC (2006) Barking up the wrong tree: are small groups rational agents? Exp Econ 9:209–222
Dufwenberg M, Kirchsteiger G (2004) A theory of sequential reciprocity. Games Econ Behav 47:268–298
Endres J, Poggenpohl C, Erben C (1999) Repetitions, warnings and video: cognitive and motivational components in preschool children's suggestibility. Legal Criminol Psychol 4: 129–146
Ericsson KA, Simon HA (1984) Protocol analysis – verbal reports as data. MIT Press, Cambridge, MA, London, 2nd edn, 1993
Falk A, Fischbacher U (2006) A theory of reciprocity. Games Econ Behav 54:293–315
Fehr E, Schmidt KM (1999) A theory of fairness, competition and cooperation. Q J Econ 14:815–868
Flood MM (1958) Some experimental games. Manage Sci 5:5–26
Frith CD, Singer T (2008) The role of social cognition in decision making. Philos Trans R Soc Lond B Biol Sci 363:3875–3886
Gächter S, Riedl A (2005) Moral property rights in bargaining. Manage Sci 51:249–263
Geng H (2004) Decision processes in ultimatum bargaining – text analysis of transcripts of an intercultural video experiment. Diploma Thesis, University of Bonn
Geng H, Hennig-Schmidt H (2007) In your shoes – a video experimental study on communication and quasi-communication in ultimatum bargaining with groups. Working Paper, Department of Economics, University of Bonn
Gilhooly K, Green C (1996) Protocol analysis: theoretical background. In: Richardson JTE (ed) Handbook of qualitative research methods for psychology and the social sciences. The British Psychological Society, Leicester, UK, pp 44–54
Gill R (1996) Discourse analysis: practical implementation. In: Richardson JTE (ed) Handbook of qualitative research methods for psychology and the social sciences. The British Psychological Society, Leicester, UK, pp 125–140
Güth W, Schmittberger R, Schwarze B (1982) An experimental analysis of ultimatum bargaining. J Econ Behav Organ 3:367–388
Hennig-Schmidt H (1999) Bargaining in a video experiment – determinants of boundedly rational behavior, vol 467, Lecture notes in economics and mathematical systems. Springer, Berlin
Hennig-Schmidt H (2002) The impact of fairness on decision making – an analysis of different video experiments. In: Andersson F, Holm H (eds) Experimental economics. financial markets. auctions. and decision making. Kluwer, Dordrecht, pp 185–210

Hennig-Schmidt H, Geng H (2005) Decision mechanisms in tacit barganing – a video experimental investigation on the power-to-take game. Working paper, Department of Economics, University of Bonn

Hennig-Schmidt H, Geng H (2006) Über Emotionen und Klugheit in impliziten Verhandlungen – Eine experimentelle Untersuchung anhand des Power-to-take Spiels. In: Scherzberg A et al (eds) Kluges Entscheiden. Disziplinäre Grundlagen und interdisziplinäre Verknüpfungen. Mohr Siebeck, Tübingen, pp 91–107

Hennig-Schmidt H, Leopold-Wildburger U (2010) The shadow of the past. Working Paper, University of Bonn and University of Graz

Hennig-Schmidt H, Yan Z (2009) Equity and prominence in asymmetric bargaining – an experimental study on aspiration formation and adaptation in Germany and P. R. China. Working paper, University of Bonn, Germany

Hennig-Schmidt H, Yi N (2009) The impact of culture on bargaining – a cross-cultural experimental study. Working Paper (in German with English abstract)

Hennig-Schmidt H, Li Zh, Yang C (2008) Why people reject advantageous offers – non-monotone strategies in ultimatum bargaining. J Econ Behav Organ 65:373–384

Hennig-Schmidt H, Geng H, Reintgen M, Walkowitz G, Yang C (2009a) Strategies and conflict potential in inter-cultural bargaining – a comparison between China and Germany. Working paper (in German with English abstract)

Hennig-Schmidt H, Kohnz S, Schlüter B (2009b) Choices, beliefs and fairness in symmetric and asymmetric ultimatum game experiments. Working paper (in German with English abstract)

Hennig-Schmidt H, Irlenbusch B, Rilke RM, Walkowitz G (2010) Asymmetry and personality in ultimatum bargaining – experimental evidence from Germany and China. mimeo

Henrich J, Boyd R, Bowles S, Camerer C, Fehr E, Gintis H, McElreath R (2001) In search of homo economicus: behavioral experiments in 15 small-scale societies. Am Econ Rev 91:73–78

Henrich J, Boyd R, Bowles S, Camerer C, Fehr E, Gintis H (eds) (2004) Foundations of human sociality. Economic experiments and ethnographic evidence from fifteen small-scale societies. Oxford University Press, Oxford

Herrmann B, Thöni C, Gächter S (2008a) Anti-social punishment across societies. Science 319:1362–1367

Herrmann B, Thöni C, Gächter S (2008b) Supporting online material www.sciencemag.org/cgi/content/full/319/5868/1362/DC1

Hertwig R, Ortmann A (2001) Experimental practices in economics: a methodological challenge for psychologists? Behav Brain Sci 24:383–451, including Open Peer Commentaries

Hofstede G (1980) Culture's consequences: international differences in work-related values. Sage, Newbury Park, CA

Hofstede G (1991) Cultures and organizations: software of the mind. McGraw Hill, London

Jacobsen E, Sadrieh A (1996) Experimental proof for the motivational importance of reciprocity. Sonderforschungsbereich 303, University of Bonn, Discussion Paper B-386

Jullien François (2004) Über die 'Zeit'. Elemente einer Philosophie des Lebens. Aus dem Französischen von Heinz Jatho, Diaphanes, Zürich, Berlin; (Original: Du 'temps'. Élements d'une philosophie du vivre, Grassiet and Fasquelle, Paris, 2001)

Kerr NL, Aronoff J, Messé LA (2000) Methods of small group research. In: Reis HT, Judd CM (eds) Handbook of research methods in social and personality psychology. Cambridge University Press, Cambridge, UK, pp 160–189

Keser C (1992) Experimental duopoly markets with demand inertia: game-playing experiments and the strategy method, vol 391, Lecture notes in economics and mathematical systems. Springer, Berlin

Klemisch-Ahlert M (1996) Bargaining in economics and ethical environments. An experimental study and normative solution concepts, vol 436, Lecture notes in economics and mathematical system. Springer, Berlin

Kocher MG, Sutter M (2005) The 'Decision Maker' matters. Individual versus team behavior in experimental 'Beauty-Contest' games. Econ J 115:200–223

Kocher MG, Sutter M (2007) Individual versus group behavior and the role of the decision making process. Empirica 34:63–88
Kohlert DA, Krumov K, Ostmann A, Schneider JF (2005) Gabe und Gegengabe. Eine kulturvergleichende Studie zum Schenken. Zeitschrift für Sozialmanagement 3(2):41–56
Kohnz S, Hennig-Schmidt H (2005) When asymmetry matters – an experimental study on asymmetric outside options in ultimatum bargaining. Working Paper, Bonn University
Komorita SS, Kravitz DA (1979) The effects of alternatives in bargaining. J Exp Soc Psychol 15:147–159
Kravitz DA (1987) Size of smallest coalition as a source of power in coalition bargaining. Eur J Soc Psychol 17:1–21
Krippendorff K (1980) Content analysis: an introduction to its methodology. Sage, Beverly Hills
Leung K, Bond MH (1984) The impact of cultural collectivism on reward allocation. J Pers Soc Psychol 47:793–804
McClintock CG, Kramer RM, Keil LJ (1984) Equity and social change in human relationships. In: Berkowitz L (ed) Advances in experimental social psychology, vol 17. Academic, New York, pp 184–228
Merten K (1995) Inhaltsanalyse, 2nd edn. Westdeutscher Verlag, Opladen
Nagel R (1995) Unraveling in guessing games: an experimental study. Am Econ Rev 85:1313–1326
Nagel R (1998) A survey on experimental "Beauty-Contest Games:" bounded rationality and learning. In: Budescu D, Erev I, Zwick R (eds) Games and human behavior. Essays in honor of Amnon Rapoport. Lawrence Erlbaum Associates, New Jersey, pp 105–142
Nisbett RE, Wilson T (1977) Telling more than we can know: verbal reports on mental processes. Psychol Rev 84(3):231–259
Orbell JM, Dawes RM, van de Kragt AJC (1988) Explaining discussion-induced cooperation. J Pers Soc Psychol 5:811–819
Ostmann A (1990) On rationality issues in the bargaining context. J Inst Theor Econ 146: 673–683
Ostmann A (1992a) On the relationship between formal conflict structure and the social field. Small Group Res 23:26–48
Ostmann A (1992b) The interaction of aspiration levels and the social field in bargaining. J Econ Psychol 13:233–261
Ostmann A (1994a) Anpassung und Einflußnahme im Verhandlungskontext – Symlog Ratings als Datenquelle in experimentellen Verhandlungen. Gruppendynamik/Zeitschrift für angewandte Sozialpsychologie 6:133–147
Ostmann A (1994b) Aspiration processing in multilateral bargaining – experiment, theory and simulation. In: Schulz U, Albers W, Müller U (eds) Social dilemmas and cooperation. Springer, Berlin, pp 469–494
Ostmann A (1996) Representing interactional judgements in multilateral bargaining. Small Group Res 27:450–470
Ostmann A (2003) Das freundliche Lächeln beim Schenken – Strategien der Beziehungsgestaltung. Zeitschrift für Sozialmanagement 1(2):81–103
Pidgeon N (1996) Grounded theory: theoretical background. In: Richardson JTE (ed) Handbook of qualitative research methods for psychology and the social sciences. The British Psychological Society, Leicester, UK, pp 75–85
Potter J (1996) Discourse analysis and constructionist approaches: theoretical background. In: Richardson JTE (ed) Handbook of qualitative research methods for psychology and the social sciences. The British Psychological Society, Leicester, UK, pp 125–140
Raab Ph, Schipper BC (2009) Cournot competition between teams: an experimental study. J Econ Behav Organ 72:691–702
Rabin M (1993) Incorporating fairness into game theory and economics. Am Econ Rev 83:1281–1302

Reis HT, Gable SL (2000) Event-sampling and other methods for studying everyday experience. In: Reis HT, Judd CM (eds) Handbook of research methods in social and personality psychology. Cambridge University Press, Cambridge, UK, pp 190–222
Reuben E, van Winden F (2008) Social ties and coordination on negative reciprocity, the role of affect. J Public Econ 42:34–53
Reuben E, van Winden F (2010) Fairness perceptions and prosocial emotions in the power to take game. Forthcoming J Econ Psychol
Robert C, Carnevale PJ (1997) Group choice in ultimatum bargaining. Organ Behav Hum Decis Process 72:256–279
Rockenbach B, Sadrieh A, Mathauschek B (2007) Teams take the better risk. J Econ Behav Organ 63:412–422
Ross L, Greene D, House P (1977) The false consensus phenomenon: an attributional bias in self-perception and social perception processes. J Exp Soc Psychol 13(3):279–301
Roth AE (1995) Introduction to experimental economics. In: Kagel JH, Roth AE (eds) The handbook of experimental economics. Princeton University Press, Princeton, NJ, pp 3–109
Roth AE, Prasnikar V, Okuno-Fujiwara M, Zamir Sh (1991) Bargaining and market behavior in Jerusalem, Ljubljana, Pittsburgh, and Tokyo: an experimental study. Am Econ Rev 81:1068–1095
Sadrieh A, Hennig-Schmidt H (1999) The Tripled Take Game – text protocols of a video experiment. Sonderforschungsbereich 303, University of Bonn, Experimental Data Documentation Series No. 6.99
Sauermann H, Selten R (1962) Anspruchsanpassungstheorie der Unternehmung. Z Gesamte Staatswiss 118:577–597
Sauermann H, Selten R (1967) Zur Entwicklung der experimentellen Wirtschaftsforschung. In: Sauermann H (ed) Beiträge zur experimentellen Wirtschaftsforschung. Mohr, Tübingen, pp 1–8
Sauermann H (ed) (1967–1978) Contributions to experimental economics/Beiträge zur experimentellen Wirtschaftsforschung, 8 volumes. Mohr, Tübingen
Selten R (1967) Die Strategiemethode zur Erforschung des eingeschränkt rationalen Verhaltens im Rahmen eines Oligopolexperimentes. In: Sauermann H (ed) Beiträge zur experimentellen Wirtschaftsforschung. Mohr, Tübingen, pp 136–168
Selten R (1978) The equity principle in economic behavior. In: Gottinger HW, Leinfellner W (eds) Decision theory. social ethics, issues in social choice. D Reidel, Dordrecht, pp 289–301
Selten R (1981) A non-cooperative model of characteristic-function bargaining. In: Aumann RJ et al (eds) Essays in game theory and mathematical economics. Bibliographisches Institut AG, Zürich, pp 131–151
Selten R (1987) Equity and coalition bargaining in experimental three-person games. In: Roth AE (ed) Laboratory experimentation in economics. Cambridge University Press, New York, pp 42–98
Selten R (1993) In search of a better understanding of economic behavior. In: Heertje A (ed) The makers of modern economics. Simon and Schuster, New York, pp 115–139
Selten R (1995) Autobiographic notes. In: Frängsmyr T (ed) Les Prix Nobel. The nobel prizes 1994. Nobel Foundation, Stockholm
Selten R (1998) Aspiration adaptation theory. J Math Psychol 42:191–214
Selten R (2000) Eingeschränkte Rationalität und ökonomische Motivation. Schriften des Vereins für Socialpolitik 274:129–157
Selten R, Leopold-Wildburger U (1982a) Equilibrium selection in a wage bargaining situation with incomplete information. Operations Research Verfahren, vol 41. Athenäum-Hain, Königstein, pp 97–101
Selten R, Leopold-Wildburger U (1982b) Gleichgewichtsauswahl in einer Gehaltsverhandlungssituation mit unvollständiger Information. In: Streißler E (ed) Information in der Wirtschaft, Schriften des Vereins für Socialpolitik, vol 126. Gesellschaft für Wirtschafts- und Sozialwissenschaften, Berlin, pp 155–161

Selten R, Schuster KG (1968) Psychological variables and coalition forming behavior. In: Borch K, Mossin J (eds) Risk and uncertainty. Proceedings of a Conference held by the International Economic Association, pp 221–240

Selten R, Schuster KG (1970) Psychologische Faktoren bei Koalitionsverhandlungen. In: Sauermann H (ed) Beiträge zur experimentellen Wirtschaftsforschung, vol 2. Mohr, Tübingen, pp 99–135

Selten R, Stöcker R (1986) End behavior in sequences of finite prisoner's dilemma supergames: a learning theory approach. J Econ Behav Organ 7:47–70

Selten R, Tietz R (1980) Zum Selbstverständnis der experimentellen Wirtschaftsforschung im Umkreis von Heinz Sauermann. Z Gesamte Staatswiss 136:12–27

Selten R, Mitzkewitz M, Uhlich GR (1988) Duopoly strategies programmed by experienced players. Sonderforschungsbereich 303, University of Bonn, Discussion Paper No. B-106

Selten R, Mitzkewitz M, Uhlich GR (1997) Duopoly strategies programmed by experienced players. Econometrica 65:517–555

Siegel S, Castellan NJ Jr (1988) Nonparametric statistics for the behavioral sciences, 2nd edn. McGraw-Hill, New York

Simon HA (1955) A behavioral model of rational choice. Q J Econ 69:99–118

Simon HA (1957) Models of man. Wiley, New York

Singer T (2006) The neuronal basis and ontogeny of empathy and mind reading – review of literature and implications for future research. Neurosci Biobehav Rev 30:855–863

Singer T, Fehr E (2005) The neuroeconomics of mind reading and empathy. Am Econ Rev 95:340–345

Smith JAN (1996) Evolving issues for qualitative psychology. In: Richardson JTE (ed) Handbook of qualitative research methods for psychology and the social sciences. Leicester, UK, The British Psychological Society, pp 189–201

Smith CP (2000) Content analysis and narrative analysis. In: Reis HT, Judd CM (eds) Handbook of research methods in social and personality psychology. Cambridge University Press, Cambridge, UK, pp 313–335

Sonderforschungsbereich 303 and Laboratory for Experimental Economics Data Documentation Series (1996–2002) University of Bonn, Laboratory for Experimental Economics

Trommsdorff V, Wilpert B (1994) Deutsch-Chinesische Joint Ventures, 2nd edn. Wiesbaden, Gabler

Twain M. (1880/2010) The Awful German Language. In: A Tramp Abroad, US-Botschaft Berlin, Public Affairs (ed.), http://usa.usembassy.de/marktwain (downloaded: 21.04.2010)

Universität Bonn (2009) Traditionell modern; Die Universität Bonn – ein Portrait. http://www.uni-bonn.tv/podcasts/ImageDeutsch.mp4/view, produced 20.05.2009 (downloaded: 21.03.2010)

van de Vijver F, Leung K (1997) Methods and data analysis for cross-cultural research. Sage, Thousand Oaks

van Winden F (2001) Emotional hazard exemplified by taxation-induced anger. Kyklos 54:491–506

Walkowitz G, Hennig-Schmidt H, Oberhammer C (2009) Experimenting over a long distance – a method to facilitate inter-cultural experiments and its application to a trust game. Working Paper, Department of Economics, University of Bonn

Walkowitz G, Hennig-Schmidt H, Geng H (2010) On the nature of fairness – experimental evidence from Germany and PR China. Working paper, University of Bonn, Germany

Zhang ZX, Yang CF (1998) Beyond distributive justice: the reasonableness norm in Chinese reward allocation. Asian J Soc Psychol 1:253–269

Chapter 11
Institutions Fostering Public Goods Provision

Ernst Fehr, Simon Gächter, Manfred Milinski, and Bettina Rockenbach

"Never do an experiment on public good provision or the ultimatum game!" This advice was given by a senior colleague to a young mathematician (BR) joining the Selten group at the Bonn Laboratory in the late 1980s. A well-meant advice to someone entering the field of experimental economics, grounded in the colleague's observation that simple games, like ultimatum, dictator or prisoners' dilemma games have already been subject to numerous experimental studies and that more complicated settings are non-tractable. For hand-run experiments the degree of complexity seemed very restricted and computerized experiments faced serious technical limitations at that time. Today, more than 20 years later, we can look back to numerous intriguing new insights that have been gained through additional public-goods and ultimatum experiments. Some of them will be reviewed in this paper and some of them are co-authored by the formerly young mathematician who did not follow the advice of the senior colleague. Undisputable, technical progress has enriched our possibilities for handling richer and more complex games and experimental settings. This, however, is at best a necessary requirement. To conduct a good experiment, it needs a clever design which allows discriminating between conflicting explanations and guides to a positive behavioral theory of human behavior. Reinhard Selten is one of the most brilliant researchers and each of the authors experienced his razor-sharp arguments, especially when it comes to the design of an experiment. More than the technical progress behavioral

E. Fehr
Institute for Empirical Research in Economics, University of Zurich, Blümlisalpstrasse 10, CH-8006 Zürich, Switzerland

S. Gächter
University of Nottingham, Sir Clive Granger Building, University Park, Nottingham NG7 2RD, UK

M. Milinski
Max Planck Institute for Evolutionary Biology, August-Thienemann-Str. 2, 24306 Plön, Germany

B. Rockenbach (✉)
Universität Erfurt, Postfach 900 221, 99105 Erfurt, Germany
e-mail: bettina.rockenbach@uni-erfurt.de

A. Ockenfels and A. Sadrieh (eds.), *The Selten School of Behavioral Economics*, 167
DOI 10.1007/978-3-642-13983-3_11, © Springer-Verlag Berlin Heidelberg 2010

economics needed personalities like Reinhard Selten to advance to an indispensable field of economics.

Reinhard Selten did not shy away from complexity. Think of his hand-run SINTO experiments (Becker and Selten 1970). He was most enthusiastic when thinking about completely new approaches of understanding human behavior. When students asked for his advice on a game or an experimental design that was "incremental" in his terms, i.e. a marginal variation either in the parameters or the game structure, they often received mild answers like "Tja, das ist auch nicht verboten, nich!"[1] or even harsher ones like "Sie sind viel zu jung, um solch einen Unsinn zu machen".[2] His advice to conduct experiments which are purely explorative in the sense that there is no theoretical benchmark available ("Why should you care for a theoretical solution, when we know in advance that it will be falsified") was not easy to take at the Bonn faculty at that time. What he did not like however, were strong prior beliefs about empirical propositions that rested in nothing but the conventions prevailing in scientific groups: I (EF) heard him once criticizing a paper at an international conference with the words: "One does not start with the assumption that cows are green, brown, black or white but one goes out observing the color of the cows". According to his opinion of how to best achieve scientific progress, the practices in theoretical economics and game theory had far too little connection to the empirical world and lacked rigorous empirical testing. This attitude allowed him a fresh view on how to study human behavior.

Having said all this, it might sound that Reinhard Selten is sometimes disrespectful on the work of others. However, the contrary is true. He was and is absolutely honest (and non-strategic) in his judgment on others' research and does not withhold highest credits to others. He has an excellent knowledge of the economic literature as well as the research in other disciplines (like psychology or biology) which aid our understanding of human behavior. For example, I (SG) remember vividly a seminar in Bonn in July 1998 where I presented the first version of our paper on punishment. Someone asked about the literature and before I had a chance to answer this question, Reinhard Selten gave a complete overview of the then-existing literature, in particular the seminal studies by the social psychologist Toshio Yamagishi (1986) and the political scientist Elinor Ostrom and her co-workers (1992). His remarkable memory is admirable and he recalled research he heard in a seminar talk decades ago. This, however, has to be qualified. I (BR) recall many research seminar talks in Bonn, which we discussed afterwards and it became clear that Reinhard Selten listened to the introduction and the first modeling attempts, while the rest of the talk evolved in his head. Inspired by the motivation, he envisioned his own modeling, which often was completely different and almost always more brilliant than what the presenter actually said.

[1]"This is not forbidden, nich".

[2]"You are much too young to do such stupid things", answer given to an advice seeking PhD-student at the summer school in Stony Brook in the early 1990s.

When talking to Reinhard Selten about voluntary contributions and public goods provision, he naturally looked at the problem from the perspective of an economist, but the political implications and the relevance from an evolutionary biology perspective were not less important. His close contacts to theoretical and evolutionary biologists opened his (economics) students' view to evolutionary perspectives. The research on public good provision, described in this paper, is inspired by this broad view on voluntary contributions.

Coming from the other side, evolutionary biology, I (MM) was deeply impressed by Reinhard Selten's knowledge and interest in evolutionary and biological problems. During his time in Bielefeld I met him several times at ESS meetings. He was eager to learn about experimental studies with animals and whether they act "economically". I was impressed by his comments on my research on sticklebacks, which I presented that time. I remember him saying something like: "Animals are more suited to test economic theories, because we know what they have to maximize: their fitness. With humans it is much more difficult, because it is not so clear what they maximize". Reinhard Selten even acted as a reviewer of my Habilitation-thesis on sticklebacks.

Voluntary Cooperation

In his seminal paper, Garrett Hardin (1968) elaborates on *the tragedy of the commons*:

> The tragedy of the commons develops in this way. Picture a pasture open to all. It is to be expected that each herdsman will try to keep as many cattle as possible on the commons. Such an arrangement may work reasonably satisfactorily for centuries because tribal wars, poaching, and disease keep the numbers of both man and beast well below the carrying capacity of the land. Finally, however, comes the day of reckoning, that is, the day when the long-desired goal of social stability becomes a reality. At this point, the inherent logic of the commons remorselessly generates tragedy. As a rational being, each herdsman seeks to maximize his gain. Explicitly or implicitly, more or less consciously, he asks, "What is the utility to me of adding one more animal to my herd?" This utility has one negative and one positive component. (1) The positive component is a function of the increment of one animal. Since the herdsman receives all the proceeds from the sale of the additional animal, the positive utility is nearly +1. (2) The negative component is a function of the additional overgrazing created by one more animal. Since, however, the effects of overgrazing are shared by all the herdsmen, the negative utility for any particular decision-making herdsman is only a fraction of −1. Adding together the component partial utilities, the rational herdsman concludes that the only sensible course for him to pursue is to add another animal to his herd. And another; and another . . . But this is the conclusion reached by each and every rational herdsman sharing a commons. Therein is the tragedy. Each man is locked into a system that compels him to increase his herd without limit – in a world that is limited. Ruin is the destination toward which all men rush, each pursuing his own best interest in a society that believes in the freedom of the commons. Freedom in a commons brings ruin to all (p. 1244).

In a very illustrative way, Hardin describes the nature of a social dilemma: the conflict between collective and individual interests. This conflict is inherent in the voluntary provision of public goods and results in their inefficient undersupply due to free-riding (Samuelson 1954). However, despite Hardin's bleak prediction, humans often manage to avoid the *tragedy of the commons* and achieve high levels of cooperation. This holds for hunter-gatherer societies to complex modern nation states which would not exist without large-scale cooperation. The fact that people vote even in anonymous situations, take part in collective actions, often do not overuse common resources, care for the environment, mostly do not evade taxes on a large scale, donate to public radio, as well as to charities, etc. suggests that the strict self-interest hypothesis is inconsistent with the degree of voluntary cooperation that we observe around us. Understanding cooperation is an important challenge across all social sciences.

Voluntary Cooperation in the Laboratory

The linear public goods game (or voluntary contribution mechanism) has proved extremely useful for testing social dilemmas in the lab. In a typical linear public goods experiment, n group members, each endowed with z "tokens" may independently decide how many tokens (between 0 and z) to contribute to a common project (the public good). All contributions are augmented by a factor $\alpha > 1$ and distributed equally among the group members, irrespective how much an individual has contributed. Thus each subject i's payoff is

$$\pi_i = z - g_i + \frac{\alpha}{n} \sum_{j=1}^{n} g_j, j = 1, \ldots, n, \alpha > 1, \alpha/n < 1.$$

Each individual benefits from the contributions of other group members, even if he or she has contributed nothing to the public good. A rational and selfish individual therefore has an incentive to keep all tokens for him- or herself. By contrast, since $\alpha > 1$, the group as a whole is best off if everybody contributes all z tokens.

Numerous experiments have falsified the prediction of complete free-riding of all subjects. Instead there exists substantial cooperation in a variety of setups (for overviews, see e.g. Ledyard 1995; Gächter and Herrmann 2009; Gächter 2007). Figure 11.1 depicts a typical finding of a public goods experiment, where the identical game is repeated ten times. Subjects play in groups of four with an initial endowment of 20 in each period. Figure 11.1 shows the resulting cooperation patterns in a "Stranger" condition, where group members change randomly from round to round, and a "Partner" condition, in which groups stay constant for all rounds.

As Fig. 11.1 illustrates subjects contribute substantially more than theoretically predicted, however, cooperation, is very fragile and tends to collapse with repeated interactions. The decay of cooperation has been replicated numerous times and has

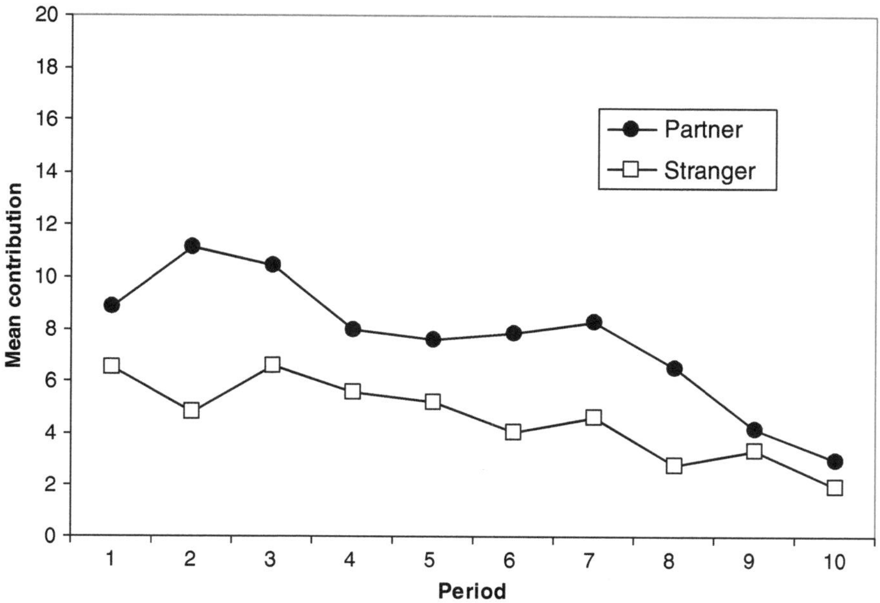

Fig. 11.1 Contributions to a public good in constant ("Partner") and randomly changing groups ("Strangers") over ten repetitions
Source: Fehr and Gächter (2000)

also been observed across a variety of participant pools (Herrmann et al. 2008). What explains this almost inevitable outcome? Reinhard Selten was very strict in correcting people when they named the observed non-rational behavior *irrational behavior*. He preferred to speak about *boundedly rational* behavior. He was also very definite in rejecting the view that deviations from rationality are caused by errors that can be wiped out by teaching or training people. Indeed, the explanation that positive contributions in public good provision are a lack of understanding free-rider incentives can be excluded. The fact that in experiments with a surprise re-start contributions start high again is inconsistent with a pure learning hypothesis (Andreoni 1988; Croson 1996; Cookson 2000).

An long held hypothesis by social psychologists (e.g., Kelley and Stahelski 1970) is that many people may be "conditional cooperators", who in principle are willing to cooperate if others do so as well, but get frustrated if others do not pull their weight. Therefore, the breakdown of cooperation is due to "frustrated attempts at kindness" (Andreoni 1995; p. 900). There is now mounting evidence from psychological and economic experiments for the importance of conditional cooperation both in the lab and the field (Gächter 2007). In experiments that elicited participants' beliefs about how much they think others will contribute contributions are indeed positively correlated with beliefs (Dufwenberg et al. 2006; Croson 2007; Fischbacher and Gächter 2010; Neugebauer et al. (2009)). A correlation does of course not establish causation and it is perfectly possible that a false consensus effect induces people to believe that others contribute the same as them (e.g., Kelley and Stahelski 1970). To circumvent this problem, Fischbacher et al. 2001 developed an experimental design closely related to Selten's strategy method (Selten 1967). Subjects have to indicate how much they contribute to the public good as a function of all possible average contribution levels of other group members. The results show that about 50% are "conditional co-operators", who increase their contributions if others contribute more, whereas about 25% are "free riders" who never contribute anything – irrespective of how much others contribute. Fischbacher and Gächter 2010 use the same method and show that the interaction of differently motivated people explains the decay of cooperation. The significance of this finding is that the decay of cooperation will occur not just because people eventually learn what is in their best interest but because frustrated conditional cooperators reduce their contributions. Thus, after some time all types *behave* like income-maximizing free riders, even though only the free-rider types are motivated by income-maximization alone.

Punishment in Public Goods Provision

Reciprocity is a likely source of conditional cooperation (Rabin 1993; Dufwenberg et al. 2006). The reason is that cooperating is a nice act towards the other group members and people may want to return the favor. By contrast, free riding is an unkind act which people may want to punish. However, in the public goods experiments described above the only way to punish free riding is to withdraw cooperation, with the consequence that other cooperators in the group get punished as well. This raises two questions: Will people be willing to punish if they could target a free rider directly? Will the possibility to punish affect cooperation? The seminal studies of Yamagishi 1986 and Ostrom et al. 1992 show that subject in a repeated public goods game with a stable group composition will indeed punish free riders. In addition, the punishment of free riders is associated with an increase in the aggregate rate of cooperation. However, subjects in these studies did not know the exact number of periods which they will stay together in a public goods game. They, therefore, left open the question whether punishment is driven by the strategic motive to induce the free-riders to contribute more to the public good in future periods or whether the punishment occurs because of social preferences for reciprocity and equity.

A typical design of most recent studies is the one implemented by (Fehr and Gächter 2000, 2002) who examined the extent to which punishment is driven by strategic motives or by social preferences. After participants have made their contribution decisions, group members are informed about how much the other group members have contributed to the public good. Each group member can then decide to punish each of the other group members. Each point assigned reduces the punished member's income, but is also costly for the punishing member. There is by now sound evidence that many subjects punish those who contribute less and that peer punishment increases and stabilizes cooperation at higher levels than without punishment (cf. Fig. 11.2). In their set up subjects are either in groups with a stable group composition for 10 periods (partner design) or groups are randomly recomposed in each of the ten periods in a stranger design (Fehr and Gächter 2000) or the subjects in the group are even ensured that they meet every other participant only once (perfect stranger design, Fehr and Gächter 2002).

It turns out that the strategic nature of interaction (repeated interaction versus one-shot interaction) matters for cooperation but not much for punishment (Fehr and Gächter 2000). Put differently, while cooperation rates are significantly and substantially higher in repeated interactions as compared to repeated one-shot interactions, people punish free riding similarly irrespective of whether it occurs in a repeated relationship or in random one-shot interactions. Remarkably, people punish even in strict one-shot games with no repetition (Falk et al. 2005; Gächter and Herrmann 2009; Gächter and Herrmann forthcoming). This suggests that the level of cooperation is influenced by strategic considerations (free riding is less likely in repeated interactions), whereas punishment is to a large part non-strategic. Punishment seems to be an impulse triggered by negative emotions (Pillutla and Murnighan 1996; Bosman and van Winden 2002; Fehr and Gächter 2002; Sanfey et al. 2003; de Quervain et al. 2004; Knoch et al. 2006; Ben-Shakhar et al. 2007; Fehr and Camerer 2007; Seymour et al. 2007; Reuben and van Winden 2008) and not much by forward-looking considerations.

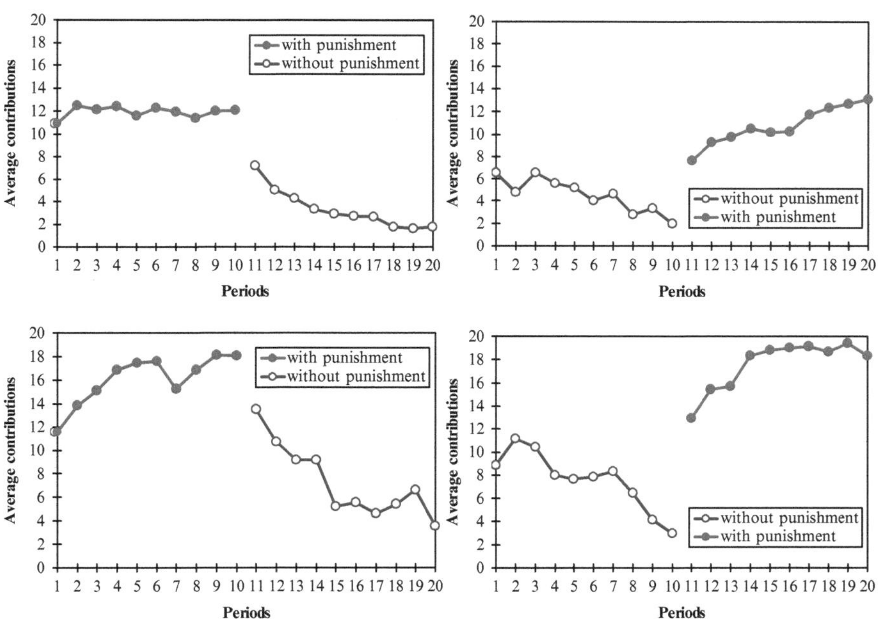

Fig. 11.2 Average contributions over time in the stranger treatment (*upper panel*) and the partner treatment (*lower panel*)
Source: Fehr and Gächter (2000)

In most experiments in which punishment has material payoff consequences, punishment turned out to be an inefficient tool to enforce cooperation because resources are destroyed. Indeed, in most experiments – which typically ran for ten periods or less – net payoffs in treatments with punishment are often lower than in treatments without punishment (e.g., Fehr and Gächter 2000; Page et al. 2005; Bochet et al. 2006; Botelho et al. 2007; Sefton et al. 2007; Egas and Riedl 2008; Herrmann et al. 2008; Masclet and Villeval 2008; Nikiforakis 2008). For instance, Herrmann et al. 2008 report public goods experiments with and without punishment conducted in 16 comparable participant pools around the world. With the exception of three participant pools the average payoff in the experiments with punishment opportunities was lower than without punishment; and in those three participant pools with higher payoffs the increase was modest and amounted to 9.1, 2.8 and 0.5%, respectively. Thus, 13 participant pools would have been better off not having had a punishment opportunity. The detrimental consequences of punishment are even more conspicuous if "counter-punishment", that is, multiple rounds of punishment, is possible (Denant-Boemont et al. 2007; Nikiforakis 2008). Gächter et al. (2008) play a public goods game for 50 periods and they compared payoffs with those in ten-period experiments. Like in previous experiments, in the ten-period experiments punishment was detrimental in terms of payoffs as compared to ten-period experiments without punishment. In the 50-period experiments the opposite conclusion holds – cooperation is high and punishment costs negligible.

Endogenous Choice in Public Goods experiments

Given the payoff reducing consequences of punishment in the short run, it has to be questioned whether a sanctioning institution would be deliberately adopted when subjects actually have the choice. This question gains further support in light of the natural resentments against punishment caused for example by its detrimental effects (Fehr and Rockenbach 2003). Gürerk et al. (2006) approach this question in a laboratory experiment that implements permanent competition between a sanctioning and a sanction-free framework through endogenous choice. This design allows studying the evolution of the communities in the different frameworks over time as well as the changes in behavior in the same individual when participating in different social settings.

A public goods game is played over 30 repetitions. Each repetition consists of three stages: An institution choice stage (S0), a voluntary contribution stage (S1), and a sanctioning stage (S2). In stage S0, the participants simultaneously and independently choose between a sanctioning institution (SI) and a sanction-free institution (SFI) in which neither positive sanctioning (rewards) nor negative sanctioning (punishment) is possible. In stage S1, each participant interacts in a public goods game with all other participants who have chosen the same institution in S0: each player is endowed with 20 money units (MUs) and may contribute between 0 and 20 MUs to a public good. Each group member equally profits from the public good, independent from his or her own contribution. The MUs not contributed to the public good are transferred to the participant's private account.

After the players have simultaneously made their contribution decisions, they are informed about the contributions of each member in the own group. In stage S2 each player in SI may positively or negatively sanction other members of SI by assigning between zero and 20 tokens to other members. Each token employed as a negative sanction costs the punished member 3 MUs and the punishing member 1 MU. Each token employed as a positive sanction yields the receiving member 1 MU and costs the employing member 1 MU. At the end of the period each participant receives detailed (but anonymous) information about each of the other participants from both institutions.

The initial choice of institution provides a clear picture: only about one third of the participants (mean 36.9%) prefer SI to SFI in the first period. Participants who initially join SI contribute significantly higher amounts (on average 12.7 MUs) than those joining SFI (on average only 7.3 MUs) in the first period. There is a high level of punishment in order to discipline low contributors and to enforce and establish a norm of high cooperation. However, the initially significantly higher contributions in SI do not result in higher payoffs. Due to immense punishment activities, average payoffs in the first period of SI are significantly lower than in SFI.

Although subjects are initially reluctant to join SI, it becomes predominant over time; in the end virtually all participants (mean 92.9%) choose SI and cooperate fully (see Fig. 11.3). Simultaneously, contributions in SFI decrease to a level of zero.

A closer look at individual behavior immediately before and after migration from one institution to the other exhibits a bipolar pattern of behavior induced by the two

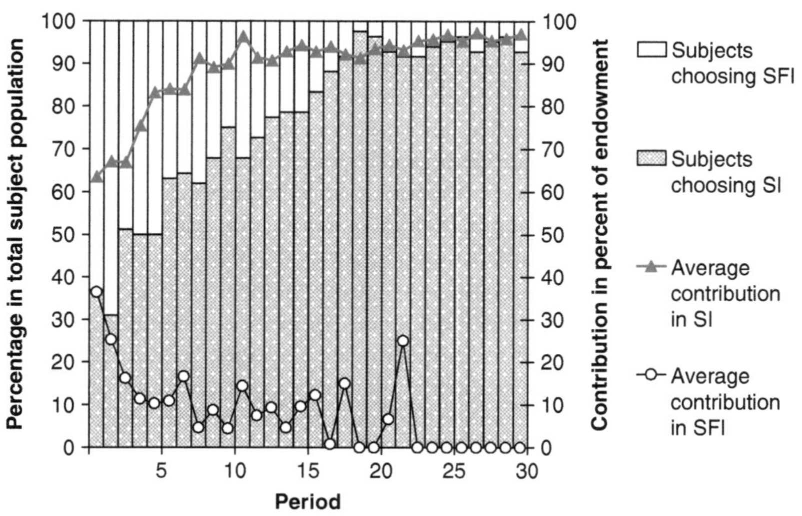

Fig. 11.3 Subjects' choice of institution and their contributions. The figure displays the average contributions in both institutions over the 30 periods of the interaction measured as the percentage of endowment contributed to the public good
Source: Gürerk et al. (2006)

institutions. In fact, 80.3% of subjects increase their contribution when migrating from SFI to SI in two consecutive periods. And 27.1% of subjects even "convert" from being a complete free-rider (contributing 0 MUs) to a full cooperator (contributing 20 MUs) when switching from SFI to SI. The migration behavior in the opposite direction, i.e. from SI to SFI, is similarly extreme. Roughly 70% of subjects reduce their contribution when switching from SI to SFI and about 20% switch from full cooperation to free-riding.

Individual payoff maximization cannot explain why new members in SI punish low contributors. The most successful behavior would be to contribute in SI (and hence avoid being punished), but refrain from the costly punishment of others. Since punishment of defectors constitutes a second-order public good (in which defection cannot be sanctioned in our setting), individual payoff maximization would rule out punishment. However, only a minority of subjects follow this payoff maximizing behavior. The overwhelming majority of 62.9% of the subjects immediately conforms to and adopts the prevailing norm of punishment in SI, i.e. they always use punishment immediately after they switch to SI. This results in a quite stable proportion of roughly 40% of subjects who both contribute highly and punish during the last 20 periods.

How can we explain the evident success of the sanctioning institution SI? Although in almost all sessions the SI communities start off with only one third of the subjects, they grow large over time and ultimately reach almost 100% of the total population. This observation leaves room for two non-exclusive explanations. The voting with one's feet choice allows for "self-selection" of the community members into the preferred institution and due to the different nature of the institutions – with and without punishment possibilities – the selection may be driven by different predispositions to cooperate. Particularly in the beginning, this selection process of like-minded people (Falk et al. 2009; Gächter and Thöni 2005) may initiate and foster a culture of high levels of cooperation in the punishment community. Later on, others with less cooperative attitudes might also be attracted simply by the success of the cooperative culture in SI. A second explanation which is independent from the "self-selection" argument is that a group that starts small and grows slowly can coordinate better than a group that already starts at "full size." Evidence pointing into this direction is presented by Weber (2006), who finds that a slow growth path improves coordination in a coordination game.

To disentangle these two explanations Gürerk et al. (2010a) conduct a new series of experiments on endogenous community choice. In the main experiment subjects may choose between a punishment and a non-punishment community. In the first control experiment, the experimenters simulate the same growth paths as they endogenously occurred in the sessions of the main experiment, but subjects were exogenously allocated to the two institutions in each period. Hence institution allocations do not emerge from self-selection, but are exogenously imposed by the experimenter. Observed contribution rates are significantly lower than in the main experiment. That shows that self-selection of subjects plays a crucial role for the superior performance of the voting with one's feet mechanism. A second control experiment eliminates the effects of an increasing growth path. Subjects are exogenously allocated into fixed-sized communities in which they remain for the entire experiment. Also in this case, contributions are significantly lower.

The results indicate that starting with a small group of subjects (and growing afterwards) does not per se foster high contributions. Instead, it seems that the group has to be composed of the "right" subjects, who initially establish a cooperative environment through high contributions and rigid punishment. Our findings underline the importance of the endogenous choice with voting with one's feet for establishing high levels of cooperation in public goods settings.

The Interaction of Reputation and Punishment

Not only the possibility to costly punish defectors has been to shown to increase cooperation rates, but also the ability to establish reputation in games (Milinski et al. 2002, 2006; Panchanathan and Boyd 2004). If the public goods game is incorporated in a broader context that promises rewards for those with a good reputation (Milinski et al. 2002), i.e. an 'indirect reciprocity game', cooperation in the public goods game can be maintained at a high level. In indirect reciprocity situations, individuals who have helped others are given support, that is, the supporters improve their reputation and are rewarded in turn (Wedekind and Milinski 2000; Bolton et al. 2005; Seinen and Schram 2006; Bshary and Grutter 2006). Subjects use withholding help in the indirect reciprocity game as a "punishment" device for defectors in the public goods game. A key difference, however, is that this device does not generate a direct efficiency reduction, like altruistic punishment does. Withholding help does not afford any monetary investment and the "punisher" actually saves money by not rewarding a defector (who does not lose his money either; instead he does not receive additional gains). A straightforward prediction is that in a social dilemma situation that allows for effective reputation building costly punishment becomes extinct because a much cheaper and equally powerful alternative to sustain cooperation is available. However, if this is true we face the following puzzle. Why is direct punishment of defectors a universal feature in all known human societies?

Rockenbach and Milinski (2006) experimentally study a public goods game with an attached indirect reciprocity game that provides effective possibilities for reputation. Before interacting subjects choose by a voting with one's feet procedure to either have an additional costly peer-punishment possibility or not. The experimental design studies a situation where reputation cannot be avoided but the punishment option can be chosen or avoided. The cost-disadvantage of punishment strongly suggests that subjects go directly for the punishment-free group because they would utilize the indirect reciprocity game for cooperation enhancement and avoid the detrimental effects of costly punishment.

Indeed, about 70% of the subjects initially chose the group without costly punishment. However, contrary to expectations, subjects significantly preferred the group with the opportunity of costly punishment in the second half of the experiment. Contributions to the public goods were significantly higher in the punishment group than in the punishment-free group.

The interaction between costly punishment and indirect reciprocity not only increased the contributions to the public good but also significantly reduced the number of punishment points allocated per member of the punishment group, compared to a control (PUN) which allowed for punishment, but not did not entail an IR game. Nevertheless, the punished free-riders in the main treatment (PUN&IR) were more heavily punished than those in the control. Thus, although fewer group members were punished, punishment of a free-rider, i.e. a non-cooperator, was even more severe in the presence of the indirect reciprocity game than without it. This is remarkable, because it could have been expected that the costly punishment acts are avoided completely at this stage and punishment is moved to the indirect reciprocity stage. Contributions to the public goods pool per allocated punishing point are about three times higher in PUN&IR than in the control. Thus, subjects were able to generate higher contributions with fewer punishment expenses.

Even if we compare the efficiency in the public goods game including all costs of punishment of the combination of both options (with and without punishment) of a treatment we find that PUN&IR is more efficient than PUN (cf. Fig. 11.4).

Thus, although subjects could choose a punishment-free option where reputation building alone would allow for high levels of cooperation in a public goods game, the great majority chose the punishment option that was also combined with reputation building. Indirect reciprocity did not simply replace punishment. Both instruments interact. In PUN&IR the free-riders were more heavily punished than the free riders in PUN and they were hurt a second time through withheld help in the indirect reciprocity game which amplified the punishment effect.

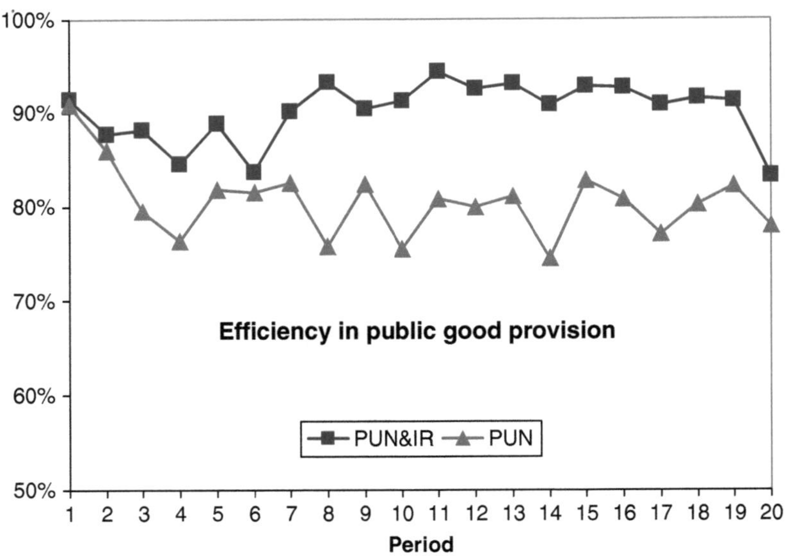

Fig. 11.4 Efficiency in public good provision. The efficiency is measure over the entire treatment efficiency, i.e. the combination of both choice alternatives (with and without punishment)
Source: Rockenbach and Milinski (2006)

Endogenous Rule Design in Public Good Provision

Talking to Reinhard Selten about the endogenous choice by voting with one's feet he was very enthusiastic. He argued that this very basic form of voting could help us to provide insights on rudimentary community building from an evolutionary perspective. In addition, he continued, it would be very intriguing to let subjects create their own rules of interaction. From scratch without any pre-defined rule of the experimenter. This is an excellent example for the introductory remark that he does not shy away from complex experimental designs. Of course, such an experiment would be very intriguing. It would allow studying the rise and fall of institutional regulations and the determinants of their success. But, how to design an experiment in which subjects repeatedly create and modify the interaction rules, while playing? After several discussions with Reinhard Selten and some hard work on our side, we made an attempt in Rockenbach and Wolff (2010).

In this paper, we endogenize the design of the institutional regulations. This approach bears a close resemblance to the studies of Axelrod (1984), Selten et al. (1997), Keser and Gardner (1999), and Keser (2000). While Axelrod asked scholars of game theory to specify complete strategies for a prisoners' dilemma, which could be refined in a second round, Selten, Mitzkewitz and Uhlich had student subjects do the same for an asymmetric Cournot duopoly over several rounds. Keser and Gardner (1999) and Keser (2000) applied the method to a common-pool resource and a public goods problem, respectively. While, in the studies mentioned, subjects were completely free in their *design of a strategy* for a given situation, Rockenbach and Wolff (2010) tackle the question of *institution design* for social-dilemma situations in this way. In other words, in this experiment, subjects – as Reinhard Selten proposed – actually act as lawmakers empowered to shape the institutional environment of the game.

The experiment was conducted over 3 months in the framework of two student seminars in which the participants' task was to design institutional regulations to overcome a social dilemma. Before designing the institutional regulations subjects gathered experience in playing the basic public goods game in an anonymous laboratory setting. Following that, they were given a week to develop a set of rules of play for this game. There was no predefined set of rules so that subjects could freely choose whatever rules they wanted to implement, as long as the incentive structure still exhibited the social-dilemma characteristics. To achieve a certain degree of external validity we attached a certain cost to each of the proposed rules according to the true (relative) costs such an institution would give rise to in common real-world settings. After this first design phase, a different set of subjects played the public goods game under these rules and the designers were rewarded according to the efficiency, i.e. sum of players' profits minus rule costs, the players achieved under their rule. The design-and-play process was repeated three times.

Three observations are noteworthy. First, punishment, in various disguises, was the initial focal point of all groups. Remarkably, however, punishment mechanisms were not designed in the form of peer punishment, but rather in the form of pre-specified rules of deduction and/or redistribution contingent on complying with provision targets. These provision targets were either fixed levels (e.g. full provision) or contingent on the other group members (e.g. not being the lowest-contributing player). A second important observation, namely the role of framing as a form of communication between "lawmakers" and their "people", is rather unexpected. This is noteworthy because subjects playing under these rule sets were experienced players, who, in their capacity as "lawmakers", have a relatively deep understanding of the logic of the game and were completely aware that the chosen frames have no connection to reality whatsoever, but are pure imagination. Nonetheless, framing was increasingly chosen and a successful means in achieving efficiency. While one of the groups opted to tell its subjects the public good was a school in Afghanistan in two of the tournaments, another group had its subjects play in a virtual neighborhood consisting of spouses and children, cats, dogs, and rat poison. The third noteworthy finding is that subjects render information on a player's fellow providers rather opaque – and that this tends to be a successful strategy. The implications are not straightforward and call for further research. Conditionally cooperative subjects are assumed to align their provision with what they believe others will contribute. Providing them with detailed information on past contributions of their peers may yield the most precise basis for the calculations. Yet, it seems that a certain degree of opaqueness by just providing the average contribution is more successful in enhancing cooperation. A possible explanation may be that the reduced information precludes individual comparisons detecting advantageous as well as disadvantageous inequalities with respect to the other players. Even if these comparisons do not lead to equilibrium predictions different from the Nash equilibrium resting on the assumption of money maximizing actors, they may be important determinants of contribution dynamics (Engel and Rockenbach 2009).

Dynamic Public Good Games

In all studies reported so far, interaction based on the repetition of a stage game. Subjects collected experience from repetition to repetition, but payoffs accumulated in the past did not have any consequences for the strategic options today. Often, reality is different. Today's contribution capabilities depend on past behavior. Financial or physical resources may be low due to past excessive unilateral cooperation. Having taken the costs of emission reduction in the heating system of one's house reduces the financial capabilities in future social dilemmas. Being hurt after showing civil courage lowers the future income possibilities during times of recovery. Ceteris paribus, having been a free-rider in past situations provides a healthy and financially well-equipped starting point for future actions. In the limit, past providers may not be *able* to contribute in the future due to excessive free-riding by others, while free-riders accumulate resources on their private accounts.

Although there is a considerable literature on cooperation in social dilemmas, surprisingly little is known about its dynamic aspects. Gürerk et al. (2010b) experimentally study a linear public-good game in which a subject's provision ability today depends on the subject's and her group members' behavior in the past. Additionally, they allow for the possibility of costly peer-to-peer punishment with a convex punishment technology that is similar to that of Fehr and Gächter (2002) for low values of assigned punishment points. The distinctive feature of the design is the endogeneity of players' contribution capabilities. Instead of providing subjects with (new) endowments in every round they play, they receive an initial endowment on their wealth account and subsequently play with whatever is currently on that account. Consequently, their payoff does not consist of the sum of period payoffs, but it is given by the final amount on their wealth account.

The structure of the game puts all the weight on the long run. Notably, this leads to incentives for cooperation even in the absence of a punishment mechanism if at least a fraction of the players is motivated by social considerations. On the other hand, the introduction of a punishment mechanism could have devastating effects if future contribution capabilities are determined by present behavior, especially because early punishment has been shown to be particularly strong in experimental studies of peer-punishment mechanisms. Alternatively, potential punishers, being aware of this hazard, might refrain from sanctioning other group-members. As a consequence, play in the game with and without the punishment mechanism might not differ.

The experimental data show that players do punish, leading to an initial disadvantage of groups with punishment possibilities as compared to groups that do not dispose of punishment. However, groups with punishment possibilities are able to keep players' contributed fractions of their current wealth at a constant level, whereas in the punishment-free environment, these fractions exhibit the typical declining trend. With punishment levels falling over time, wealth levels in the groups having punishment opportunities are able to catch up with those in the groups without. In contrast to the latter, average wealth levels in the former exhibit an increasing growth path, such that significantly higher wealth and, consequently, contribution levels seem to be a question of an extension of the time horizon by a small number of rounds.

Thus, punishment enhances cooperation even in a dynamic setting. In this sense, the results are a reassuring sign of robustness for public-good studies on punishment. At the same time, they underline the fact that peer-punishment will not be a suitable solution of social dilemmas for all groups: in a dynamic setting, its double-edged character clearly asserts itself: in some instances, the enhancement of cooperation comes at too high a price, leading the respective society to end up worse than it might have in the absence of sanctioning opportunities.

Some Final Words

Yes, we did! Unlike the advice initially quoted we conducted experiments on public goods provision reaching for a better understanding on institutions fostering cooperation in social dilemmas. The inspiring discussions with Reinhard Selten were most valuable for designing experiments aiding our understanding of human behavior. We are very grateful that we had the opportunity to benefit from Reinhard Selten's academic advice.

References

Andreoni J (1988) Why free ride – strategies and learning in public-goods experiments. J Public Econ 37:291–304

Andreoni J (1995) Cooperation in public-goods experiments – kindness or confusion? Am Econ Rev 85:891–904

Axelrod R (1984) The evolution of cooperation. Basic Books, New York

Becker O, Selten R (1970) Experiences with the management game SINTO-Market. In: Sauermann H (ed) Contributions to experimental economics, vol. 2. J.C.B. Mohr (Paul Siebeck), Tübingen, pp 136–150

Ben-Shakhar G, Bornstein G, Hopfensitz A, van Winden F (2007) Reciprocity and emotions in bargaining using physiological and self-report measures. J Econ Psychol 28:314–323

Bochet O, Page T, Putterman L (2006) Communication and punishment in voluntary contribution experiments. J Econ Behav Organ 60:11–26

Bolton GE, Katok E, Ockenfels A (2005) Cooperation among strangers with limited information about reputation. J Public Econ 89:1457–1468

Bosman R, van Winden F (2002) Emotional hazard in a power-to-take experiment. Econ J 112:147–169

Botelho A, Harrison GW, Costa Pinto LM, Rutström EE (2007) Social norms and social choice. Working paper no. 05-23, Economics Department, University of Central Florida

Bshary R, Grutter AS (2006) Image scoring and cooperation in a cleaner fish. Nature 441:975–978

Cookson R (2000) Framing effects in public goods experiments. Exp Econ 3:55–79

Croson R (1996) Partners and strangers revisited. Econ Lett 53:25–32

Croson R (2007) Theories of commitment, altruism and reciprocity: evidence from linear public goods games. Econ Inq 45:199–216

de Quervain DJF, Fischbacher U, Treyer V, Schellhammer M, Schnyder U, Buck A, Fehr E (2004) The neural basis of altruistic punishment. Science 305:1254–1258

Denant-Boemont L, Masclet D, Noussair CN (2007) Punishment, counterpunishment and sanction enforcement in a social dilemma experiment. Econ Theory 33:145–167

Dufwenberg M, Gächter S, Hennig-Schmidt H (2006) The framing of games and the psychology of strategic choice. CeDEx Discussion Paper 2006-20, University of Nottingham

Egas M, Riedl A (2008) The economics of altruistic punishment and the maintenance of cooperation. Proc R Soc B Biol Sci 275:871–878

Engel C, Rockenbach B (2009) We are not alone: the impact of externalities on public good provision. Max Planck Institute for Research on Collective Goods 29, Bonn

Falk A, Fehr E, Fischbacher U (2005) Driving forces of informal sanctions. Econometrica 73:2017–2030

Falk A, Gächter S, Fischbacher U (2009) Living in two neighborhoods – social interaction effects in the lab. CEDEX Discussion Paper 1

Fehr E, Camerer C (2007) Social neuroeconomics: the neural circuitry of social preferences. Trends Cogn Sci 11:419–427

Fehr E, Gächter S (2000) Cooperation and punishment in public goods experiments. Am Econ Rev 90:980–994

Fehr E, Gächter S (2002) Altruistic punishment in humans. Nature 415:137–140

Fehr E, Rockenbach B (2003) Detrimental effects of sanctions on human altruism. Nature 422:137–140

Fischbacher U, Gächter S (2010) Social preferences, beliefs, and the dynamics of free riding in public good experiments. Am Econ Rev 100:541–556
Fischbacher U, Gächter S, Fehr E (2001) Are people conditionally cooperative? Evidence from a public goods experiment. Econ Lett 71:397–404
Gächter S (2007) Conditional cooperation: behavioral regularities from the lab and the field and their policy implications. In: Frey BS, Stutzer A (eds) Psychology and economics. The MIT Press, Cambridge
Gächter S, Herrmann B (2009) Reciprocity, culture, and human cooperation: previous insights and a new cross-cultural experiment. Philos Trans R Soc B Biol Sci 364:791–806
Gächter S, Herrmann B (forthcoming) The limits of self-governance when cooperators get punished – experimental evidence from urban and rural Russia. European Economic Review
Gächter S, Thöni C (2005) Social learning and voluntary cooperation among like-minded people. J Eur Econ Assoc 3:303–314
Gächter S, Renner E, Sefton M (2008) The long-run benefits of punishment. Science 322:1510
Gürerk Ö, Irlenbusch B, Rockenbach B (2010) Community choice in social dilemmas. Mimeo
Gürerk Ö, Rockenbach B, Wolff I (2010) The effects of punishment in dynamic public-good games. Mimeo
Gürerk Ö, Irlenbusch B, Rockenbach B (2006) The competitive advantage of sanctioning institutions. Science 312:108–111
Hardin O (1968) The tragedy of the commons. Science 162:1243–1248
Herrmann B, Thöni C, Gächter S (2008) Antisocial punishment across societies. Science 319:1362–1367
Kelley H, Stahelski A (1970) Social interaction basis of cooperators' and competitors' beliefs about others. J Pers Soc Psychol 16:190–219
Keser C (2000) Strategically planned behavior in public good experiments. CIRANO, Montreal
Keser C, Gardner R (1999) Strategic behavior of experienced subjects in a Common Pool Resource Game. Int J Game Theory 28(2):241–252
Knoch D, Pascual-Leone A, Meyer K, Treyer V, Fehr E (2006) Diminishing reciprocal fairness by disrupting the right prefrontal cortex. Science 314:829–832
Ledyard JO, (1995) Public Goods. A Survey of Experimental Research. The Handbook of Experimental Economics. J.H. Kagel und A.E. Roth. Princeton, NJ, Princeton University Press: 111–194
Masclet D, Villeval MC (2008) Punishment, inequality, and welfare: a public good experiment. Soc Choice Welf 31:475–502
Milinski M, Semmann D, Krambeck H-J (2002) Reputation helps solve the 'tragedy of the commons'. Nature 415:424–426
Milinski M, Semmann D, Krambeck H-J, Marotzke J (2006) Stabilizing the Earth's climate is not a losing game: supporting evidence from public goods experiments. Proc Natl Acad Sci USA 103:3994–3998
Neugebauer T, Perote J, Schmidt U, Loos M (2009) Self-biased conditional cooperation: on the decline of cooperation in repeated public goods experiments. J Econ Psychol 30:52–60
Nikiforakis N (2008) Punishment and counter-punishment in public good games: can we really govern ourselves? J Public Econ 92:91–112
Ostrom E, Walker JM, Gardner R (1992) Covenants with and without a sword – Self-governance is possible. Am Polit Sci Rev 86:404–417
Page T, Putterman L, Unel B (2005) Voluntary association in public goods experiments: reciprocity, mimicry, and efficiency. Econ J 115:1032–1052
Panchanathan K, Boyd R (2004) Indirect reciprocity can stabilize cooperation without the second-order free rider problem. Nature 432:499–502
Pillutla M, Murnighan KJ (1996) Unfairness, anger, and spite: emotional rejections of ultimatum offers. Organ Behav Hum Decis Process 68:208–224
Rabin M (1993) Incorporating fairness into game-theory and economics. Am Econ Rev 83:1281–1302

Reuben E, van Winden F (2008) Social ties and coordination on negative reciprocity: the role of affect. J Public Econ 92:34–53
Rockenbach B, Milinski M (2006) The efficient interaction of indirect reciprocity and costly punishment. Nature 444:718–723
Rockenbach B, Wolff I (2010) Institution design in social dilemmas: how to design if you must? Mimeo
Samuelson P (1954) The pure theory of public expenditures. Rev Econ Stat 36(4):387–389
Sanfey AG, Rilling JK, Aronson JA, Nystrom LE, Cohen JD (2003) The neural basis of economic decision-making in the ultimatum game. Science 300:1755–1758
Sefton M, Shupp R, Walker JM (2007) The effect of rewards and sanctions in provision of public goods. Econ Inq 45:671–690
Seinen I, Schram A (2006) Social status and group norms: indirect reciprocity in a repeated helping experiment. Eur Econ Rev 50:581–602
Selten R (1967) Die Strategiemethode zur Erforschung des eingeschränkt rationalen Verhaltens im Rahmen eines Oligopolexperiments. In: Sauermann H (ed) Beiträge zur Experimentellen Wirtschaftsforschung. J. C. B. Mohr, Tübingen, pp 136–168
Selten R, Mitzkewitz M, Uhlich GR (1997) Duopoly strategies programmed by experienced players. Econometrica 65(3):517–555
Seymour B, Singer T, Dolan R (2007) The neurobiology of punishment. Nat Rev Neurosci 8:300–311
Weber R (2006) Managing growth to achieve efficient coordination in large groups. Am Econ Rev 96(1):114–126
Wedekind C, Milinski M (2000) Cooperation through image scoring in humans. Science 288:850–852
Yamagishi T (1986) The provision of a sanctioning system as a public good. J Pers Soc Psychol 51:110–116

Chapter 12
Social Behavior in Economic Games

Gary E. Bolton, Werner Güth, Axel Ockenfels, and Alvin E. Roth

The Solidarity Game: What Drives Social Behavior in Economic Games? (Prepared by Axel Ockenfels)

In 1993, I attended Reinhard Selten's class on "bounded rationality". The class polarized the students: some were immediately struck by his unorthodox view on economic behavior. Others were more skeptical for the same reason. I remember vividly, for example, when Selten explained to us how people buy a new house. He spent maybe 15 minutes illustrating in great detail how people form aspirations and how these aspirations get adjusted during the search process in non-compensatory ways. Then his lesson reached an unexpected climax: He explained how, after all these boundedly rational adaptations, the house searchers might come across a house of "outstanding attractiveness" and would then buy it – regardless of whether it fits their aspirations or not. Well, I must admit that, having studied the elegant theory of rational decision making in Bonn, I was not well-prepared for this perspective on economic behavior. However, over the years Selten taught me that, in his own words, it is not human behavior that is anomalous – rational behavior is the anomaly. Nonetheless, I concentrated my studies on rational choice

G.E. Bolton
Laboratory for Economic Management and Auctions, Smeal College of Business, Pennsylvania State University, 334 Business Building, University Park, PA 16802, USA

W. Güth
Strategic Interaction Group, Max Planck Institute of Economics, Kahlaische Straße 10, D-07745 Jena, Germany

A. Ockenfels (✉)
Universität zu Köln, Albertus-Magnus-Platz, D-50923 Köln, Germany
e-mail: ockenfels@uni-koeln.de

A.E. Roth
Harvard Business School, Baker Library 441, Soldier's Field Road, Boston, MA 02163, USA

A. Ockenfels and A. Sadrieh (eds.), *The Selten School of Behavioral Economics*,
DOI 10.1007/978-3-642-13983-3_12, © Springer-Verlag Berlin Heidelberg 2010

and general equilibrium theory, and eventually began a diploma thesis in this field. But Selten had seeded enough doubts with his class, and I increasingly challenged what I was doing. Eventually, I decided to ask him for advice on how to reduce all the cognitive dissonance that he had implanted in my brain. He listened to me and, after several discussions, convinced me to start my diploma thesis from scratch on boundedly rational behavior.

At that time, following the seminal paper by Güth et al. (1982) on ultimatum games, other-regarding behavior was an increasingly central topic in experimental economics. The leading, unifying approach to ultimatum game and other non-equilibrium behaviors was Roth and Erev's (1995) reinforcement learning model. The model is very successful in capturing behavior in various repeated games, including games where standard rational theory succeeds and where it fails: clearly, *adaptation matters* [see Section "Adaptation (Prepared by Al Roth)"]. Another interesting development in the early 1990s were one-shot dictator experiments, where (almost by definition) social behavior cannot be organized by purely materially based adaptation but seemed to call for models of social motivation such as altruism.

Against this background, Selten proposed that I should write my diploma thesis either on learning models, or on social motivation in dictator-like experiments. He recommended the experiment and I agreed. As it turned out, Selten's advice became the tipping point of my academic life.

Reinhard Selten, Karim Sadrieh, who at that time was a research assistant to Selten, and I then designed a variant of the dictator game (although my role was rather small in the process), the solidarity game: Each subject in the experimental (one-shot) game participates in exactly one three-person-game in which each of the subjects independently wins either DM 10.00 with probability 2/3 or zero with probability 1/3. Before the random draws subjects have to decide, how much in the case of their winning they are willing to give to a single loser – who is the only one in the group and so can be helped by two winners – and to each of two losers. Of course, no gifts are made if there is no loser in the group. The pledge to help a loser is conditioned on winning and on the presence of losers.[1]

The solidarity game creates a situation in which subjects can show solidarity in the sense that they are willing to help others who by chance came to a much worse position than they themselves. To some extent solidarity is similar to reciprocity, a motivation which urges you to give something in exchange for something you have received, even if you are not compelled to give anything. However, solidarity is different. Gifts are made but not reciprocated. They are made to recipients who presumably, if one were in need oneself, would have made a gift to oneself. Solidarity aims at a reciprocal relationship, but a more subtle one than giving after one has received.

[1]The experiment was conducted in the students' restaurant employing an elaborate double blind procedure. The set-up was later applied to many other experiments.

The solidarity game produced a couple of interesting results, indicating what a theory of social behavior in economic games needs to organize. Maybe the most important observation is that only few (25 out of 120 subjects) were not willing to leave anything to losers conditional on being a winner. Rather, a majority of subjects chose a systematic pattern that we later called "fixed total sacrifice". They gave the same total amount in both loser conditions; e.g. they kept DM 5 for themselves, in case of winning, and gave DM 5 to a single loser and DM 2.50 to each of two losers (which is one of the most frequently chosen patterns). This kind of behavior suggests that subjects traded-off two competing goals: on the one hand they wanted to keep a certain share of their winning (DM 10) and at the same time they were helping losers in their group. Keeping half of the DM 10 seemed to be a fair way of resolving the trade-off. In fact, it has been shown that a concern for relative (or fair) payoffs organizes a large and seemingly disparate set of behavioral patterns in a wide range of games: *motivations matter* [see Section "Motives (Prepared by Gary Bolton)]".

However, observe how this simple heuristic to resolve the trade-off plays out in the solidarity game. If everybody chooses to fix the total sacrifice, regardless of the number of losers, then these subjects give double as much to a single loser than to one of two losers. Suppose that everyone in a group behaves in this way, all with the same gift parameters. Then a single loser receives four times as much as one of two losers. That is, the needs of the other players or the reduction of inequality do not seem to be the guiding considerations of these subjects. For example, it can be shown that the behavior cannot be the result of 'pure' altruism (under some straightforward assumptions). Similarly, it is easy to see that a fixed total sacrifice does not easily lend itself to an interpretation by inequality aversion. Inequality aversion would imply something closer to what can be called 'fixed gift to loser'; e.g., a conditional gift of DM 10/3, regardless of the number of losers, by all three players would yield the egalitarian solution, independent of the outcome of the chance move.[2]

Thus, fixed total sacrifice seems to require a different explanation for how the trade-off between selfish and social motivation is resolved, one that is not relying on utility maximization. One plausible heuristic is that subjects in the solidarity game first fix the total amount to be sacrificed for solidarity and then distribute it among losers regardless of their number. In fact, this idea turned out to be quite similar to one proposed by Bolton et al. (1998b) for the interpretation of behavior in which one dictator can make gifts to many potential recipients. Bolton et al. (1998a, b) observed that in such situations the dictator often distributes his sacrifice quite unevenly among the anonymous recipients. This could not result from altruistic utility maximization under the assumptions made above. Bolton et al. (1998a, b) suggested that the decision process has two parts. In the first part, the amount to be distributed is fixed and in the second part the distribution is decided upon. In our

[2]In fact, 16% of the non-egoistical subjects chose a fixed gift to loser pattern, with about half of them choosing a gift of DM 10/3.

case there is no analogue to the second part because, in the case of two losers, we forced the subjects to give the same amount to both.[3]

Summing up, the evidence in the solidarity game suggests that, beyond motivation, *cognition matters* (see Section "Cognition and Bounded Rationality (Prepared by Werner Güth)"; see also Selten 1978a, 1983, 1987 for the role of fairness in models of boundedly rational behavior). In fact, the solidarity game produced additional evidence for the important role of cognition. Applying a theory of the prominence structure of the decimal number system developed by Albers and Albers (1983) and Selten (1987), assigning prominence levels to numbers, one can achieve information about the exactness of the decision processes in the solidarity game. This information can then be used in order to describe and organize the rounding processes underlying the subjects' behavior.[4]

In 1994, while I was writing my diploma thesis on the solidarity game, Selten received the Nobel Prize for his work in game theory. One could have expected that Selten would now have less time for his students. Quite to the opposite, shortly after I submitted my diploma thesis, he suggested that we could write a joint paper on the solidarity game. Needless to say that I was thrilled. The result can be found in Selten and Ockenfels (1998). It stimulated more research on solidarity games. For instance, it has been shown that solidarity systematically and robustly depends on gender, education and culture (Ockenfels and Weimann 1999; Brosig et al. 2010), but not on measures of empathy-driven pro-social behavior used in social science (Büchner et al. 2007), that people are not only motivated by different varieties of outcome fairness (Bolle et al. 2008), but also by procedural fairness concerns in solidarity game-like environments (Trhal and Radermacher 2008; Bolton et al. 2005, Bolton and Ockenfels 2010, but see also Bolle and Costard 2009),[5] that solidarity may be subject to framing effects (Bischoff and Frank 2009), among other effects.

In the same year when the solidarity game paper came out, Selten (1998) published a paper on "Features of experimentally observed bounded rationality" based on his extensive work and on Simon's seminal work in 1957. There he explained his approach to investigating bounded rationality with its three roots, adaptation, motivation and cognition. In the next sections, this classification is taken up and explored.

---

[3]In this connection, I first came across the work of Gary Bolton. Selten brought the two of us together in Bonn, while I was still writing my diploma thesis. Gary then invited me to spend time at Penn State University. Again, Selten had a substantial positive influence on the path of my academic life.

[4]Moreover, applying the same method to a related dictator experiment, I later found that with hypothetical choices subjects put less cognitive effort in resolving the trade-off but rather restricted themselves to fewer, more prominent gifts (Ockenfels 1999). In Mussweiler and Ockenfels (2010), we employ procedural priming techniques from psychology in order investigate the cognitive underpinnings of social preferences.

[5]Not giving can be considered fair in the sense that everybody has the same chance of winning DM 10, and so everybody has the same expected payoff.

Cognition and Bounded Rationality (Prepared by Werner Güth)

In his Presidential Address on "Features of experimentally observed bounded rationality" Reinhard Selten says "that *cognition* stands for all the reasoning processes in the human mind, regardless of whether they are fully conscious or not" (Selten 1998). Clearly, the "or not" part renders this a very general definition of "cognition". For reinforcement learning [see Section "Adaptation (Prepared by Al Roth)" below] it would mean that its underlying assumption that "what was good in the past will be good in the future" describes "cognition", even when not consciously perceived. But then all our behavior obeying to the law of effect would have to rely on that "cognition". Here we, however, want to distinguish between adjustment without reasoning and "cognition" in the sense of forward looking deliberation. Although one usually can argue about definitions, we therefore take the liberty to exclude the "or not" part. Thus, in the following, cognition stands for all the conscious reasoning processes in the human mind and will be used without citation marks.

When, as actually observed, Reinhard Selten takes his umbrella when hiking in the desert with no rain possible, the adjustment interpretation would be: he has developed the habit of always taking his umbrella without anymore reasoning about it. A cognitive interpretation would be: he views it as a weapon against dangerous animals and humans or to protest against too much sunshine. We will stick to justifications of the latter kind.[6]

Rational choice theory has very little to say about human cognition. Orthodox decision and game theory essentially view some of the most crucial aspects of decision making like

- the preference relations or utility functions
- the beliefs concerning the decision environment, including others behavior
- the choice sets
- the adequate mental representation of the decision environment

as exogenously given. But to determine what one should achieve, which choices one wants to consider, and how to predict what these choices imply for one's goals, are crucial aspects of human cognition. Furthermore, human cognition must pay attention to our limited cognitive abilities and the relevance of what is at stake. In game theory, we usually impose further common knowledge and consistency

[6]Similar explanations are possible for the desire and ability to learn driving the car. Reinhard Selten and John Harsanyi were not only closely collaborating but also very close friends. John Harsanyi never drove a car and Reinhard Selten has learnt it rather late in life. In fact, when John and Ann Harsanyi came for their first yearly visit to Bielefeld, I (Werner Güth) began to commute by car once or twice a week between Bielefeld and Münster where I did all my career work. Ann Harsanyi then told me: "Werner, I fear you will never become a game theorist – you drive too well!"

requirements. This altogether justifies that rational choice theory is only absorbable by rational (wo)men whose genetical distance to homo sapiens exceeds by far our genetical distance to our primate relatives in the animal kingdom.

But then Reinhard Selten is not only one of the pioneers of orthodox rationality theory (Selten 1965, 1975; Harsanyi and Selten 1972, 1988) but also a leading scholar in evolutionary game theory (Selten 1988; Hammerstein and Selten 1994). Here nobody reasons at all: one is borne with a given inherited behavioral trait which Mother Nature has selected in the light of its relative fitness in the past. Don't we now throw out the baby with the bath water? In our view, we should not recommend psychotherapy for Reinhard Selten to cure him from schizophrenia in science but rather assume a multiple-selves interpretation of Reinhard Selten. He himself has done this by the different roles of his Nancy Schwartz Lecture (Selten 1991):

- When refining equilibria (Selten 1965, 1975) or selecting among them (Harsanyi and Selten 1972, 1988) his alter ego is that of a philosopher reasoning about common(ly known) rationality and how it can be justified as an approximation of rationality in a world where no possible behavior can be excluded.
- When further developing evolutionary (game) theory, his alter ego is that of an evolutionary biologist studying what would evolve when everything is only path dependent.

In our view, when actually predicting what will happen in the real world, Reinhard Selten will trust neither of these two alter egos. Rather he would assume his – at least – third alter ego, namely the one of a behavioral social scientist who denies

– neither conscious reasoning in boundedly rational ways how behavior affects goal achievement
– nor path dependence in the sense of learning from past experiences and inherited traits.

It is this alter ego of Reinhard Selten who very early on (Sauermann and Selten 1962) has faced the challenge of developing the theory of bounded rationality, based on conscious deliberation how goal achievement, expressed by so-called aspirations, can be achieved via searching successively for adequate plans and how the success of such search may result in aspiration adaptation and/or extended search.

When trying to generalize such ideas to stochastic decision environments and games, it is important not to import ideas of orthodox or evolutionary (game) theory without examining whether they are suitable for an at best bounded rationality approach. This may concern the clear distinction of means and ends which may not apply to satisficing (Simon 1957) since aspirations could be both, ends and (choices of) means.

Here we do not want to discuss this in full generality but only focus on one crucial aspect, namely that bounded rationality should be developed without necessarily importing Bayesianism in the sense of probabilistic reasoning. Since one should not discard a well accepted concept without offering an alternative, here is our idea of a more broadly defined process of decision making:[7]

1. Rather than considering the universe CxS_{-i} of all possible constellations (c, s_{-i}) of chance events $c \in C$ and behavioral constellations $s_{-i} = (s_j)_{j \neq i} \in S_{-i} = \underset{j \neq i}{x} S_j$ of i's interaction partners, decision maker i will rather focus on a few relevant scenarios $(c, s_{-i}) \in Sc_i$ with $\emptyset \neq Sc_i \subset CxS_{-i}$.
2. For each scenario $(c, s_{-i}) \in Sc_i$ decision maker i then forms an aspiration level $A_i(c, s_{-i})$ and thus altogether an aspiration profile $A_i = (A_i(c, s_{-i})_{(c,s_{-i}) \in Sc_i})$.
3. Based on the – in all likelihood simplified – cognitive representation $C_i(R)$ of the actual situation R decision maker i successively develops action plans s_i in order to check whether or not they are satisficing A_i. If $U_i^{C_i(R)}(s_i(c, s_{-i}))$ measures i's perceived goal achievement when using s_i and expecting (c, s_{-i}), according to i's cognitive representation $C_i(R)$ of R the choice s_i satisfices A_i if $U_i^{C_i(R)}(s_i(c, s_{-i})) \geq A_i(c, s_{-i})$ for all $(c, s_{-i}) \in Sc_i$.Note that this satisficing definition does not require intrapersonal payoff aggregation as assumed by expected utility maximization (see Güth and Kliemt 2010, for a more general discussion of intrapersonal and interpersonal payoff comparisons).
4. Especially, when very easily finding a satisficing action plan s_i, decision maker i may return to step (1) or (2) and develop a more ambitious aspiration profile before deciding. If, however, it is difficult to find an action plan s_i satisficing A_i, one might search more or return to (1) or (2) in order to form more moderate aspirations before searching again.

This, of course, provides at best only a general framework of boundedly rational decision making as a process model. A more complete process model would have to specify the cognitive representation part $C_i(R)$ about which we so far know very little. It is especially here where cognitive psychologists and behavioral economists should join forces.

Even without specifying the mental representation part $C_i(R)$, one can, however, test the formally defined satisficing hypothesis which does not require but also does not exclude probabilistic reasoning and allows for optimal satisficing as an – for complex tasks – unlikely border case (in the sense that one cannot increase the aspiration for one scenario without having to reduce those for other scenarios; see

[7]Reinhard Selten was not always fully convinced that I am the right person for a formal approach. When, in 1970, I (Werner Güth) got my diploma in economics, I was interested in game theory. So I went to Berlin to consult Reinhard Selten, the famous German game theorist and a former colleague of my boss at the University of Münster, Jochen Schumann. Reinhard Selten's first comment: "With your background, you should not work in game theory, maybe you could do empirical studies." My reaction: I visited him again in the same year. Reinhard Selten's second comment: "We can talk already." After regularly commuting to Bielefeld, I also engaged in empirical research by becoming an experimentalist. I also accepted the advice of John Harsanyi and Reinhard Selten to study mathematics parallel to my career work. Like my daughter Sandra, I was very fond of studying mathematics. But as Sandra said: "There are better genes than ours for studying mathematics!"

Fellner et al. 2009, and Güth et al. 2009). The main trick of doing so is to avoid the "curse-of-revealed- aspirations" method by directly eliciting the "belief sets" Sc_i and the aspiration profiles $A_i = (A_i(c, s_{-i})_{(c,s_{-i}) \in Sc_i})$ in addition to the choice behavior rather than by inferring them from observed choices.

Adaptation (Prepared by Al Roth)

I first met Reinhard Selten in 1978, at the Fourth International Conference on Game Theory at Cornell University. I recall that he was accompanied by several young German game theorists, who had clearly been advised to refer to him, in the American manner, as "Reinhard" when they spoke to him in English. It was evident that they found this difficult, not only when speaking to him, but even when speaking about him when he was not present.

I met him again when he came to Pittsburgh around 1986, for the conference I hosted that eventually led to the volume of papers that included Selten (1987). I recall that during a coffee break I sat with him and he asked me something like "what else are you working on?" I briefly told him of a very early-stage idea I had been thinking about. He listened carefully, and finally pronounced "That's the wrong approach."

In 1999, Reinhard and I had both been invited to speak in Chicago to the Society for Quantitative Prediction of Behavior. (He spoke about direction learning, and I spoke about my work with Ido Erev on reinforcement learning, and on what eventually became the Equivalent Number of Observations (ENO) measure for evaluating the predictive power of a theory, Erev et al. 2007).[8] Afterwards, we took a walk along the lake, and I heard, not for the first time, about the various ways in which Reinhard had been an outsider in economics, and still regarded himself as one. I remarked to him that this sounded different to me now that he had won the Nobel Prize. He seemed to take this under consideration, and hasn't since mentioned to me that he's an outsider, but I can't tell if this only affected his conversations with me.

The last of these anecdotes (and perhaps the first) suggests that Selten adapts his behavior based on his experience, at least to some extent. (The second anecdote may show that he hoped I would do the same.) But of course Selten has been one of the leading theorists of adaptive behavior and learning (see not only his work on direction learning, but also the famous Chain Store Paradox (Selten 1978b), which, along with the iconic example he invented, also contains a model of learning). Selten's view of learning and adaptation is nuanced. In what follows I'll briefly mention some of the ways in which my colleagues and I have also found a need for nuance.

---

[8]For direction learning, see e.g. Selten and Stoecker (1986); the later developed concept of impulse balance equilibrium is based on a simple principle of ex-post rationality similar to Selten's learning direction theory (see, e.g., Ockenfels and Selten 2005).

Learning Models and Experimental Data

Ido Erev and I were drawn to models of learning for some of the same reasons that Gary Bolton and then also Axel Ockenfels were drawn to models of other-regarding preferences. We wanted to be able to explain why the behavior observed in some experiments looked very much as if it were influenced more by considerations of fairness than by conventional notions of equilibrium, while in other games, sometimes with similar, unfair equilibria, equilibrium was a good predictor of behavior, after players had even a little experience.

In this we were very much influenced and inspired by earlier discussions with Werner Güth, who had been a pioneer in framing these issues, and who thought that perhaps it would be more fruitful to approach different kinds of games separately. His idea, I think, was that notions of fairness were activated by the presence or absence of relevant information about payoffs, or by certain kinds of activity such as bargaining, but not by other activities, such as participation in markets. We wanted to see, instead, how far we could get with a simple theory that would encompass both kinds of games, without depending on the information available to the participants beyond what they could acquire through their own direct experience. This is what led to our exploration of models of reinforcement learning.

We began (in Roth and Erev 1995) by looking at the data from the four-country experiment on ultimatum and market games that I conducted with Hiro Okuno-Fujiwara, Vesna Prasnikar, and Shmuel Zamir (Roth et al. 1991), and with the earlier (but published later) experiment, Prasnikar and Roth (1992) on ultimatum and best-shot games. We followed that up (in Erev and Roth 1998) with an analysis of repeated play of games in strategic form with unique equilibria in mixed strategies. In both papers we found that simple learning models tracked the data well when initialized from early period play, and predicted the data quite well even when initialized with random initial play.

Of course we could reject the hypothesis that the data were generated by the same distributions being produced by our learning models. That is what led us to develop the statistical notion of the Equivalent Number of Observations (ENO) of a model, a measure of how many experimental observations you would need of behavior in a particular game to predict future behavior more accurately than the prediction of a given theoretical model. This helped us formalize what we meant when we said that a particular model was a useful approximation to, and useful for predicting, the observed behavior.

These investigations showed that even simple models of reinforcement learning (much simpler than anyone supposes a human being is, much as models of perfect rationality are much more rational than any human) do well as models of repeated play of games with small effective strategy sets (e.g. games against different other players, or zero-sum games even against a fixed player) in which personal history of play doesn't matter or isn't known. This is so even when the actual players have, and can use, much more information than the reinforcement models of learning use:

the simplest models employ only a player's knowledge of his own past choices and payoffs.

Learning theories are harder to use in repeated games against a fixed opponent, or where personal history is known and matters, since the relevant strategy spaces explode. To use learning models to predict behavior as successfully as those models work in simpler environments may depend on a breakthrough in how to handle large strategy sets, or in how to select appropriate strategy subsets without first looking at the data.

In the meantime, for repeated games it may be possible to use learning theories in a qualitative way (much as we use equilibrium theories for comparative statics predictions when we don't know all the relevant parameters needed to make quantitative equilibrium predictions). This is what we tried to do in Bereby-Meyer and Roth (2006), following up on the experiments concerning repeated prisoner's dilemma games reported by Selten and Stoecker (1986). Like them we considered a repeated prisoner's dilemma, in which players learn over time to cooperate in the early periods and defect in the late periods. We showed that when payoff variance is increased, which slows learning, this can entirely disrupt the learning of cooperation. That is, reinforcement learning is slower when payoff variance increases, and the repeated prisoner's dilemma is an environment in which this effect is magnified, since what one player learns in early periods depends on what other players are learning (and hence doing), so that when early learning was slowed, early cooperation was not reinforced, and learning to cooperate did not occur.

Another place where learning theories don't help is in predicting first-period play. This is one of the important contributions of theories of fairness and other-regarding preferences, particularly in games in which players have good information about one another's payoffs. It would be tempting to try to combine a model of fairness with a learning model. Right now the most successful models of each kind are quite different: the learning models have barely-rational (non-equilibrium) players who are motivated only by their own payoffs, while the fairness models have ideally rational players who play equilibrium strategies that are sensitive to their own and others' other-regarding motivations. My guess is that a well articulated learning model with other-regarding preferences would have to take into account that notions of fairness can change with experience. That is, my reading of the evidence is that peoples' ideas about fairness are often clear, but can change fairly rapidly in response to certain kinds of experience. So I think a promising direction for research would be to investigate learning models in which initial propensities of play are influenced by initial other-regarding preferences, and in which preferences as well as propensities to play particular strategies adapt to experience.

Learning in Market Design

One reason I personally am interested in theories of learning is that I am interested in practical market design. While traditional mechanism design in the theory literature compares mechanisms by comparing their equilibria, practical market design at least sometimes involves thinking about new markets in which all users are new, and about established markets in which at least some users may be new. So it is far from enough to have good equilibria; markets also have to be safe to participate in when they are new, and when new users are present in established markets. Both of these things may mean that we won't always be observing equilibrium behavior, but may have to think about how the market will operate when at least some participants are still learning (see Roth 2008 for a range of examples).

When I began studying bidding behavior on eBay with Axel Ockenfels, and later also with Dan Ariely, it became clear that this was an issue of considerable importance. There is a lot of very late bidding ("sniping") in eBay auctions, particularly among experienced bidders. At least part of this seems to be in response to the fact that less experienced bidders sometimes bid incrementally, and engage in bidding wars if provoked, something that experienced bidders seek to avoid.

These were issues we could study in field data (Roth and Ockenfels 2002), and theoretically (Ockenfels and Roth 2006) and in the laboratory (Ariely et al. 2005). In this respect this work is very much in the spirit of Selten, who has never been a captive of particular methods, but has sought to study the questions he addresses with whatever tools are most appropriate.

Motives (Prepared by Gary Bolton)

In the summer of 1993, I gave a talk in the Stony Brook Summer Festival on Game Theory. That year featured sessions on experimental research. I was a junior faculty then, and I gave a talk on work I was doing with Rami Zwick and Elena Katok on ultimatum and dictator games (Bolton and Zwick 1995; Bolton et al. 1998b). My Ph.D. thesis had argued that shrinking pie bargaining games, such as the ultimatum game (Güth et al. 1982), could be explained by preferences for fairness. But the theory didn't fit with data from dictator games and much of my talk was taken up with hypotheses and tests aimed at resolving that inconsistency (something that would turn out to take quite a few years). At the time, research on ultimatum and especially dictator games, while a growing enterprise, sparked contention among economists. The sessions were well attended. As I panned the room during my talk, my eye could not help but note two especially senior scholars, both of whom would subsequently win the Nobel Prize. One sat towards the back. About mid-way he got up and left, making what appeared to me to be a dismissive wave in my direction. The other sat up front, and upon completion of my talk, immediately approached me and asked whether I'd send the paper to a journal he was editing. No guarantees of course, but he thought the paper would have a decent chance.

That was how I met Reinhard Selten – sitting up front as he always does at talks – and it was the first of a great deal of generous support I would receive from him over the years. Selten actively facilitated the early work on social preference theory. I have learned and adopted a great deal from Selten and his work. At the same time, there have been points of disagreement, and some quite central. In particularly, Selten has always expressed strong reservations about the basic social utility approach, grounded in his doubts about the efficacy of utility functions to explain behavior. Selten is that rare scholar, supremely passionate in his own views but tolerant of those who see it another way, even to the point of helping them advance their view - albeit forcefully debating them all the way.

Selten has always championed an approach to descriptive theory that starts small, explaining a handful of selective empirical facts, and then broadening out the theory through successive revision. This is the approach I took in my Ph.D. thesis. As a graduate student, I helped Jack Ochs and Al Roth run a series of experiments on multi-stage shrinking pie bargaining games. The data produced were inconsistent with what we would expect if people were both self-interested and fully rational. In the subsequent paper, Ochs and Roth (1989) argued that a satisfactory theory of their data would explain five particular regularities they observed. They also noted that a propensity for comparative equity was a promising starting point. I was captivated by all of this and began working on a formal model. The model retained the rationality assumption but assumed that people were willing to trade off some self interest to obtain a more equitable portion of the bargaining pie– for themselves; so individuals cared about fairness for themselves as well as their money pay out. This comparative bargaining model as published in Bolton (1991) fits with all of Ochs and Roth's regularities. Still, the comparative model could be criticized along certain dimensions.

For one, the predominant assumption among economists has been that people are purely self interested, and this model broke with that, introducing fairness as a motive. This was not entirely novel to the study of games; Maschler's (1963) theory of the power of coalitions and Selten's (1987) theory of equal division payoff bounds are early examples of game theoretic models based on equity considerations. Both models were introduced to explain data culled from experiments on coalition bargaining games. Many economists find fairness a slippery concept, with questions starting at the most fundamental level: How do we define 'fair'? Viewed broadly, fairness is a multi-faceted concept, applied to particular circumstances in different ways by different people. The thing to remember is that, approached the same way, self interest is also a slippery concept. The multi-faceted nature of self interest is famously and popularly stated by Gordon Gekko, a character in the movie Wall Street: "Greed . . . captures the essence of the evolutionary spirit. Greed, in all of its forms - greed for life, for money, for love, knowledge - has marked the upward surge of mankind." We have a self interest in many facets of our lives, with different people weighting each facet differently. Economists, however, have traditionally defined self interest more narrowly in terms of material wants and needs. This has proven productive, it seems to me, for two reasons: First, it is well defined, in a compact manner suitable for theorizing. Second, it is a motive that has proven to be a good guide to a lot of market behavior, particularly competitive market pricing. Social preference models strive to do the same with regard to the fairness motive: accept a narrow characterization of the motive in trade-off for relatively strong explanatory power.

A second issue was how the motive should be represented in the theory. Clearly, fairness implies some sort of comparison and so the theory need adopt some normative standard to which payoffs can be compared. The earlier coalition theories adopted such normative standards but not as part of the utility functions. Selten (1987) explicitly eschews utility, arguing out that the maximizing subjective utility approach "fails to do justice to the structure of human decision processes." I have always been sympathetic to this argument. The primary reason I inserted fairness into the utility function was simply that I wanted a noncooperative model, and within this framework I could find no other place to put it (the earlier coalition models were cooperative models, solved with stability conditions that do not stipulate individual maximization). A limitation then of noncooperative social preference models is that they do not provide a statement of the decision process in which norms of fairness are employed (or as Werner Güth pithily puts it, the models are the product of "the neoclassical repair shop"). At the same time, the models can be useful in predicting outcomes of the process, under the important condition that the normative standard of comparison employed by the decision maker is stable (Bolton et al. 2005).

A third issue raised by the comparative model was generality, and this for me was the stickiest issue. My approach was start small, and use experimental data as a hothouse to grow the model, revising the model as data demanded. The model did well explaining comparative statics and other ordinal characteristics of shrinking pie game behavior. While it wasn't clear how to generalize the comparative model to a larger set of games, perhaps with more effort a revised model could. Economists have tended to favor models that start out as very general (later modifying them as data demands). When it works, it is the faster approach. But not all problems lend themselves to the broad theoretical vision necessary for it to work. The start small with experimental evidence approach offers an alternative, albeit a more time consuming one. As Selten argues "Pure speculation is an unreliable guide when descriptive theories are concerned. Unfortunately, the development of successful descriptive theories is a slow process that must be guided by experimental evidence."[9]

This is where things stood in the summer 1993 – it wasn't clear how to generalize the comparative model to a larger set of games – and it is where things still stood in the summer of 1995 when I and Elena Katok (we were recently

[9]The start small approach has been associated with the theory of games for a long time. In the introduction to their famous book, Von Neumann and Morganstern (1944) that the theory they would introduce would not solve practical economic problems right away but there was reason to show patience: "The great progress in every science came when, in the study of problems which were modest as compared with ultimate aims, methods were developed which could be extended further and further. The free fall is a very trival physical phenomenon, but it was the study of this exceedingly simple fact and its comparison with the astronomical material, which brought forth mechanics."

married) arrived in Bonn at Selten's invitation for a 6 month visit at his research center. There I began work with Selten's students Klaus Abbink, Karim Sadieh and Fang Fang Tang on an experiment aimed at clarifying the relationship between social preferences and reinforcement learning in ultimatum game behavior (Abbink et al. 2001). Al Roth and Ido Erev had recently published a paper (1995) showing that a reinforcement learning model could capture much of the pattern in the ultimatum game data. In their framework, it was unclear what, if any, role social preferences might have. Al and Ido's paper was an important stimulus to the later development of social preference theory because it was widely seen as offering an alternative explanation to social preferences, and one with the potential to generalize to a larger set of games.

It was towards the very end of my Bonn visit that I met Axel Ockenfels. He came into my office and told me that Professor Selten thought it would be a good idea if he visited me at Penn State, with the aim of collaboration. Axel was a graduate student then, and I knew about his work with Selten on the solidarity game. But beyond this, I really didn't know what Axel or Selten had in mind. At any rate, Axel and I began working on the ERC model almost immediately after his arrival at Penn State in the fall of 1996.

'ERC' is the acronym for equity, reciprocity and competition, the patterns of behavior that the model reconciles. The basic idea behind the model is to fashion a normative comparative standard in the utility function so that it can explain two important observations concerning fairness: why, in the ultimatum game, second movers turn down money to gain relative payoff (all receiving nothing is an equal split) and why players in the dictator game give money. Then apply this model to explain other observations about fairness, reciprocity as well as competition in bargaining, social dilemma and market games. The model can be estimated from ultimatum and dictator data to provide quantitative as well as ordinal predictions. Bolton and Ockenfels (1998, 2000) provide full accounts for the model. See DeBruyn and Bolton (2008) for an empirical fit of the model to an array of shrinking pie bargaining games.

The model was first presented at the Bonn Conference on Bounded Rationality hosted by Reinhard Selten in the summer of 1997. Immediately after I presented it, Ernst Fehr got up and presented a model he had been working on with Klaus Schmidt (Fehr and Schmidt 1999) that turned out to be similar in approach and scope to ERC. Cooper and Kagel (2010) provide an excellent overview of the experimental and theoretical literatures these models have spawned.

Let me mention three of the big messages from social preference theory. To begin, negotiation literature likes to distinguish between issues of economics and issues of principle, depending on whether bargainers are willing to compromise on the issue or not. Social preference models argue that, in its empirical form, fairness is an economic issue for most people. This contrasts with the philosophical work where fairness is often studied in its principle form (i.e., Rawls 1971). The empirical nature of fairness may have important implications for social welfare. Second, fairness is a critical component of reciprocity. "What is fair" is a judgment one must make in deciding whether to reciprocate as well as how much to reciprocate. I say that fairness is a "critical" component because the social preference models of Bolton and Ockenfels as well as Fehr and Schmidt show that these judgments are sufficient to trigger reciprocal acts (absent any judgment of intentionality). Third, preferences for fairness are compatible with the seemingly self interested behavior observed in competitive markets. In social preference models, people compete because it advances both pecuniary and fairness interests; the latter because not competing can greatly raise the probability of getting less than ones fair share. The fact that economists have traditionally given most of their attention to markets and price formation may explain their reluctance to admit fairness as a motive in other settings such as bilateral monopoly and public goods games – fairness preferences are effectively invisible in competitive pricing. See Bolton and Ockenfels (2008) for a controlled experiment on eBay that makes this point.

Yet there are other facets of market behavior where social preferences are likely to be visible. For instance, trust and trustworthiness typically play a role in markets – even markets where price formation is very competitive. Bolton et al. (2008) reports on an engineering study done on eBay's feedback system, essentially a reputation mechanism to enforce trust and trustworthiness in the promised quality and shipping of goods (see also Ockenfels et al. 2010). A key observation in the study is that the production of reputation information involves reciprocal behavior that can help the market, in terms of facilitating the provision of reputation information, but can also hurt the market, in terms of distorting the information provided. This behavior shows more than a passing resemblance to the reciprocity we observe in the lab and that can be accounted for by current social preference models. Nevertheless, a full theoretical accounting will no doubt require another round of model revision. These descriptive models have come a long way, but still have a ways to go.

Summary

According to Reinhard Selten (1998), there are three interacting roots of behavior: motivation (the driving force), adaptation (routine adjustment without reasoning) and cognition (reasoning). This article surveys our research endeavors along these lines, and how they were influenced not only by Selten's research in the field but also by our personal relationships to him. We did not always take approaches that Selten was happy with. Maybe this is because Reinhard Selten, as it turned out to be often the case, is years ahead of research. Yet, it was his research and his support that, directly or indirectly, shaped much of our views and that brought us together, in different subgroups, in numerous fruitful joint research projects with the goal of developing descriptively sound theories of human behavior – both as fundamental research, and increasingly as an input for economic engineering science and practical economic advice.

Acknowledgments Financial support from the German Science Foundation (DFG) is gratefully acknowledged

References

Abbink K, Bolton GE, Sadrieh K, Tang FF (2001) Learning versus punishment in ultimatum bargaining. Games Econ Behav 37:1–25

Albers W, Albers G (1983) On the prominence structure of the decimal system. In: Scholz RW (ed) Decision making under uncertainty. Elsevier Science, Amsterdam, pp 271–287

Ariely D, Ockenfels A, Roth AE (2005) An experimental analysis of ending rules in internet auctions. Rand J Econ 36(4):891–908

Bereby-Meyer Y, Roth AE (2006) Learning in noisy games: partial reinforcement and the sustainability of cooperation. Am Econ Rev 96(4):1029–1042

Bischoff I, Frank B (2009) Zwei unterschwellige Einflüsse auf die Entscheidungen im Solidaritätsspiel. Mimeo

Bolle F, Breitmoser Y, Heimel J, Vogel C (2008) Multiple motives- evidence from the solidarity game. Mimeo

Bolton GE (1991) A comparative model of bargaining: theory and evidence. Am Econ Rev 81:1096–1136

Bolton GE, Ockenfels A (1998) Strategy and equity: an ERC-analysis of the Güth-van Damme game. J Math Psychol 42:215–226

Bolton GE, Ockenfels A (2000) ERC: a theory of equity, reciprocity and competition. Am Econ Rev 90:166–193

Bolton GE, Ockenfels A (2008) Does laboratory trading mirror behavior in real world markets? Fair bargaining and competitive bidding on Ebay. University of Cologne, Working Paper Series in Economics, 2008, No. 36

Bolton GE, Ockenfels A (2010) Betrayal aversion: Evidence from Brazil, China, Oman, Switzerland, Turkey, and the United States: Comment. American Economic Review 100:628–633

Bolton GE, Zwick R (1995) Anonymity versus punishment in ultimatum bargaining. Games Econ Behav 10:95–121

Bolton GE, Brandts J, Ockenfels A (1998a) Measuring motivations in the reciprocal responses observed in a dilemma game. Exp Econ 1:207–219

Bolton GE, Katok E, Zwick R (1998b) Dictator game giving: rules of fairness versus acts of kindness. Int J Game Theory 27:269–299

Bolton GE, Brandts J, Ockenfels A (2005) Fair procedures: evidence from games involving lotteries. Econ J 115(506):1054–1076

Bolton GE, Greiner B, Ockenfels A (2008) Engineering trust: reciprocity in the production of reputation information. University of Cologne. Working paper

Bolle F, Costard J (2009) Solidarity, responsibility, and group identity. Working paper

Brosig J, Helbach C, Ockenfels A, Weimann J (2010) Still different after all these years: solidarity in eastern and western Germany. University of Cologne. Working paper

Büchner S, Coricelli G, Greiner B (2007) Self centered and other regarding behavior in the solidarity game. J Econ Behav Organ 62(2):293–303

Cooper D, Kagel J (2010) Other-regarding preferences: a selective survey of experimental results. In: Kagel J, Roth A (eds) The handbook of experimental economics, vol. 2. Princeton University Press, Princeton, NJ

DeBruyn A, Bolton GE (2008) Estimating the influence of fairness on bargaining behavior. Manage Sci 54:1774–1791

Erev I, Roth AE (1998) Predicting how people play games: reinforcement learning in experimental games with unique, mixed strategy equilibria. Am Econ Rev 88(4):848–881

Erev I, Roth AE, Slonim RL, Barron G (2007) Learning and equilibrium as useful approximations: accuracy of prediction on randomly selected constant sum games. Econ Theory 33:29–51, Special issue: Behavioral Game Theory Symposium
Fehr E, Schmidt K (1999) A theory of fairness, competition and cooperation. Q J Econ 114:817–868
Fellner G, Güth W, Maciejovsky B (2009) Satisficing in financial decision making – a theoretical and experimental approach to bounded rationality. J Math Psychol 53:26–33
Güth W, Kliemt H (2010) (Un)Bounded rationality in decision making and game theory – back to square one? Games 1:53–65
Güth W, Schmittberger R, Schwarze B (1982) An experimental analysis of ultimatum bargaining. J Econ Behav Organ 3:367–388
Güth W, Levati VM, Ploner M (2009) An experimental analysis of satisficing in saving decisions. J Math Psychol 53:265–272
Hammerstein P, Selten R (1994) Game theory and evolutionary biology. In: Aumann RJ, Hart S (eds) Handbook of game theory with economic applications, vol. 2. Elsevier, Amsterdam, pp 929–993
Harsanyi JC, Selten R (1972) A generalized nash solution for two-person bargaining games with incomplete information. Manage Sci 18(5): Part 2, 80–106
Harsanyi JC, Selten R (1988) A general theory of equilibrium selection in games. MIT-Press, Cambridge, MA
Maschler M (1963) The power of a coalition. Manage Sci 10:303–309
Mussweiler T, Ockenfels A (2010) How similarity affects beliefs and social preferences: a Procedural Priming Study. Work in progress
Neumann, John von, Morgenstern O (1944) Theory of Games and Economic Behavior, Princeton University Press
Ochs J, Roth AE (1989) An experimental study of sequential bargaining. Am Econ Rev 79:355–384
Ockenfels A (1999) Fairness, Reziprozität und Eigennutz. Tübingen: Mohr Siebeck
Ockenfals A, Selten R (2005) Impulse balance equilibrium and feedback in first price auctions. Games Econ Behav 51:155–170
Ockenfels A, Roth AE (2006) Late and multiple bidding in second-price internet auctions: theory and evidence concerning different rules for ending an auction. Games Econ Behav 55:297–320
Ockenfels A, Weimann J (1999) Types and patterns – an experimental East-West-German comparison of cooperation and solidarity. J Public Econ 71(2):275–287
Ockenfels A, Sliwka D, Werner P (2010) Bonus payments and reference point violations. University of Cologne. Working paper
Prasnikar V, Roth AE (1992) Considerations of fairness and strategy: experimental data from sequential games. Q J Econ 107:865–888
Rawls J (1971) A theory of justice. Harvard University Press, Cambridge MA, Reprint 1999
Roth AE (2008) What have we learned from market design? Econ J 118:285–310
Roth AE, Erev I (1995) Learning in extensive-form games: experimental data and simple dynamic models in the intermediate term. Games Econ Behav 8:164–212
Roth AE, Ockenfels A (2002) Last-minute bidding and the rules for ending second-price auctions: evidence from eBay and Amazon auctions on the internet. Am Econ Rev 92(4):1093–1103
Roth AE, Prasnikar V, Okuno-Fujiwara M, Zamir S (1991) Bargaining and market behavior in Jerusalem, Ljubljana, Pittsburgh, and Tokyo: an experimental study. Am Econ Rev 81(5): 1068–1095
Sauermann H, Selten R (1962) Anspruchsanpassungstheorie der Unternehmung. Z Gesamte Staatswiss 118:577–597
Selten R (1965) Spieltheoretische Behandlung eines Oligopolmodells mit Nachfrageträgheit. Zeitschrift für die gesamte Staatswissenschaft 121:301–324, 667–689
Selten R (1975) Reexamination of the perfectness concept for equilibrium points in extensive games. Int J Game Theory 4(1):25–55

Selten R (1978a) The equity principle in economic behavior. In: Gottinger HW, Leinfellner W (eds) Decision theory, social ethics, issues in social choice. D. Reidel, Dordrecht, pp 289–301
Selten R (1978b) The chain store paradox. Theory Decis 9(2):127–159
Selten R (1983) Equal division payoff bounds for three-person characteristic function experiments. In: Tietz R (ed) Aspiration levels in bargaining and economic decision making, vol 213, Lecture notes in economics and mathematical systems. Springer, Berlin, pp 255–275
Selten R (1987) Equity and coalition bargaining in experimental three-person games. In: Roth A (ed) Laboratory experimentation in economics. Cambridge University Press, Cambridge, pp 42–98
Selten R (1988) Evolutionary stability in extensive two-person games, correction and further development. Math Soc Sci 16(3):47–70
Selten R (1991) Evolution, learning, and economic behavior. Games Econ Behav 3(1):3–24
Selten R (1998) Features of experimentally observed bounded rationality. Eur Econ Rev 42:413–436
Selten R, Ockenfels A (1998) An experimental solidarity game. J Econ Behav Organ 34:517–539
Selten R, Stoecker R (1986) End behavior in sequences of finite prisoner's dilemma supergames: a learning theory approach. J Econ Behav Organ 7(1):47–70
Simon HA (1957) Models of man: social and rational. Wiley, New York
Trhal N, Radermacher R (2008) Bad luck vs. self-inflected neediness – an experimental investigation of gift giving in a solidarity game. J Econ Psychol 30(4):517–526

Part IV
Organizational Behavior

Chapter 13
The Equity Principle in Employment Relationships

Sebastian J. Goerg and Sebastian Kube

If you think of Reinhard Selten, you might first think of his outstanding contributions to game-theory which were honored with the Prize in Economic Sciences in Memory of Alfred Nobel in 1994. [1] At the same time, you might think of his devotion to the field of bounded rationality and his continual effort to promote the use of experiments in the Economic discipline (interestingly, his first journal publication in 1959 was not a theoretical, but an experimental paper featuring Cournot oligopoly models). [2] By jointly using theoretical and empirical methods, he wants *"[...] to build up a descriptive branch of decision and game theory which takes the limited rationality of human behavior seriously."* [3] This aim is of utmost importance, because only then will we be able to develop worthwhile economic models that enable us to understand and predict how people behave in certain economic situations.

Ideally, these models should consider bounded rationality, and the fact that human behavior includes additional traits like unbounded willpower and unbounded selfishness. Progress has been made in all three directions, but particularly the latter has received much attention during the last decade(s). For example, it has been pointed out that people frequently compare themselves to other persons; they judge situations, outcomes and procedures in terms of fairness, they care about social norms and they even spend considerable amounts of their personal resources

[1] Of course, if you know him better different things might come to your mind first, e.g., Esperanto, cats, or hiking in the Siebengebirge.

[2] Sauermann and Selten (1959).

[3] From Selten's biography as reported in: Les Prix Nobel. The Nobel Prizes 1994, Editor Tore Frängsmyr, [Nobel Foundation], Stockholm, 1995.

S.J. Goerg
MPI for Research on Collective Goods, Kurt-Schumacher-Str. 10, D-53113 Bonn, Germany
S. Kube (✉)
Department of Economics, Institute for Empirical Research in Economics, University of Bonn, Adenauerallee 24-42, 53113 Bonn, Germany
e-mail: kube@uni-bonn.de

A. Ockenfels and A. Sadrieh (eds.), *The Selten School of Behavioral Economics*,
DOI 10.1007/978-3-642-13983-3_13, © Springer-Verlag Berlin Heidelberg 2010

to (re-)establish fair allocations and to obey social norms. (Not only) based on these findings, interesting alternative models of human behavior have emerged (e.g., Fehr and Schmidt (1999), Bolton and Ockenfels (2000), Dufwenberg and Kirchsteiger (2004), Falk and Fischbacher (2006); among many others).

Most of these models assume that players use specific distributional norms (e.g., equal sharing) as a reference point to evaluate outcomes and ultimately to guide their behavior. The basic idea underlying these models and experiments is, of course, not a new one in Economics. For example, as early as 1972, Selten formulated two desirable features of economic models, namely that they should be easy to compute and that they should consider distributive norms. He stated: *"The importance of distributive norms has the consequence that the influence of individual utility functions on the behavior of the players is overshadowed by the prescriptions of norms, which do not refer to utilities, but directly observable variables such as money payoffs."*[4] Moreover, already back in 1978, Selten applied the equity principle from social psychology to economic behavior.[5] He described the principle as a *"[...] very reasonable [...] normative rule which can be applied by decision-makers without extraordinary capabilities of logical analysis and computation."*[6]

The equity principle can be operationalized as a simple "norm" or "formula" which states that the relation between an individual's "input" and his "output" should be equal for all individuals in a reference group (if this is not the case, individuals experience distress and seek to re-establish equity). For example, in Selten (1978) it is used to allocate rewards between members of a group: equitable reward combinations are formed given a standard of comparison, which assigns a weight w_i to the reward r_i of each group member *I*, with $i \in \{1,..,n\}$. An equitable reward combination then satisfies the condition:

$$r_1/w_1 = r_2/w_2 = \ldots = r_n/w_n.$$

Selten gives the example of sharing a dollar between two experimental subjects (Nydegger and Owen 1974) and of results from duopoly experiments (Selten and Berg 1970). While dividing a dollar between two experimental subjects might be a very simple task, the underlying normative rule for distributing the dollar is applicable on a wide range of situations. The same principle can be applied on equitable cost combinations or equitable quota combinations. In particular, it is applicable to employment relationships – an area in which much laboratory research has been conducted during past years. This includes our own research, too. Thus, in the following we will present a short selection of papers

[4]Selten (1972)

[5]cp. Selten (1978), Homans (1961), Adams (1963, 1965)

[6]Selten's article from 1978 also nicely illustrates the way he likes to work. He creates theories that explain actual economic behavior (i.e., experimental data), while being inspired by findings from surrounding areas of research (i.e., Biology, Psychology,...).

that experimentally investigate the influence of distributive norms in stylized employer–employee relationships.

Bilateral Relationship Between Employer and Employee

Usually the employment relationship between an employer and an employee is set in an environment characterized by incomplete contracts. Employers pay a fixed, unconditional wage and employees have significant discretion over their effort level. The employer's profit typically increases with the employee's effort due to the goods that are produced, while it decreases with the wage payment. The employee's payoff increases with the wage, while increasing effort lowers the payoff due to individual costs of effort exertion. Thus, under the classic assumptions of egocentric money-maximization, an employee should not be expected to provide effort above the minimal level, and an employer anticipating this should not pay more than the market-clearing wage.

Of course, this prediction runs counter to what we observe every day in actual employment relationships. Employers often pay fixed wages, and still workers frequently exert effort under these contracts. This form of behavior has been termed "gift exchange". In recent years, a vast body of literature has stressed the importance of gift exchange for incomplete contracts (e.g., Akerlof 1982, Akerlof and Yellen 1990): since many agents repay the employer's gift (taking the form of higher wages) by providing higher efforts, high effort can be elicited even under incomplete contracts. Robust evidence for this stems from experimental labor markets with bilateral "gift-exchange games" between a principal and a single agent (e.g., Fehr et al. 1993, Fehr, Gächter, and Kirchsteiger 1997, Fehr and Falk 1999, Fahr and Irlenbusch 2000, Charness 2004, and many more).

How does the gift-exchange game work? Let us look at Chmura et al. (2010) who use the classical setup of a gift-exchange experiment to investigate the wage-effort relationship in China. First, the employer chooses the wage from a given set of possible wages. Later, this wage is deducted from the employer's total earnings. In this specific experiment, the employer has an initial endowment of 20 Yuan and selects the wage w from the closed interval of 11 Yuan (minimum wage) and 20 Yuan (maximum wage). Given this wage the employee then decides on his effort level e. Effort is either chosen from a range of possible levels, or it is a binary decision, as in this experiment. Chmura et al. (2010) apply the strategy method (Selten 1967) to the employee's decision, i.e., for every possible wage the employee decides whether to work normal ($e = 0$) or hard ($e = 1$).

In the stylized employer–employee relationship of the gift-exchange game, higher effort leads to higher costs for the employees, while at the same time earnings of the employers increase. Here, if the employee works normal, effort costs of 5 Yuan occur. If he works hard these costs are doubled, i.e., he has to bear effort costs of 10 Yuan. A employee who works normal produces earnings of 5 Yuan for the employer, while a hard-working one produces four times higher earnings of 20 Yuan. Typically in gift-exchange experiments, the increase of the employer's earning from exerting effort is higher than the resulting cost for the employee. Therefore, an employment relationship that leads to higher effort by the employee increases total efficiency.[7]

In bilateral gift-exchange experiments, usually a positive wage-effort relationship is observed. Figure 13.1 from Chmura et al. (2010) is representative for this finding. For each possible wage payment, the figure shows the corresponding percentage of employees that choose to work hard. Obviously, the higher the wage, the more likely it is that the employee provides a high effort. This suggests that most employees care about fairness along a *vertical dimension*: a high wage is considered fair and this "gift" is repaid by a high effort. Moreover, this gift-exchange works even in one-shot situations, like the one investigated in Chmura

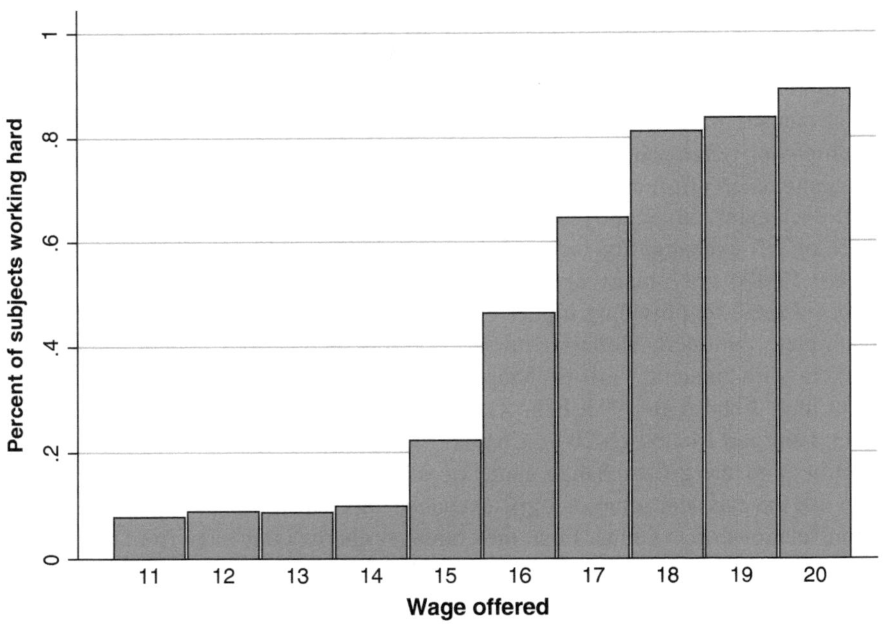

Fig. 13.1 Gift-exchange behavior in Chmura et al. (2010)

[7]Efficiency here refers to the sum of payoffs. The formal payoff functions are as follows:
Employer:

$$\pi = 20 + 5 + 15e - w, \quad \text{with} \quad 10 < w \leq 20$$

Employee:

$$\pi = w - 5 - 5e$$

et al. (2010). Thus, the results are not driven by the attempt of building up a positive reputation, because no future monetary gains are to be expected in one-shot situations (though this could certainly matter in repeated interactions).

In many papers, the relationship between increasing wage and increasing effort is explained by *reciprocity*, a concept that incorporates intentions into the decision process (cp. the models by, e.g., Rabin 1993, or Dufwenberg and Kirchsteiger 2004). Yet, if the increasing effort provision were solely driven by motives of intention, effort in gift-exchange games should not increase if the wage is determined by a lucky draw from an urn or by a third (unrelated) person. In Charness (2004), wages are either chosen by the employer or randomly determined. In both treatments, however, effort increases with the size of the wage. Therefore, not only reciprocal but also distributional concerns (between the employer and the employee) seem to play a major role in eliciting high efforts in gift-exchange settings.[8]

Is it possible that those distributional concerns are reflected by the equity principle? Let us apply the principle to the gift-exchange situation. In real-world employer–employee relationships, costs do not share the same unit, but one might still apply the equity principle on the ratio of outcomes to inputs (e.g., Mowday 1991): the input of the employer is the provided wage and the input of the employee is his effort level. If workers act in line with the equity principle their effort should increase with the wage offered by the employer. Thus, the prediction based on the simple equity principle is also reconcilable with the observed behavior in the gift-exchange situation. Moreover, the principle can even account for behavior observed in situations in which intentions are absent, as it is the case in Charness (2004) where employers do not select the wage.

Gift Exchange in Multi-agent Environments

In the preceding section, we reported results from bilateral relationships. We saw that many agents provided higher efforts under higher wage payments. Therefore, effort could be elicited even under incomplete contracts – and even in the absence of reputation, i.e., in one-shot situations where no future monetary gains were to be expected.

Frequently, however, relationships in actual organizations are characterized by multiple agents working side-by-side rather than by bilateral relationships. As (not only) Abeler et al. (2010) note, in this environment, agents are likely to judge the principal's actions not only in terms of their individual wage payment, but also how their own payment compares to their co-workers' payment. Thus, it becomes an important question how to treat agents relative to each other. On the one hand,

[8]cp., e.g., Falk and Fischbacher (2006) for a model of reciprocity that incorporates intentions *and* distributional concerns.

agents might care about horizontal wage *equality*. Differential pay of co-workers could reduce their work motivation and ultimately lower performance – in particular by causing resentment and envy within the workforce.[9] If workers care foremost about equality, a wage scheme that guarantees equal wages for co-workers should lead to an efficiency-enhancing gift-exchange relation. On the other hand, as we have already pointed out above, agents' behavior might be driven by the *equity principle*. In a multi-agent work environment, the equity principle or "equity norm" demands that a person who exerts higher effort should receive a higher wage compared to his co-workers. Of course, when performances of co-workers are the same, equity and equality coincide. Yet, whenever workers differ in their performance, horizontal wage equality violates the equity principle since a higher effort is not rewarded with a higher wage.[10] This might have important implications for the success of gift-exchange relationships.

Abeler et al. (2010) study this paradigm in a laboratory experiment, i.e., they explore the effectiveness of gift exchange as a contract enforcement device in multi-agent relationships.[11] In their experiment, one principal and two agents play a (reversed) gift-exchange game. In the first stage, agents decide simultaneously and independently how much effort they want to provide. Exerting effort is costly for the agents, with effort choices ranging from 1 to 10. After observing their efforts, the principal pays them a wage.

In one treatment, the principal can choose the level of the wage, but he is obliged to pay the same wage to both agents (*equal wage treatment* or *EWT*). In the second treatment, the principal can wage discriminate between the two agents (*individual wage treatment* or *IWT*). In both treatments, neither efforts nor wages are contractible. As in the bilateral gift-exchange game, if all players are rational and selfish the principal will not pay anything to the agents since wage payments only reduce his monetary payoff. Anticipating this, both agents will provide the minimal effort of one in the first stage. Neither the reversed move order nor the presence of additional agents changes this standard prediction.

However, in the case of bilateral gift-exchange relationships (reported above), we saw that gift exchange typically takes place, i.e., that efforts and wages exceed the smallest possible value and are positively correlated. Therefore, one might naively expect the same to happen in the multi-agent setup – but a fundamental prerequisite for the functioning of gift-exchange relations is that workers perceive

---

[9]See, e.g., Pfeffer and Langton (1993), Bewley (1999). In addition, consider that wage equality is also often referred to in employer-union bargaining as being a cornerstone of a fair wage scheme. Moreover, equal-wage payment is one of the most prevalent payment modes (e.g., Baker et al. 1988, Medoff and Abraham 1980).

[10]Indeed, in real-life work relations, this is likely to happen quite frequently and it is likely to matter. Thus, if equity is important, the often-heard slogan "equal pay for equal work" implies "unequal pay for unequal work".

[11]There are, of course, other studies on this topic. For example, Charness and Kuhn (2007) use a setting where agents differ in their productivity. Gächter et al. (2008) investigate the effects of pay comparison information and effort comparison information in a multi-agent firm.

their wage as fair. The fairness of a wage payment, however, may not only be evaluated in absolute terms, but also *relative* to the wages of other members in a worker's reference group. The multi-agent setup allows exploring if such horizontal fairness considerations play a crucial role. Moreover, one can shed light on the question which requirements a wage scheme must fulfill to be considered fair by the agents. If agents care foremost about wage equality, there should be no treatment difference: principals in the individual wage treatment might anticipate this and simply pay the same wage to both agents. However, if equity considerations are more important, one should observe that the EWT elicits lower effort levels than the IWT.

In fact, this is what Abeler et al. (2010) observe. In both treatments, the agent's individual effort and his own wage are positively correlated; which in bilateral situations was sufficient to establish a successful gift-exchange relationship. The monetary incentives, i.e., the average size of the wages paid for a given effort level, are also similar in both treatments. Nevertheless, agents' behavior differs significantly between treatments (cp. Fig. 13.2). Agents who are paid equal wages exert significantly lower efforts than agents who are paid individually. Effort levels are nearly twice as high under individual wages, and efforts decline over time when equal wages are paid.

The strong differences in actual efforts and especially the high frequency of low effort choices in the equal wage treatment suggest that the relative treatment of agents indeed plays an important role in the multi-agent environment. More precisely, equal wages are apparently not reconcilable with agents' horizontal fairness considerations. As the authors show, the frequent violations of the equity principle in the equal wage treatment are able to explain the effort differences between the treatments. In both treatments, agents who exert a higher effort and

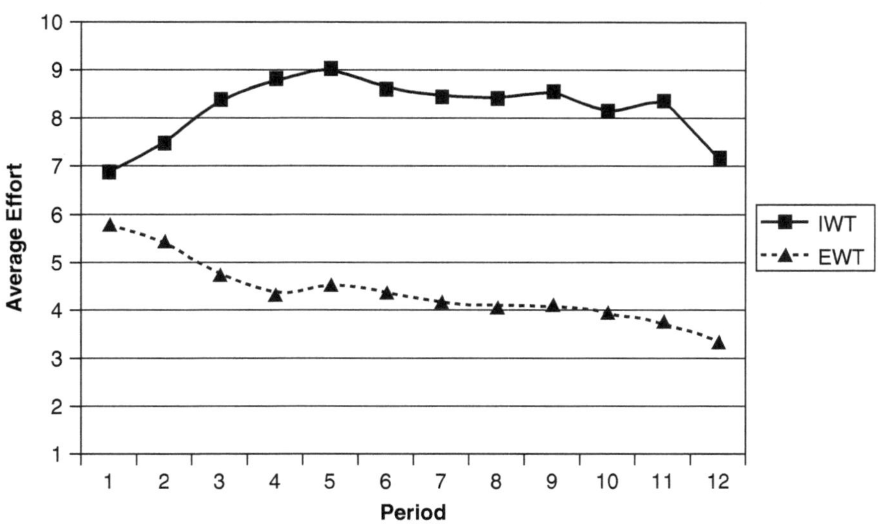

Fig. 13.2 Effort over time in Abeler et al. (2010)

earn a lower payoff than their co-worker strongly decrease their effort in the subsequent period.[12] However, the norm of equity is violated much more frequently under equal wages (whenever agents' effort levels and thus their effort costs differ). Principals in the individual wage treatment seem to understand the mechanisms of equity quite well. When efforts differ, they do pay different wages, rewarding the harder-working agent with a higher payoff in most cases – and it is in these cases that successful gift-exchange relations between principal and agents are established.

(Asymmetric) Reactions to Norm Violations

The observed effort dynamics in Abeler et al.'s experiment are likely to be caused by an asymmetry of agents' reactions to the violations of the equity norm. As Abeler et al. show, those agents who experience a disadvantageous norm violation (e.g., who work harder than their co-worker but do not earn more) subsequently reduce their effort. At the same time, their co-worker, who by design experiences an "advantageous" norm violation (e.g., they work less, but earn more), subsequently increases his effort. This suggests that agents change their effort provision in the direction that makes a violation of the equity norm less likely to occur in the next period. The strengths of these reactions are found to be asymmetric. The reactions to a disadvantageous norm violation are stronger than reactions to an advantageous one.

This asymmetry in reaction to norm violations is mirrored in other studies as well.[13] For example, Kube et al. (2010) conduct a field experiment in which they hire workers to catalogue books for a library. Their experimental design features three treatments. In the first treatment, workers are paid exactly the *projected* wage that was advertised in the recruiting poster (baseline treatment). In the second treatment, workers' wage is increased right before they start to work (treatment wage-raise). In a third treatment, workers' wage is cut right before they start to work (treatment wage-cut). The electronic measurement of the number of books entered by the worker during the subsequent working period provides a proxy for workers' effort. It allows exploring whether workers behave differently between the three treatments.

Although it is not the focus of their paper, one might apply the norm of equity to this situation. To see what the norm would imply, we need to interpret the wage payment as the input of the employer and the work effort as the input of the employee. The output for the employer might be what the worker "produces" for the employer, e.g., a function of the number of catalogued books. The output for the

[12]Remember that a higher effort implies higher effort costs and thus under equal wages translates into a lower payoff.

[13]e.g., Loewenstein et al. (1989), Mowday (1991), Thöni and Gächter (2008).

worker is likely to be the received wage minus the cost of effort, i.e., a function that depends on a worker's effort level. Compared to the projected wage, a wage increase thus constitutes an advantageous norm violation for the worker, while a wage cut is a disadvantageous norm violation for the worker. In light of the evidence reported above, one would therefore expect workers to react to the norm violation by increasing (decreasing) their work effort in response to a wage raise (wage cut). Moreover, one would expect the reactions to a disadvantageous norm violation to be stronger than the reactions to an advantageous one.

The findings in Kube et al. are basically in line with these predictions (cp. Fig. 13.3, which reports the average number of books logged during the entire working period in the respective treatment). The reactions to the advantageous norm violations are very weak and statistically insignificant when compared to the baseline treatment.[14] By contrast, the effort reactions in response to the wage cut, i.e., to the disadvantageous norm violation, are strong and highly significant. The average number of books logged during the entire working period in treatment wage-cut drops by 21% when compared to treatment baseline. It seems that workers indeed suffer from disadvantageous norm violations, feel dissatisfied and subsequently try to restore equity by adjusting their effort downwards.[15] In our view, this

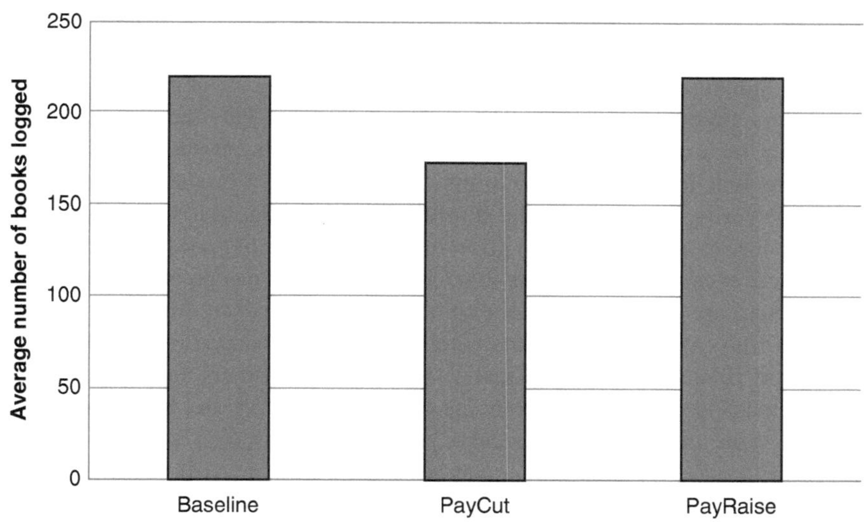

Fig. 13.3 Average performance per treatment in Kube et al. (2010)

[14]The size of the baseline wage is already rather high in relation to what is usually paid for comparable jobs. Thus, workers might perceive both treatments, the baseline and the wage-raise, to be an advantageous norm violation – which might explain why there are no significant effort differences between these two treatments.

[15]Of course, this is not to say that no other motives exist which are reconcilable with the observed behavior. See also our corresponding discussion in the last section.

indicates that the importance of the norm of equity which is observed in the "artificial" employment relationship in the laboratory carries over to natural work environments, too.

Inevitable Norm Violations

Up to now, we have reported behavior from settings in which workers directly interact with the wage-setting principal. Moreover, these settings always included actions that made it possible to fulfill the equity norm. But how will workers behave if norm violations are inherent in the wage scheme that they face and that are inevitable to them?

Two of the treatments (345COM and 345SUB) reported in Goerg et al. (2010) feature such a situation.[16] In their experiments, three subjects form a workgroup and work on a joint project. Each worker individually decides whether to exert effort or not. Effort exertion is costly to the workers, and their total sum of effort determines the number of some goods produced for a "virtual" principal (i.e., a principal who is neither explicitly modeled nor part of the experiment) under a given production technology (which is manipulated between treatments). The payoff of a worker is given by the total number of produced goods multiplied by worker's individual reward per unit produced, minus the cost of effort exertion.

Workers are designed to be perfectly symmetric, i.e., they have identical constant costs for exerting effort and their effort is equally productive. The actual production function can take two forms. It can either be a production function of complementarity, in which case the number of goods produced is strictly increasing in the number of agents who exert effort (e.g., the number of goods produced if 0/1/2/3 workers exert effort equals 20/40/65/100 units). Or it can be of complementarity, in which case the number of goods produced is strictly decreasing in the number of agents who exert effort (e.g., the number of goods produced if 0/1/2/3 workers exert effort equals 20/55/80/100 units). Although the workers are designed to be perfect clones in the sense that they are equally productive and have equal effort costs, in some treatments they receive unequal rewards, i.e., within a working group, one worker receives a wage payment of 3T (low reward), the other one of 4T (medium reward) and the last one of 5T (high reward), with T being the total number of goods produced by the working team.

With regard to the norm of equity, one could interpret a worker's effort cost to be his input and the individual wage payment to be his output. The norm of equity would demand the ratio of worker's input and output to be equal among workers. However, due to the specific design and parameterization chosen for the treatments

[16]Coincidentally, the experiments in Georg et al. not only shed light on the question at hand in this section, but furthermore they were funded by a research grant of Reinhard Selten and Eyal Winter. The experiments were actually designed to test the implications of a model by Winter (2004).

345COM and 345SUB in Goerg et al., workers have no action available that would make it possible to fulfill this norm. Moreover, norm violations are inevitable for their workers because by design there is no employer in the experiment who might adapt the wage scheme.

Goerg et al. observe that under these circumstances, the majority of workers behave in line with the predictions derived under the standard assumption of self-centered money-maximization. In 345COM, there exists a unique Nash equilibrium in which all agents exert effort, irrespective of their individual reward. As can be seen on the left side of Fig. 13.4, most workers indeed exert effort in 345COM. By contrast, in 345SUB, there exists a unique Nash equilibrium in which the high- and medium-reward agents exert effort, while the low-reward agent does not.[17] This is reflected in the data displayed on the right side of Fig. 13.4. Only 22% of all decisions by low-reward agents are to exert effort. This is not only significantly lower than that of the other two co-workers in 345SUB. The average rate of effort provision is also significantly lower than the corresponding rate of the same reward type in 345COM.

Although the observed behavior closely resembles the predictions derived under the assumption of self-centered money-maximization, it would be premature to conclude that equity considerations did not matter at all for the workers. It could

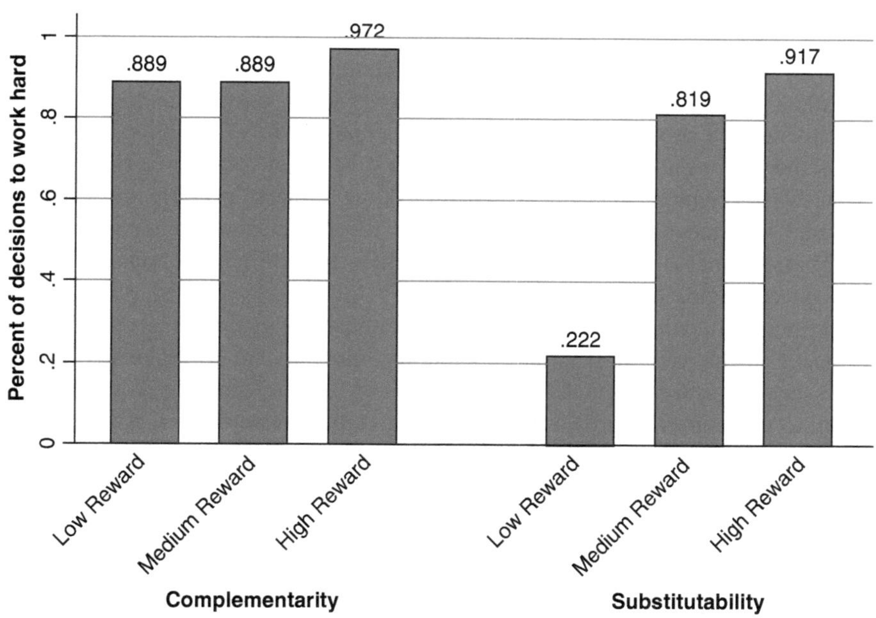

Fig. 13.4 Effort decisions in Goerg et al. (2010)

[17]Due to space constraints, we would like to refer the interested reader to Winter (2004) or Goerg et al. (2010) for the details.

still be that subjects were dissatisfied with the situation at hand and disliked the inevitable inequity (but chose to make the best out of it). In the next section, we present supportive evidence for this claim, which stems from the exciting new field of Neuroeconomics.

Supportive Evidence in Neuroeconomics

Recent years have seen the emergence of several fascinating studies that use neuroimaging techniques in combination with behavioral experiments to explore economic behavior. Of these studies, Fliessbach et al. (2007), and Dohmen et al. (2010) are of particular interest for us because they study social comparisons in a context that is related to the questions at hand here.[18]

In Fliessbach et al. (2007) and Dohmen et al. (2010), subjects repeatedly work on simple guessing tasks. They can either succeed or fail in a task. If subjects do (do not) solve a task correctly, they are (are not) rewarded with a certain amount of money. After completing the task, subjects are informed about their own performance and reward size as well as about the performance and reward size of a randomly assigned other subject. The experiment is designed such that in case both subjects were successful in their task, the size of their individual rewards is randomly varied in terms of absolute and relative levels. Afterwards, the next task starts. Since the reward sizes vary between tasks, during the course of the experiment each subject experiences several situations in which his performance equals the performance of his "co-worker", but he is rewarded by a higher (or a lower) wage payment than his co-worker. While subjects perform the guessing tasks and experience the situations where they receive unequal wage payments for equal performance, their brains are permanently scanned in a functional Magnetic Resonance Imaging (fMRI) scanner.

The authors find that subjects' payment affects blood oxygenation level-dependent responses in the ventral striatum – a brain region known to be related to the processing of rewards (including the valuation of basic rewards like food delivery as well as monetary rewards). Most interestingly, the data shows that the activation in this area is not only sensitive to the size of a worker's own wage (the higher the own wage, the higher the activation in this reward-related area), but also to the co-worker's wage. In particular, the activation in the ventral striatum is significantly lower when someone is paid less than his co-worker (controlling for the absolute size of the individual's wage).

Thus, to put it simple, the results imply that the "pleasurable experience" of receiving my own wage is lowered if at the same time my co-worker receives a higher wage for the same performance. Furthermore, the situation closely resembles several of the situations described in the previous sections (e.g., where workers'

[18]See also, e.g., Tricomi et al. (2010).

wage-payments differed although their effort costs and productivity did not). Thus, in some sense the brain-imaging studies reported here constitute a first neurophysiological foundation of the equity principle – although we admit that this might be far-fetched. Still, the results of the brain-imaging studies provide suggestive evidence for the importance of equity considerations (not only) in employment relationships.

Conclusions

This essay reviewed a selection of papers that experimentally investigate stylized employment relationships. While the original motivation of most of these papers is different, we reinterpreted the data in light of the equity principle – a concept from social psychology which Reinhard Selten applied to economic behavior as early as 1976. The equity principle can be operationalized as a simple norm or "formula" which states that the relation between an individual's "input" and his "output" should be equal for all individuals in a group (if this is not the case, an individual experiences displeasure and seeks to (re-)establish equity). As we saw, the evidence reported in the above papers is reconcilable with the idea of subjects' behavior being guided by such a principle.

Of course, there are limitations to this approach. In particular, interpreting existing empirical evidence can often be difficult because important aspects such as the cost of effort or the relevant reference group are ambiguous and the predictions are thus subject to interpretation (cp. Mowday 1991). Moreover, observed behavior is not exclusively compatible with a model based on the equity principle. There are different models that predict the same effects, e.g., models of reciprocity, guilt aversion, or social image concerns (to name only a few). In fact, this multiplicity of potential models constitutes a challenge to the research agenda of understanding how people really behave in certain economic situations and why they do so. A cause for this challenge is that the alternative models of human behavior usually (and often necessarily) introduce additional degrees of freedom. Thus, a certain pattern of observed behavior is reconcilable with different models using different parameterizations.[19] A possible work-around is to estimate the parameters of models from one dataset and apply them to different other datasets; implicitly assuming stability between situations. However, corresponding attempts with existing models recently demonstrate that this is problematic (cp., e.g., Brosig et al. 2007, or Blanco et al. 2008). Stability across situations is not necessarily given, since small changes in the game description influence subjects' behavior

[19]Which is why Selten usually tries to use as few free parameters as possible in his models – arguing (and joking) that with four parameters, he is already able to draw an elephant.

(cp., e.g., Tversky and Kahneman 1981, Abbink and Hennig-Schmidt 2006, Goerg and Walkowitz 2010). Alternatively, the evidence might suggest that the current models are missing factors that play an important role for individuals' behavior. Either way, it shows that much more work is needed until the discipline might finally settle on a sufficiently precise model (or set of models) of human behavior – but we are optimistically looking forward to future developments in this area.

References

Abbink K, Hennig-Schmidt H (2006) Neutral versus loaded instructions in a bribery experiment. Exp Econ 9(3):103–121

Abeler J, Altmann S, Kube S, Wibral M (2010) Gift exchange and workers' fairness concerns – when equality is unfair. J Eur Econ Assoc 8(6)

Adams JS (1963) Wage inequities, productivity and work quality. Ind Relat 3:9–16

Adams JS (1965) Inequity in social exchange. In: Berkowitz L (ed) Advances in experimental social psychology, vol 2. Academic, New York, pp 267–299

Akerlof GA (1982) Labor contracts as partial gift exchange. Q J Econ 97(4):543–569

Akerlof GA, Yellen JL (1990) The Fair Wage-Effort Hypothesis and unemployment. Q J Econ 105:255–283

Baker GP, Jensen MC, Murphy KJ (1988) Compensation and Incentives: practice vs. theory. J Finance 43:593–616

Bewley TF (1999) Why wages don't fall during a recession. Harvard University Press, Cambridge, MA

Blanco M, Engelmann D, Normann H-T (2008) A within-subject analysis of other-regarding preferences. Working Paper, Available at SSRN: http://ssrn.com/abstract=934700

Bolton GE, Ockenfels A (2000) ERC – a theory of equity, reciprocity and competition. Am Econ Rev 90(1):166–193

Brosig J, Riechmann T, Weimann J (2007) Selfish in the end?:an investigation of consistency and stability of individual behavior. FEMM Working Paper No. 05, Feb 2007

Charness G (2004) Attribution and reciprocity in an experimental labor market. J Labor Econ 22(3):1–25

Charness G, Kuhn P (2007) Does pay inequality affect worker effort? Experimental evidence. J Labor Econ 25:693–723

Chmura T, Goerg SJ, Pitz T (2010) Gift-exchange in China: wage discrimination across provinces. Mimeo

Dohmen T, Falk A, Fliessbach K, Sunde U, Weber B (2010) Relative versus absolute income, joy of winning, and gender: brain imaging evidence. Mimeo

Dufwenberg M, Kirchsteiger G (2004) A theory of sequential reciprocity. Games Econ Behav 47:268–298

Fahr R, Irlenbusch B (2000) Fairness as a constraint on trust in reciprocity: earned property rights in a reciprocal exchange experiment. Econ Lett 66:275–282

Falk A, Fischbacher U (2006) A theory of reciprocity. Games Econ Behav 54(2):293–315

Fehr E, Falk A (1999) Wage rigidity in a competitive incomplete contract market. J Polit Econ 107(1):106–134

Fehr E, Gächter S, Kirchsteiger G (1997) Reciprocity as a contract enforcement device: experimental evidence. Econometrica 65(4):833–860

Fehr E, Schmidt KM (1999) A theory of fairness, competition and cooperation. Q J Econ 114:817–868

Fehr E, Kirchsteiger G, Riedl A (1993) Does fairness prevent market clearing? An experimental investigation. Q J Econ 108(2):437–459
Fliessbach K, Weber B, Trautner B, Dohmen T, Sunde U, Elger CE, Falk A (2007) Social comparison affects reward-related brain activity in the human ventral striatum. Science 318:1305–1308
Gächter S, Nosenzo D, Sefton M (2008) The impact of social comparisons on reciprocity. IZA Discussion paper, 3639
Goerg S. Walkowitz G (2010) On the prevalence of framing effects across subject-pools in a two-person cooperation game. Journal of Econ Psych
Goerg S, Kube S, Zultan R (2010) Treating equals unequally – incentives in teams, workers' motivation and production technology. J Labor Econ
Homans GC (1961) Social behavior: it's elementary forms. Routledge and Kegan Paul, London
Kube S, Maréchal MA, Puppe C (2010) Do wage cuts damage work morale? Evidence from a natural field experiment. Working Paper, Institute for Empirical Research in Economics, University of Zurich
Loewenstein G, Thompson L, Bazerman M (1989) Social utility and decision making in interpersonal contexts. J Pers Soc Psychol 57:426–441
Medoff JL, Abraham KG (1980) Experience, performance, and earnings. Q J Econ 95:703–736
Mowday RT (1991) Equity theory predictions of behavior in organizations. In: Steers RM, Porter LW (eds) Motivation and work behavior. McGraw-Hill, New York, pp 111–131
Nydegger R, Owen G (1974) Two-person bargaining: an experimental test of the Nash axioms. Int J Game Theory 3(4):239–249
Pfeffer J, Langton N (1993) The effect of wage dispersion on satisfaction, productivity, and working collaboratively: evidence from college and university faculty. Adm Sci Q 38:382–407
Rabin M (1993) Incorporating fairness into game theory and economics. Am Econ Rev 83(5):1281–1302
Sauermann H, Selten R (1959) Ein Oligopolexperiment. Z Gesamte Staatswiss 115:427–471
Selten R (1967) Die Strategiemethode zur Erforschung des eingeschränkt rationalen Verhaltens im Rahmen eines Oligopolexperiments. In: Sauermann H (ed) Beiträge zur experimentellen Wirtschaftsforschung. J.C.B. Mohr (Paul Siebeck), Tübingen, pp 136–168
Selten R (1972) Equal share analysis. In: Sauermann H (ed) Beiträge zur Experimentellen Wirtschaftsforschung, vol 3. J.C.B. Mohr (Paul Siebeck), Tübingen, p 163
Selten R (1978) The equity principle in economic behavior. In: Gottinger HW, Leinfellner W (eds) Decision theory social ethics, issues in social choice. D. Reidel, Dordrecht/Holland, pp 289–301
Selten R, Berg C (1970) Drei experimentelle Oligopolspielserien mit kontinuierlichem Zeitablauf. In: Sauermann H (ed) Beiträge zur experimentellen Wirtschaftsforschung, vol 2. J.C.B. Mohr (Paul Siebeck), Tübingen, pp 162–221
Thöni C, Gächter S (2008) Kinked conformism' in voluntary cooperation. Mimeo
Tricomi E, Rangel A, Camerer CF, O'Doherty JP (2010) Neural evidence for inequality-averse social preferences. Nature 463:1089–1091
Tversky A, Kahneman D (1981) The framing of decisions and the psychology of choice. Science 211(4481):453–458
Winter E (2004) Incentives and discrimination. Am Econ Rev 94(3):764–773

Chapter 14
The Analysis of Incentives in Firms: An Experimental Approach

Christine Harbring, Bernd Irlenbusch, Dirk Sliwka, and Matthias Sutter

"Behaviour cannot be invented in the armchair. It has to be observed." This statement by Reinhard Selten (1998, p. 414) is particularly true and relevant when designing organisations and incentive schemes with the aim to motivate employees and facilitate coordination and cooperation in firms. Experiments are a powerful tool to observe behaviour under controlled conditions and thereby to draw inferences about causal influences on behaviour.

In this paper we present three experimental studies investigating particular aspects of incentives schemes. We start by experimentally comparing two prominent compensation schemes in firms, i.e., fixed wages and output dependent variable pay. Traditional economic analysis suggests that the second scheme is better suited to motivate employees to exert higher effort compared to the first scheme. We discover that in a repeated setting the opposite can be true since variable pay may cause that individuals focus too strongly on short term bonuses crowding out motivation for acting cooperatively. Our second study focuses on the question of how far tournament incentive schemes might induce destructive activities that

C. Harbring
Lehrstuhl für Human Resource Management, Fakultät für Wirtschaftswissenschaften, Karlsruhe Institute of Technology, Waldhornstr. 27, 76131, Karlsruhe, Germany
e-mail: christine.harbring@kit.edu

B. Irlenbusch (✉)
Seminar für Allgemeine Betriebswirtschaftslehre, Unternehmensentwicklung und Wirtschaftsethik, Universität zu Köln, Herbert-Lewin-Str. 2, 50931 Köln, Germany
e-mail: bernd.irlenbusch@uni-koeln.de

D. Sliwka
Seminar für Allgemeine Betriebswirtschaftslehre und Personalwirtschaftslehre, Universität zu Köln, Herbert-Lewin-Str. 2, 50931 Köln, Germany
e-mail: dirk.sliwka@uni-koeln.de

M. Sutter
Institut für Finanzwissenschaft, University of Innsbruck, Universitätsstrasse 15, A-6020 Innsbruck, Austria
e-mail: matthias.sutter@uibk.ac.at

A. Ockenfels and A. Sadrieh (eds.), *The Selten School of Behavioral Economics*,
DOI 10.1007/978-3-642-13983-3_14,

reduce the output of competitors in the scheme. We are particularly interested whether the problem of such sabotage activities is more severe if the production technology is subject to increased uncertainty. The third study looks at a potential reason why firms often employ teams to get things done. By means of a simple coordination game we show that teams are much better in coordinating with other teams than individuals coordinate with each other.

Fixed Wages, Piece Rates, and Motivation Crowding Out

In this section we aim at contributing to the understanding of crowding out effects in ongoing principal agent relations.[1] We report an experiment to explore a new possible explanation for why extrinsic incentives might crowd out effort (Deci 1971; Frey 1997; Deci et al. 1999; Gneezy and Rustichini 2000; Bénabou and Tirole 2003; Heyman and Ariely 2004; Sliwka 2007; Ellingsen and Johannesson 2008; Bowles 2008). In particular we are interested in the influence the compensation scheme of piece rates may have on the agents' cognitive perception of the situation. A possible conjecture would be that the availability of piece rate incentives leads agents to follow a short term, individual maximisation behaviour, whereas fixed wages may cause them to pursue a more cooperative, longer term orientation.

A Simple Model of Incentives Provided by Fixed Wages and Piece Rates

The basis of our experimental analysis is a simple principal-agent model, consisting of two stages. In the first stage, the agent is offered a wage by the principal. The agent then determines his effort-level e within the second stage. The effort causes costs of $c(e) = \frac{c}{2}e^2$ for the agent. The effort level e results in an output of ke for the principal. The first experimental condition (*fixed wage* condition) gives the principal the opportunity to offer the agent a fixed wage α which is determined in the first stage and paid directly to the agent. In the second stage the agent then chooses his effort-level e. The payoff for the principal is:

$$\Pi_P^F = k \cdot e - \alpha \tag{14.1}$$

The payoff for the agent accounts for:

[1]The results presented in this section are part of a larger research project, see Irlenbusch and Sliwka (2005). Excellent overviews about theoretical and empirical findings on various incentive schemes are provided by Gibbons (1998), Lazear (1999), and Prendergast (1999).

$$\Pi_A^F = \alpha - \frac{c}{2}e^2. \tag{14.2}$$

The second experimental condition (*choice* condition) offers the principal to choose between a fixed wage and an incentive wage, i.e., a variable wage dependent on the output. If she wants to pay a fixed wage, she sets the amount α of the fixed wage as in the first experimental condition. If she prefers to pay a variable wage, the principal decides on the fixed component α as well as on the variable rate β. This means the agent receives a fixed wage α from the principal plus a payment of $\beta \cdot k \cdot e$ according to his effort exerted. If she chooses this possibility, the principal has to pay measuring costs f for determining the level of actual effort provided by the agent. It is assumed that α and β are non-negative, i.e. the agent is protected by limited liability. Offering an incentive wage yields the following payoff for the principal:

$$\Pi_P^I = (1 - \beta)k \cdot e - \alpha - f. \tag{14.3}$$

The agent receives:

$$\Pi_A^I = \alpha + \beta \cdot k \cdot e - \frac{c}{2}e^2. \tag{14.4}$$

As a benchmark for the behaviour observed in the experiment in the following we examine the one-shot situation predicted by game theory with the assumption of actors who are only interested in their own payoff. The analysis of the *fixed wage* condition is quite simple. During the second stage the agent does not contribute any effort because he has already received his wage and each positive effort level would cause him unnecessary costs. The principal anticipates this behaviour in the first stage and is consequently not willing to pay any wage. We now consider the second experimental condition. The principal has the possibility to choose between a fixed wage and an incentive wage. The analysis of the fixed wage is identical to the first experimental condition described above. However, the analysis differs if the principal chooses to pay an incentive wage. For chosen values of α and β the agent maximises his payoff according to (14.4). The first order condition determines the optimal effort level:

$$e = \frac{\beta \cdot k}{c}. \tag{14.5}$$

The principal anticipates the agent's behaviour in the first stage and sets the payment by deciding on α and β. Analogous to the argumentation above the principal does never choose any positive amount α of the fixed wage component because no effort incentives are induced by the fixed component. Thus, the principal maximises her payoff (14.3) under the constraint (14.5):

$$\max_{\beta} (1 - \beta)k \cdot \frac{\beta \cdot k}{c} - f. \tag{14.6}$$

This first order condition is solved for the variable component β and it results that the principal chooses $\beta = \frac{1}{2}$, independent of the parameter values k, f und c. Given the agent's behaviour as derived above it is optimal for the principal to hand over half of the output to the agent. In the equilibrium the agent thus chooses an effort level of[2]:

$$e* = \frac{k}{2c}. \tag{14.7}$$

In case the principal decides to pay the incentive wage, his equilibrium payoff is:

$$\Pi_P^I* = \frac{k^2}{4c} - f. \tag{14.8}$$

It is optimal for the principal to choose the incentive wage, if $\frac{k^2}{4c} > f$. In this case, the principal would not choose the pure fixed wage during the second experimental condition (*choice* condition). Moreover, the effort level and the principal's payoff should be considerably higher in the second experimental condition (*choice* condition) compared to the first experimental condition (*fixed wage* condition). In the experiment described below the discussed one-shot situations are each repeated during a finite amount of periods T. By applying backward induction (Selten 1965) no change in the game theoretical prediction for both experimental conditions occurs.

Experimental Design and Procedure

The experimental design closely follows the two settings introduced above. The employed parameters are provided in Table 14.1. In each of the two treatments we had 42 participants, half of them were assigned the role of the principal and half of them had the role of an agent. The experiment was conducted in the Laboratory of Experimental Economics at the University of Erfurt and the games were played for 20 rounds with fixed matching. After a session payoffs were converted to € and paid in cash with an exchange rate of 6€ for 100 Talers.

[2]Since we implemented the constraint of a non-negative fixed-wage component α, it is not possible to "sell the shop", i.e., to set up contracts that sell the working output to the agent. Therefore the here presented optimal contract does not induce the "first-best" working effort, i.e., the effort level that would maximize the sum of the principal's and agent's payoff: $e^{eff} = k/c$.

Table 14.1 Experimental design and procedure

	Fixed wage condition	Choice condition
# independent observations	21	21
Endowment capital for principals	100	100
Endowment capital for agents	100	100
# rounds	20	20
Integer effort levels	{0, ..., 20}	{0, ..., 20}
Integer fixed wages	{0, ..., 40}	{0, ..., 40}
Measurement costs	–	2
Variable rates	–	{10%, 20%, ...,100%}

Table 14.2 Average effort levels and payoffs

Treatment	Fixed wage	Choice		
		Totally	Pure fixed wage	Incentive wage
Effort levels	9.15	6.25	4.65	7.77
Principals' payoff	2.98	0.69	0.68	0.71
Agents' payoffs	4.71	5.33	5.08	5.56
Total payoffs	7.68	6.02	5.76	6.27

Results

In this section we first report the findings of each of the two experimental conditions. In line with the statistical rigour propagated by Reinhard Selten, all non-parametrical tests are based on the values of single independent observations. Table 14.2 provides an overview of the average payoffs and effort-levels chosen under both conditions.

The Fixed Wage Condition

We are particularly interested in the level of effort chosen by the agent. As described in the model, fixed wages should not induce any effort according to the standard theoretical analysis. The experimental data, however, draws a different picture. In return for an average fixed wage of 11.36 Talers the agents select an average effort level of 9.15 Talers. Figure 14.1 illustrates the average effort levels chosen in each round. It becomes clear that, under the fixed wage condition in rounds 1 to 19, the effort levels are constantly high until they drop in the last round.

The Choice Condition

During the second experimental condition an average effort level of 6.25 was exerted by the agents. This average effort level for each round is also plotted in Fig. 14.1. Within this experimental condition the principal has the choice between a merely

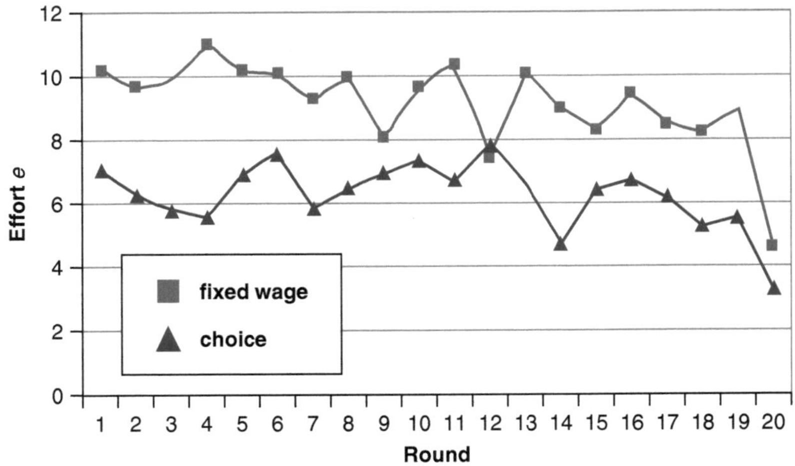

Fig. 14.1 Average effort levels over 20 rounds in both treatments

fixed wage and a variable wage. As discussed above in the subgame perfect equilibrium the principal chooses piece rate wages with a share of 50% from the working output for the agent. Following this prediction she does not choose the option of a fixed wage component. However, the experimental data are not in clear support of this benchmark: Within the *choice* condition principals on average choose fixed wages in 9.71 of the 20 rounds. This share of the fixed wages is relatively constant in each of the 20 rounds. In the cases where the principal decides to pay a mere fixed wage, the average wage is 8.63 Talers. On average the agents choose an effort level of 4.65 Talers. For the cases of a piece rate wages, the fixed wage components reach a notable positive amount of 6.27 Talers on average. The principals set the agents' share of the output at 40.88% on average if piece wages are paid.

The payoffs of principals under pure fixed wages and piece wages are surprisingly similar. This is likely to be one reason, why both wage options are chosen similarly often. Even though the difference in payoffs is considerably smaller as predicted, a comparison of the effort levels induced by both wage options qualitatively confirms the theoretical estimation: if principals choose to offer a piece-wage, agents choose significantly higher effort than under pure fixed wages (1% level, Wilcoxon rank-sum test, two-sided). Figure 14.2 shows the average effort level during the 20 rounds for the options of the incentive wage and the pure fixed wage.

The Comparison of Conditions

Surprisingly, while within the *choice* condition, incentive wages lead to higher efforts than fixed wages, the efforts exerted in the *fixed wage* condition, which simply excluded the possibility to choose variable wages, are (weakly) significantly lower than in the *choice* condition (10% level ($p = 0.061$), two-tailed,

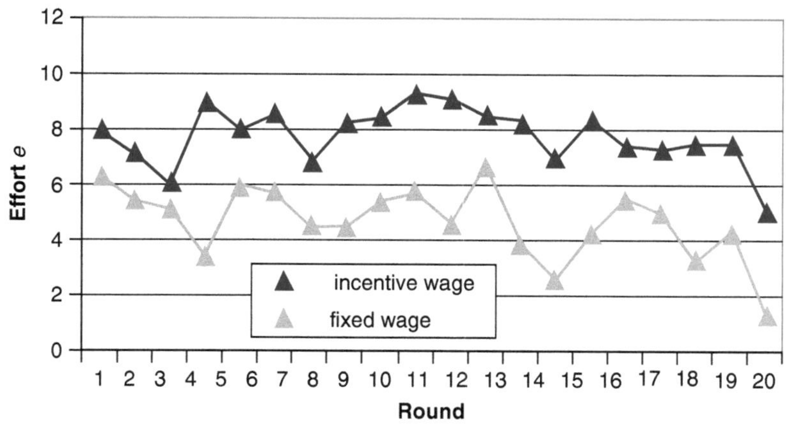

Fig. 14.2 Average effort levels over 20 rounds under the *choice* condition

Mann–Whitney U-test). This indicates that we indeed observe a "crowding out" effect, i.e., the *additional* option of introducing incentive wages seems to reduce effort. The difference in efforts is also reflected in payoffs of principals. In the *fixed wage* condition on average principals earn 2.98 Talers which is significantly higher than the average of 0.69 which they earn on average in the *choice* condition. (1% level, two-tailed, Mann–Whitney U-test). Also efficiency (total payoffs of agents and principals) is higher in the *fixed wage* setting than in the *choice* condition (7.68 versus 6.02, 1% level, two-tailed, Whitney–Mann U-test, see Table 14.2).

Insights from a Post-experimental Questionnaire

The comparison between treatments indeed suggests that participants perceive the two situations differently. This impression is supported by an analysis of a questionnaire the participants had to fill in at the end of the experiment: agents qualitatively give rather different reasons for their effort choice in the two settings. In the *fixed wage* condition agents typically argue:

- *"Type A player should earn almost – or even the same – amount as I do."*
- *"My partner should roughly make a similar profit as I do."*
- *"The other player and I should approximately yield the same number of points."*
- *"I wanted to keep him in a good mood."*

Agents seem to take care of the principals' interests in the *fixed wage* setting. They apparently are concerned not to disappoint the other player – for instance because they hope for high fixed wages in the future. The perception of the situation seems to be different in the *choice* setting. Here representative statements are:

- *"Maximal profit for me"*
- *"Depending on the percentage offered"*

– *"I aimed at profit maximisation."*
– *"Depending on the percentage share that player A gave me."*

These answers suggest that the primary focus of the agents in the *choice* setting lies in (short-term) profit maximisation without paying noteworthy attention to the interests of the other player. In fact in the *fixed* wage setting agents mention the well-being of the principals significantly more often as a reason for their effort choice than agents do in the *choice* setting (5% level, Fisher Test, two-tailed).

Discussion

We observe a *crowding out effect* in the sense that if the principals are given the *additional* possibility to set up an incentive scheme instead of a pure fixed wage regime this leads to lower efforts, lower principals' profits and also overall welfare losses. It seems that agents perceive the two situations differently. In the presence of pure fixed wages agents seem to take the interest of the principals into account to a larger extent than when the principals have the possibility to "force" agents to exert effort by implementing incentive wages. This finding lends new support for Kohn's (1993) informal discussion of the disadvantages of incentive plans: "'Do this and you'll get that´ [..] focuses attention on the 'that´ instead of the 'this´. [..] Do rewards motivate people? Absolutely. They motivate people to get rewards."

Tournament Incentives and Sabotage

Reward schemes in which remuneration is based on relative rather than on absolute performance are widely recognised as an important component in the toolbox of incentive system designers for modern organisations (see e.g., Prendergast 1999). Tournaments are often implemented in the form of internal promotion tournaments or as relative evaluation schemes such as 'force distribution systems'. A major problem of such relative remuneration schemes is seen in its incentives to harm others as only relative performance is evaluated and not the absolute performance level. Since the tendency to harm others is hardly observable in the field, experiments are an appropriate empirical tool to analyze sabotage behaviour.

We are particularly interested in the predictive power of the classical tournament model. In what follows, we focus on the influence of the volatility of an individual random shock on behaviour in a framework in which principals may endogenously decide on implementing tournament wages.[3] In standard tournament models the

[3]For experimental investigations on the influence of tournament size, prize structure, or the prize spread, see Harbring and Irlenbusch (2008, 2010).

individual performance level is usually distorted by an individual noise factor, which resembles, for example, measurement errors, technical difficulties etc. We vary the size of the interval from which the individual random term is drawn across two experimental treatments. Standard tournament theory predicts that effort and also sabotage activities should be lower if the potential influence of the random term is increased.

A Model of Tournaments with Sabotage

We employ a simple two-stage game with four players, three agents and one principal. As shown in Fig. 14.3 the principal selects the wage spread in the first stage, and in the second stage agents can exert two activities: productive effort and sabotage.

The total sum of wages amounts to 300. In the simplest case the principal selects full wage compression, i.e., a fixed wage of 100 for each agent. If unequal wages are specified we assume that the three agents compete in a tournament for a winner prize M. The two losing agents receive a loser prize m with $0 < m < M$. We denote the wage spread $(M - m)$ by Δ with $(3\,m + \Delta) = 300$, i.e., the sum of winner and loser prizes equals the wage sum.

A strategy of an agent i is constituted by a pair (e_i, s_i) where $e_i \in [0, \ldots, 100]$ denotes effort and $s_i \in [0, \ldots, 50]$ is the sabotage activity which reduces the output of the two other agents. Exerting effort and sabotage is costly for each agent i. The costs are assumed to be symmetric and are described by the functions $C_e(e_i) = e_i^2/70$ and $C_s(s_i) = s_i^2/20$, respectively.[4] The output y_i of agent i is determined by the following production function

$$y_i = e_i + \varepsilon_i - \sum_{j \neq i} s_j$$

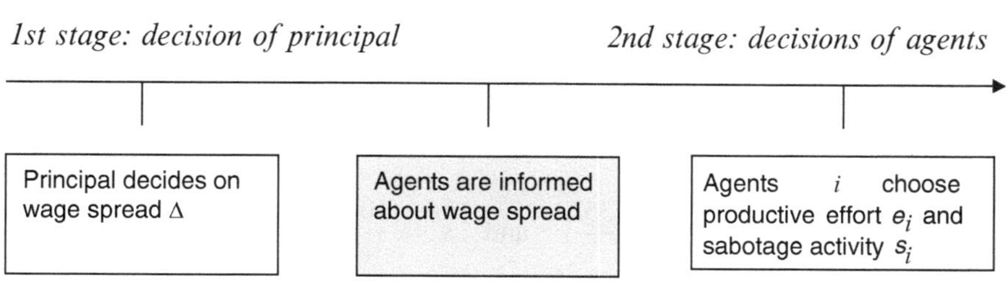

Fig. 14.3 Sequence of the game

[4]We assume that sabotage is costlier than productive effort resembling the fact that in the workplace an agent usually has to exert some extra effort to conceal the destructive activity at least in front of the employer. Agents are modeled symmetrically. For a discussion of the impact of sabotage when agents are asymmetric see Harbring et al. (2007).

with ε_i a being a random variable which is uniformly distributed over an interval $[-(\bar{\varepsilon}/2), +(\bar{\varepsilon}/2)]$ and assumed to be i.i.d. for each agent i. The random component, ε_i, resembles production luck or measurement error of output. The expected payoff for agent i is given by

$$E\Pi_i(e_i, e_{-i}, s_i, s_{-i}) = f^w(e_i, e_{-i}, s_i, s_{-i})M + [1 - f^w(e_i, e_{-i}, s_i, s_{-i})]m - C_e(e_i) - C_s(s_i)$$

with $f^w(e_i, e_{-i}, s_i, s_{-i})$ denoting the probability for agent i to receive the winner prize if the other two agents choose effort levels e_{-i} and sabotage activities s_{-i}.

To provide a benchmark for behaviour in the experiment let us have a look at the equilibrium prediction of the one-shot setting. For simplicity we assume that all players are rational, risk-neutral, and purely money-maximizing. The expected payoff of an agent i can be written as

$$E\Pi_i(e_i, e_{-i}, s_i, s_{-i}) = m + f^w(e_i, e_{-i}, s_i, s_{-i})\Delta - e_i^2/70 - s_i^2/20.$$

If the principal chooses full wage compression $\Delta = 0$ (*fixed wages*) agents should neither exert effort nor sabotage. For positive prize spreads the first-order conditions are given by

$$\frac{\partial f^w(e_i, e_{-i}, s_i, s_{-i})}{\partial e_i}\Delta = \frac{2e_i}{70} \quad \text{and} \quad \frac{\partial f^w(e_i, e_{-i}, s_i, s_{-i})}{\partial s_i}\Delta = \frac{2s_i}{20}.$$

Provided our assumptions it is straightforward to show that in a symmetric equilibrium the marginal probabilities of winning only depend on the size of the interval from which the random component in the production function is drawn (for a detailed exposition see, for example, Orrison et al. 2004, Harbring and Irlenbusch 2008), i.e., one can show that

$$\frac{\partial f^w(e_i, e_{-i}, s_i, s_{-i})}{\partial e_i} = \frac{\partial f^w(e_i, e_{-i}, s_i, s_{-i})}{\partial s_i} = \frac{1}{\bar{\varepsilon}}$$

with $\bar{\varepsilon}$ denoting the size of the interval from which each ε_i is drawn. Thus, our first-order conditions for the one-shot setting reduce to

$$e^* = \frac{35\Delta}{\bar{\varepsilon}} \quad \text{and} \quad s^* = \frac{10\Delta}{\bar{\varepsilon}}.$$

These one-shot equilibrium predictions support standard conjectures about tournament incentives, i.e., that effort and sabotage increase with the wage spread. It is important to note that in our model an additional unit of effort has the same effect on improving the own position in the ranking as has one additional unit of sabotage. The reason is that an additional unit of effort increases own output by one unit while an additional unit of sabotage reduces the output of *all other competitors* by one unit. Thus, in equilibrium the marginal costs of the two activities have to be equal.

To ensure that an interior solution exists and that agents have no incentive to deviate to activities of zero, the expected gain of an agent must not be lower than his cost, i.e., $\Delta/3 \geq C_e(e^*) + C_s(s^*)$. Additionally, the equilibrium effort and sabotage levels must lie in the feasible interval, i.e., $e^*<100$ and $s^*<50$.

We assume that the principal's expected payoff increases in the total output of the agents

$$E\Pi_P(e,s) = \tau\left[E\left(\sum_i y_i\right)\right] = \tau\left(\sum_i e_i - 2\sum_i s_i\right).$$

$\tau > 0$ indicates how much the principal values one unit of output.[5] Thus, in the symmetric equilibrium of the one-shot tournament setting the principal receives the following expected payoff depending on her choice of the prize spread Δ:

$$E\Pi_P(e^*, s^*) = \frac{45\tau\Delta}{\bar{\varepsilon}}.$$

This reflects the standard tournament result that the principal's payoff increases with the prize spread. If the principal anticipates the derived behaviour of the agents and aims at maximizing her payoff, she chooses the highest possible wage spread.

Experimental Design and Procedure

Table 14.3 summarises the design alternatives for the wage contracts. The principal may decide on the distribution of the wage sum $W = 300$ by determining one of the five prize spreads Δ_j with $j = 0, \ldots, 4$. A prize spread of zero is denoted by Δ_0, i.e., all players receive the same fixed wage of 100 irrespective of the output they have achieved. Additionally, we allow the principal to offer no contract at all, which results in a payoff of zero for the principal as well as all agents.

Table 14.3 Design alternatives for wage contracts

Fixed wages	$\Delta_0 = 0$	100 for each agent	
	Prize spread Δ_j	2 loser prizes m	1 winner prize M
Tournament incentives	$\Delta_1 = 48$	84	132
	$\Delta_2 = 96$	68	164
	$\Delta_3 = 144$	52	196
	$\Delta_4 = 192$	36	228

[5] A lump sum of 90 Taler is deducted from the principal's payoff in each round which models part of the cost of the implemented wage scheme.

Table 14.4 One-shot equilibrium predictions for each prize spread

	Fixed wages	Tournament incentives, prize spread $\Delta =$			
	No prize spread	48	96	144	192
Effort					
LOW	0	14	28	42	56
HIGH	0	7	14	21	28
Sabotage					
LOW	0	4	8	12	16
HIGH	0	2	4	6	8
Output					
LOW	0	6	12	18	24
HIGH	0	3	6	9	12

In the experiment, the value τ of one unit of output for the principal is set to $\tau = 3$. Table 14.4 provides the corresponding effort and sabotage equilibrium predictions for the one-shot setting as well as the resulting expected output. We vary the interval $\bar{\varepsilon}$ from which the random term is drawn over two treatments: In treatment *LOW* it is set to $\bar{\varepsilon} = 120$ and in our second treatment *HIGH* it is twice as high with $\bar{\varepsilon} = 240$.

The effort and sabotage equilibrium predictions are derived for one-shot tournaments assuming that agents are risk neutral and are only interested in maximizing their own payoff.

The experiment was conducted in the *Laboratory for Experimental Research* at the University of Bonn and the *Laboratory for Experimental Research* at the University of Erfurt. All sessions were computerised and the software was developed by using RatImage (Abbink and Sadrieh 1995). Recruitment was done by ORSEE (Greiner 2004) with a randomised allocation of subjects to treatments. In total, 80 students of different disciplines were involved. Each candidate was allowed to participate in one session only. To resemble the fact that tournaments are often repeatedly conducted with the same contestants – think of, for example, 'seller-of-the-year' contests – and to allow participants to gain some experience with the situation and familiarise themselves with the relatively complex strategic interaction, a session consisted of 30 repetitions (rounds) of the same tournament setting with fixed matching and roles (one principal and three agents).

After the instructions were read to the participants they were randomly and anonymously divided into groups of four. Within a group, roles as principal and agents were also randomly allocated. Each group constitutes a statistically independent observation. We collected 12 observations with 48 subjects for *LOW* and 8 observations with 32 subjects for *HIGH*. After each round each player observed his own payoff, the output of each of the three agents and the principal's payoff. Each session lasted for about 2 h and participants earned 14.50€ on average. In the experiment the payoffs were given in 'Talers', all subjects received an endowment of 1,200 Talers and at the end Talers were converted into Euro by a previously known exchange rate of 200 Talers per 1€. All subjects were paid anonymously and privately.

Results

According to the theoretic prediction effort and sabotage activities should be lower in our treatment with the random term being drawn from the smaller interval compared to the larger interval, i.e., activities should be much higher in *LOW* than in *HIGH*. Figure 14.4 depicts the average effort and sabotage activities in both treatments per prize spread.

Averages of effort and sabotage in *LOW* and *HIGH* seem to be surprisingly similar and we do not find any differences between averages per prize spread that are significant at a conventional level applying the Mann–Whitney U test. The output is generated by combining effort and sabotage. Output should be twice as high in *LOW* compared to *HIGH*. However, we do not find a significant difference between the treatments. Thus, contrary to our theoretic prediction agents' behaviour does not differ between the treatments.

Figure 14.5 depicts the frequency of contracts in both treatments. Note that the principal could also choose not to offer a contract at all resulting in zero payoffs in this particular round for everyone in the group. This option was chosen in 10.8% of all cases in *LOW* and in 8.3% in *HIGH*. From a theoretic perspective a principal should always decide to implement a tournament with the largest prize spread in both treatments as output is highest in this case.

However, Fig. 14.5 depicts a difference between the treatments showing that fixed wages seem to be more often chosen in *HIGH* than in *LOW*. The difference between the treatments is not significant, though (Mann–Whitney U test, $p = 0.1887$, two-tailed). Taking a closer look at the frequency of contracts within each treatment we find that in *LOW* incentive contracts with 70.8% are significantly more often chosen than fixed wages with 18.3% (Wilcoxon-Signed Rank test, 1% level, two-tailed). This seems to be in line with economic intuition in the sense that

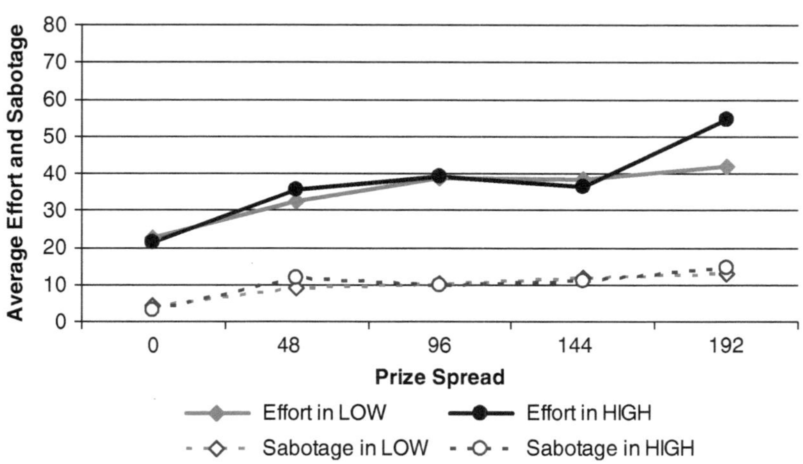

Fig. 14.4 Decisions of agents – average effort and sabotage in both treatments

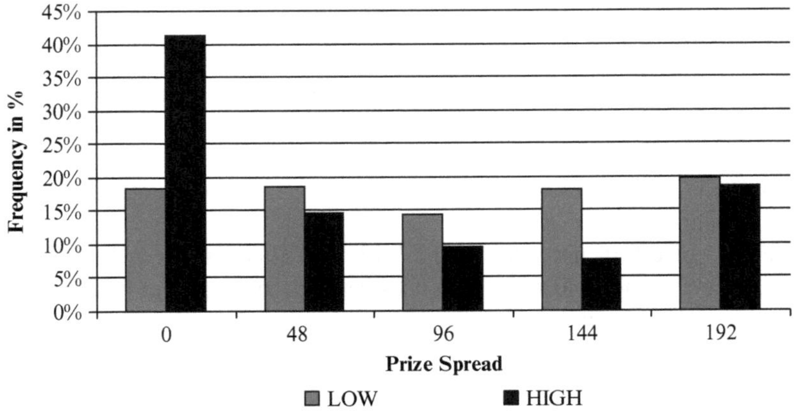

Fig. 14.5 Decisions of principals – frequency of prize spreads in both treatments

if performance measurement is difficult incentives tend to be avoided. In *HIGH*, however, fixed wages with 41.3% are similarly often chosen as incentive contracts with 50.6%. This observation indicates that the principal has a stronger preference for fixed wages compared to tournament wages if the influence of luck is high.

Discussion

Contrary to standard tournament theory we find no difference in effort and sabotage between the treatments. Thus, agents' decisions seem widely unaffected by the influence of the random term in our setting. However, principals choose fixed wages strikingly more often than tournament wages if the random component is high. This observation may indicate that principals perceive the influence of the random term as relatively important and, therefore, decide to "insure" agents by offering more fixed wages.

Team-Based Coordination

Coordination problems are ubiquitous in firms. Coordination is required for aligning the activities between individual employees, but also between groups of employees, like different teams or units in a firm. For example, a firm's success may depend on the efficient interaction of its production unit, marketing unit, and finance unit (Camerer and Knez 1997). Due to the imminent importance of successful coordination for the functioning of firms and industries a large body of research in economics is devoted to coordination games that are characterised by multiple, and typically Pareto-ranked, pure strategy equilibria. Since the determinants for coordination failure or success can more easily be identified and controlled for in laboratory studies than in field studies, most of the work on coordination has relied on controlled experiments, starting with the seminal papers by van Huyck et al. (1990, 1991) and Cooper et al. (1990, 1992). Strikingly, though, when examining the determinants of coordination failure or success all experiments up to now have exclusively focused on individual decision-making. In the reality of firms, however, it is often teams that have to coordinate with other teams. So far, very little is known whether teams might be better or worse in coordinating on efficient actions. Therefore, we investigate whether individual or team decision-making has any influence on coordination failure or success.

An Average Opinion Game

In this section we focus on a coordination situation in which it is essential that a player meets the median action within a group of players.[6] Such a situation is modelled by average opinion games proposed by van Huyck et al. (1991). Table 14.5 displays the payoffs of the average opinion game we employ. Players select a number between one and seven. If a player meets the median choice he earns a positive payoff, if not the payoff is zero. The strategy combinations with all players choosing identical numbers constitute the pure strategy equilibria. Equilibria are Pareto-ranked and using the concept of payoff-dominance (Harsanyi and Selten 1988) as an equilibrium selection device would lead to symmetric choices of “7”.

Experimental Design and Procedure

We have set up two treatments. In the “*Individuals*”-treatment we let five individuals interact in the respective game for 20 periods, and this is common knowledge.

Table 14.5 Payoffs in the average-opinion game

Own number	Median number chosen in the group						
	7	6	5	4	3	2	1
7	130	0	0	0	0	0	0
6	0	120	0	0	0	0	0
5	0	0	110	0	0	0	0
4	0	0	0	100	0	0	0
3	0	0	0	0	90	0	0
2	0	0	0	0	0	80	0
1	0	0	0	0	0	0	70

[6]This section is part of a larger research project on the impact of team decision making on efficient coordination in coordination games. See Feri et al. (2010).

In each period, each individual has to choose a number from the feasible interval. Individuals choose independently from each other. At the end of a period, individuals are informed about own payoffs and about the median number chosen, i.e., the full distribution of chosen numbers is not revealed.

The "*Teams*"-treatment is, in principle, identical to the "*Individuals*"-treatment, except that a group of players consists now of five teams – instead of five individuals – making decisions in the coordination game. We apply partner matching across the 20 periods, allowing us to treat each group of five individuals or teams as an independent unit of observation. Each team consists of three randomly chosen members.[7] They can communicate via an electronic chat where they can discuss anything except for revealing their identity. They are requested to arrive at a joint team decision by agreeing on a single number to be chosen by all team members. Each team member has to enter the team's decision individually on his computer screen. If different numbers are entered, team members can chat again and enter a decision once more. Only if the second attempt fails this team receives no payment and does not take part in the coordination game in this period. Note that if a team agrees on a specific number, the resulting payoff is to be understood as a per-capita payoff for each team member. This approach is taken to hold the marginal incentives constant across the *Individuals*- and *Teams*-treatment. The feedback in the *Teams*-treatment is identical to the one for *Individuals*, meaning that own payoffs (per team member) and the median in the group of five teams are announced at the end of each period.

In total, 120 subjects (30 subjects in the *Individuals* treatment and 90 subjects in the *Teams* treatment) participated in the computerised experiment (using zTree by Fischbacher 2007, as experimental software and ORSEE by Greiner 2004, for online recruiting), and it was run at the University of Innsbruck in the academic year 2006/2007. No subject was allowed to participate in more than one session.

Results

Figure 14.6 plots the averages for single periods. The first-period data are particularly interesting because choices in the first period cannot have been influenced by any history of the game. In this respect, the first-period data are an indicator of "genuine" differences in coordination behaviour between individuals and teams. In the first period the average numbers of teams are already higher than those of individuals (*Individuals*: 5.33, *Teams*: 6.43; significant at the 5% level, two-sided robust rank order test). This is also true for the average over all periods and the

[7]We use the term "group" to refer to the entity of players who interact with each other, whereas "team" refers to three subjects who have to choose one number. The "group size" in our experiment is always five, i.e., the coordination game is played either by five individuals or by five teams.

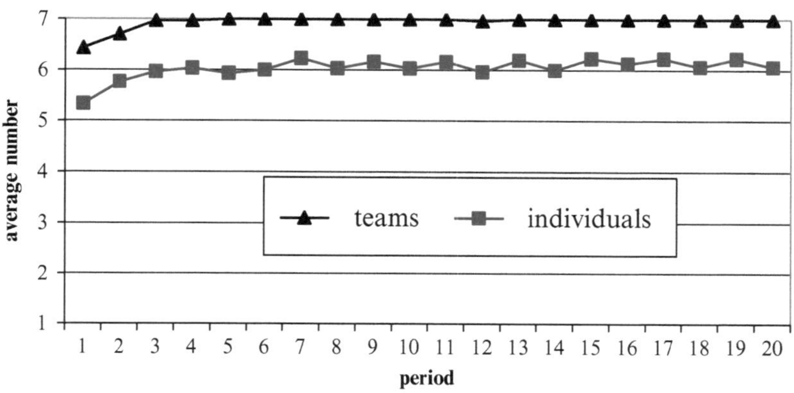

Fig. 14.6 Average numbers over periods

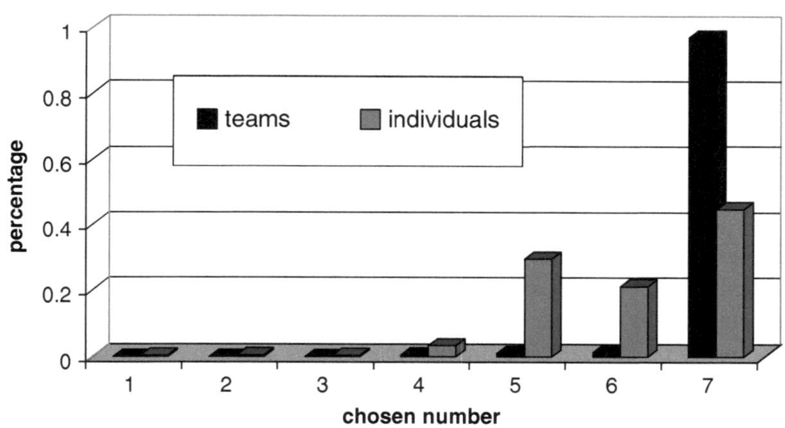

Fig. 14.7 Relative frequencies of chosen numbers

differences are again significant (*Individuals*: 6.04, *Teams*: 6.95; significant at the 5% level, two-sided robust rank order test). Note that for testing we can use all first-period choices, i.e., we can take *all five* values from each single group, because first-period choices are independent. This is true also for the high statistical standards propagated by Reinhard Selten. When examining the average data across all 20 periods, we treat each group (with five decision-makers) as *one* independent unit of observation. The number of independent observations in both treatments is six.

This observation is also clearly supported by Fig. 14.7 which shows the relative frequencies of choosing a particular number. The distribution of numbers is shifted to the right with team decision-making. Turning from the chosen numbers to the actually resulting median number within groups, we see that teams succeed in coordinating on higher median numbers (*Individuals*: 5.99, *Teams*: 6.98; significant at the 5% level, two-sided robust rank order test). This leads to substantially and significantly higher total payoffs for team members than for individuals (*Individuals*: 103.3, *Teams*: 127.1; significant at the 5% level, two-sided robust rank order test).

The superiority of teams with respect to payoffs is not only driven by coordinating on equilibria with higher payoffs, but also by a significantly smaller amount of miscoordination. Miscoordination can be measured as the absolute deviation of chosen numbers from the actual median in a given period. This indicator is significantly larger for individuals (*Individuals*: 0.25, *Teams*: 0.03; significant at the 5% level, two-sided Mann–Whitney U-test). Hence, individuals fail more often to coordinate on the same number. Perfect coordination, i.e., the fraction with which all five group members choose the same number, is indeed higher in the *Teams*-treatment than in the *Individuals*-treatment (*Individuals*: 0.61, *Teams*: 0.92; significant at the 5% level, two-sided Mann-Whitney U-test). The second indicator (adjustment) measures the absolute differences between a decision-maker's number in period t and the median in period t–1. There is a higher level of adjustment activity in the *Individuals*-treatments than in the *Teams*-treatments, meaning that teams settle quicker for an equilibrium (*Individuals*: 0.34, *Teams*: 0.02; significant at the 10% level, two-sided Mann-Whitney U-test).

Discussion

Our results show that teams are persistently and remarkably better in coordinating on efficient outcomes than individuals are. This finding adds one important cornerstone to the recent literature on the conditions for successful and efficient coordination (among individuals; see Brandts and Cooper 2006; Weber 2006; Blume and Ortmann 2007; Manzini et al. 2009) and is likely to be one of the reasons why teams are so frequently used in firms (Katzenbach and Smith 1993).

Conclusion

The examples above demonstrate that laboratory experiments provide valuable insights into the suitability and the effectiveness of incentives and thereby serve as a convincing informative basis for design issues in firms. It cannot be emphasised enough that the internal validity of results achievable in the laboratory is very high and can generally not be beaten by studies using field data. The main reason is that in the laboratory one has full control over decision situations and therefore razor-sharp conclusions can be drawn by means of appropriate experimental designs. Having said this, however, one also has to admit that the external validity of laboratory experiments on incentives in firms is often questioned. Indeed, there might be good reasons to assume that people in firms with their experience and expertise and being embedded in corporate cultures behave differently than student subjects. Thus, ideally one would like to complement laboratory results with experiments in corporate environments and sometimes go back to the lab in order to test new specific hypotheses that have emerged from the experimental findings in firms. Additionally, complementary evidence from experiments in corporate environments would certainly increase the acceptance of managers for policy recommendations.

Occasionally quasi-experiments emerge in which independent variables are manipulated, for example, due to policy changes of companies. Such changes can often be exploited to infer causal effects. Lazear (2000), for example, found a natural experiment in Safelite, a windshield manufacturing company that had just switched from paying hourly wages to piece rates. With the available data Lazear was able to show that half of the 44% productivity increase was due to turnover (unmotivated employees were replaced with people who wanted to earn more) while the other half resulted from increasing effort of continuous employees.

So far it is very rarely the case that researchers are able to convince business people to implement different treatments within a company to infer causal effects. A notable exception is a study by Bandiera et al. (2009) who engineered an exogenous change in managerial incentives, from fixed wages to bonuses based on average productivity of the workers managed. It turned out that the wage schemes had a clear impact on who the managers favoured. When they were paid fixed wages they favoured workers to whom they were socially connected irrespective of the worker's ability. This was less the case when they were paid performance bonuses which as a result increased the firm's overall performance. As is quite evident from this example, firms as well as researchers could benefit enormously from experiments within companies.

It seems, however, that firms are quite reluctant to agree to such collaborations. Reflecting on his experience Ariely (2010) argues "Companies pay amazing amounts of money to get answers from consultants with overdeveloped confidence in their own intuition [..] And yet, companies won't experiment to find evidence of the right way forward." Ariely conjectures that one reason for companies' reluctance to experiment is that experiments require to accept short-term losses (arising from the experimental treatments that from hindsight turn out not to work well) for long-term gains (evidence based recommendations) and humans are quite bad in making these trade-offs. A second reason according to Ariely is that we as humans tend to value answers over questions (which would lead to experimental designs) because answers allow us to take action while questions keep us thinking. Thus, it seems to be easier to listen to (intuitive) expert advice of consultants than to think of suitable experiments that usually could guide us to well-founded answers to arising questions.

One of the many things one should learn from Reinhard Selten is to go for approaches that one has identified as essential even if so far they are not developed much and have not reached the status of an accepted standard in the research community. Selten followed this maxim at least two times in his academic career and thereby became a co-founder of two very successful streams of research in economics. When he started working on non-cooperative game theory very few researchers saw its potential. Large parts of the community believed only in cooperative game theory. Selten often told us the story that during his early research stay at the University of California, Berkeley people left the room when they realised that his seminar talk was about non-cooperative games. Potentially even more important was the second trend Selten helped to set: Experimental economics. When he started to conduct his first experiments together with Heinz Sauermann in Frankfurt very few economists had the impression that what they were doing was serious economic research, let alone that they anticipated the impact experimental economics has today. Convincing businesses to experiment might be a promising and fruitful new research approach that is worth to follow today.

Acknowledgment The authors are grateful for the delicious cakes and the inspiring atmosphere provided by Café Jansen (Cologne), Café Sturm (Bonn), Café Zum Roten Turm (Erfurt), and Café Sacher (Innsbruck). Following the example of Reinhard Selten the authors spent a considerable number of hours in these cafés where many of the ideas discussed here were born.

References

Abbink K, Sadrieh A (1995) RatImage – research assistance toolbox for computer-aided human behavior experiments. SFB Paper B-325, Bonn

Ariely D (2010) Why businesses don't experiment. Harv Bus Rev April: 34.

Bandiera O, Barankay I, Rasul I (2009) Social connections and incentives in the workplace: evidence from personnel data. Econometrica 77(4):1047–1094

Bénabou R, Tirole J (2003) Intrinsic and extrinsic motivation. Rev Econ Stud 70:489–520

Blume A, Ortmann A (2007) The effects of costless pre-play communication: experimental evidence from games with Pareto-ranked equilibria. J Econ Theory 132(1):274–290

Bowles S (2008) Policies designed for self-interested citizens may undermine 'The Moral Sentiments': evidence from economic experiments. Science 320(5883):1605–1609

Brandts J, Cooper DJ (2006) A change would do you good ... an experimental study on how to overcome coordination failure in organizations. Am Econ Rev 96(3):669–693

Camerer CF, Knez MJ (1997) Coordination in organizations: a game theoretic perspective. In: Shapira Z (ed) Organizational decision making. Cambridge University Press, Cambridge, pp 158–188

Cooper R, DeJong DV, Forsythe R, Ross TW (1990) Selection criteria in coordination games: some experimental results. Am Econ Rev 80(1):218–233

Cooper R, DeJong DV, Forsythe R, Ross TW (1992) Communication in coordination games. Q J Econ 107(2):739–771

Deci EL (1971) Effects of externally mediated rewards on intrinsic motivation. J Pers Soc Psychol 18:105–115

Deci EL, Koestner R, Ryan RM (1999) A meta-analytic review of experiments examining the effects of extrinsic rewards on intrinsic motivation. Psychol Bull 125:627–668

Ellingsen T, Johannesson M (2008) Pride and prejudice: the human side of incentive theory. Am Econ Rev 98(3):990–1008

Feri F, Irlenbusch B, Sutter M (2010) Efficiency gains from team-based coordination: large-scale experimental evidence. Am Econ Rev, (in press)

Fischbacher U (2007) z-Tree: Zurich toolbox for ready-made economic experiments. Exp Econ 10(2):171–178

Frey BS (1997) Not just for the money – an economic theory of personal motivation. Edward Elgar, Cheltenham

Gibbons R (1998) Incentives in organizations. J Econ Perspect 12(4):115–132
Gneezy U, Rustichini A (2000) Pay enough or don't pay at all. Q J Econ 115:791–810
Greiner B (2004) An online recruiting system for economic experiments. In: Kremer K, Macho V (eds) Forschung und wissenschaftliches Rechnen 2003. GWDG Bericht 63, Ges. Für Wiss. Datenverarbeitung, Goettingen, pp 79–93
Harbring C, Irlenbusch B (2008) How many winners are good to have? On tournaments with sabotage. J Econ Behav Organ 65:682–702
Harbring C, Irlenbusch B (2010) Sabotage in tournaments – evidence from a laboratory experiment. Manage Sci (in press)
Harbring C, Irlenbusch B, Kräkel M, Selten R (2007) Sabotage in corporate contests – an experimental analysis. Int J Econ Bus 14:367–392
Harsanyi J, Selten R (1988) A general theory of equilibrium selection in games. MIT Press, Cambridge, MA
Heyman J, Ariely D (2004) Effort for payment. Psychol Sci 15(11):787–793
Irlenbusch B, Sliwka D (2005) Incentives, decision frames, and motivation crowding out – an experimental investigation. IZA Discussion Paper 1758
Katzenbach JR, Smith DK (1993) The wisdom of teams – creating the high-performance organization. Harvard Business School Press, Boston, MA
Kohn A (1993) Why incentive plans cannot work. Harv Bus Rev Sept–Oct:54–63
Lazear EP (1999) Personnel economics: past lessons and future directions. J Labor Econ 17(2):199–236
Lazear EP (2000) Performance pay and productivity. Am Econ Rev 90(5):1346–1361
Manzini P, Sadrieh A, Vriend NJ (2009) On smiles, winks and handshakes as coordination devices. Econ J 119(537):826–854
Orrison A, Schotter A, Weigelt K (2004) Multiperson tournaments: an experimental examination. Manage Sci 50:268–279
Prendergast C (1999) The provision of incentives in firms. J Econ Lit 37:7–63
Selten R (1965) Spieltheoretische Behandlung eines Oligopolmodells mit Nachfrageträgheit. Z Gesamte Staatswiss 12:301–324
Selten R (1998) Features of experimentally observed bounded rationality. Eur Econ Rev 42(3–5):413–436
Sliwka D (2007) Trust as a signal of a social norm and the hidden costs of incentive schemes. Am Econ Rev 97(3):999–1012
Van Huyck JB, Battalio RC, Beil RO (1990) Tacit coordination games, strategic uncertainty, and coordination failure. Am Econ Rev 80(1):234–248
Van Huyck JB, Battalio RC, Beil RO (1991) Strategic uncertainty, equilibrium selection, and coordination failure in average opinion games. Q J Econ 106(3):885–910
Weber RA (2006) Managing growth to achieve efficient coordination in large groups. Am Econ Rev 96(1):114–126

Part V
Risky-Choice Behavior

Chapter 15
Georges-Louis Leclerc de Buffon's 'Essays on Moral Arithmetic'

John D. Hey, Tibor M. Neugebauer, and Carmen M. Pasca

We offer a translation into English of the original French version published in 1777 of Buffon's *Essai d'Arithmetique Morale*. In this classic work, Buffon discusses degrees of certainty, probability, the moral value of money, the different evaluations of gains and of losses; moreover, he proposes repeated experiments to determine the moral value of a game. Our hope is that this remarkable work, which anticipates and relates to many aspects discussed in more recent literature, reaches a broader audience than the French original reaches today. Our belief is that we are first to translate this classic work completely. Remaining errors are ours.

Buffon (1707–1788) was acknowledged, and is mainly remembered, for his opus *Natural History*, of which he finalized thirty-six volumes during his lifetime. Another eight volumes were published posthumously.

His scientific work was generally based on the methods of empirical observation and experiment. Based on his evidence, he suggested that the origin of organisms resulted by spontaneous generation from smallest particles and the development and diversity from climatic changes. Differing from the common view at his time, he believed that the first life developed in the sea, and that the stepwise development of the species took long periods. Preceding Charles Darwin (1809–1882), he advanced the view of common descent of the species, discussing the relation between apes and man. By experimental evidence he showed that the earth was older than the 6,000 years calculated by theologians based on biblical data. This evidence led to a conflict with the church including the burning of his writings. He had to officially revoke his statement, but raised the point again in the *Essays*.

J.D. Hey
Department of Economics and Related Studies, University of York, Heslington, York YO10 5DD, UK

T.M. Neugebauer (✉)
Faculty of Law, Economics and Finance (LSF), University of Luxembourg, 4-rue Albert Borschette, L-1246 Luxembourg, Luxembourg
e-mail: Tibor.Neugebauer@uni.lu

C.M. Pasca
LUISS Guido Carli, Viale Pola, 12 - 00198 Rome, Italy

A. Ockenfels and A. Sadrieh (eds.), *The Selten School of Behavioral Economics*, DOI 10.1007/978-3-642-13983-3_15,

Before dedicating himself to natural history, Buffon studied and contributed to mathematics during the 1720s and 1730s. He communicated, and maintained correspondence with, Gabriel Cramer (1704–1752), professor of analysis and geometry at the University of Geneva, from 1727 (Weil 1961). So, while it is likely that he started editing the *Essays* in the 1760s, his interest in the topic and many of his considered ideas date back to the early 1730s.

The *Essays on Moral Arithmetic* contains 25 articles, the first two being introductory. As pointed out in them, the work is dedicated to the measurement of, and, more generally, to the valuation of uncertainty. Buffon defines levels of uncertainty, or contrariwise, levels of certainty, by available evidence observed in nature, or more generally if the causes of a certain effect are unknown; if effects repeat constantly, he asserts, then the effect becomes certain; after a long sequence of repeated observations effects become physically certain. Conversely, if an effect has constantly failed to occur, he suggests, it will come to be refuted. He also points out that a change in an assumed constant effect surprises us; such effects have been brought to the attention of a great audience through the recent literature (Taleb 2007). Buffon illustrates the concept of certainty levels by his example of the experience of the sunrise. He reflects on how an ignorant man sees the sunrise, the sunset and learns to reinforce his belief and simultaneously to decrease his doubt by repeated experience, finally reaching certainty about the return of the sunrise (see articles III–VI).[1] He refers to physical certainty as an "almost infinite" probability level to which he assigns a relationship as of one to $2^{2,189,999}$.[2] Note the power term represents the number of days following the first day in 6,000 years, as this number of years represented the accepted time of existence of man and the earth during Buffon's lifetime; the base term, 2, represents the two possible effects either the sun returns on the next day or it does not return.

If one has only a limited number of observations of a constant effect so that physical certainty cannot be inferred, Buffon argues that uncertainty can be so much removed if only the number of experiences is sufficiently large; one can achieve moral certainty about the effects. Moral certainty can be considered as a boundedly-rational judgment level, sufficiently great to draw conclusions on the certainty of constant effects. Given the limitations of time and resources on the number of experiences (observations), moral certainty is a substitute for physical certainty that enables one to make a boundedly-rational judgment about the effects of nature even without understanding its causes. In Buffon's thinking, moral certainty implies a

[1]Reinforcement learning is an algorithm that tries to maximize payoffs under uncertainty by successively increasing the weight of better experiences (for further references see Gigerenzer and Selten 2001). In contrast, Buffon's belief learning focuses only on judgment and not on strategy.

[2]So, we should warn readers at the outset that Buffon uses the word certainty not in the absolute sense that we currently use. We think of a certain event as one that is bound to happen, that is, an event that has probability 1 of happening. However, and rather schizophrenically, we also say that one event is more certain than another. This latter sense is that of Buffon. He refers to certainties of different orders and hence uses the word certainty as a synonym for probability or likelihood.

lower degree of certainty than physical certainty, and can be determined either by evidence following a constant sequence, or by analogical reasoning based on testimonies of a constant sequence. The probability level that Buffon assigns to a complete level of moral certainty is one ten-thousandth (see articles VII–IX). This level is based on the argument that a 56 year old man is fearless about dying during any given day, an event which according to his mortality tables, reported elsewhere (Buffon 1777a), occurred with the corresponding relative frequency.

Buffon refers also to different levels of probability for the case when the underlying effect is not constant. Discussing such chance effect in more detail, Buffon focuses on the 50-50 game, where an effect occurs as often as it fails to occur. He suggests that, even in these games, the observation of a large number of results from the risk device can give us an advantage in the game, if the risk device is biased (see articles X–XI). Thus he indicates that underlying probabilities can also be learned from long series of observations, and can deviate from *a priori* probabilities.

Generally, Buffon condemns participation in games of chance, judging them as generally harmful to overall well-being in society (article XII). In particular, he refers to the contemporary popular game of *Pharaoh*, apparently a forerunner of *Poker*. He argues that people behave dishonestly and irrationally with respect to these games of chance. His judgment is based on: (1) a value function approach; and (2) a classification of individual income as either *necessary* or *superfluous* (articles XIII–XIV).

1. His value function exhibits *loss aversion*, as it punishes losses more than it rewards gains. Gains are valued relatively to the *ex post* income including the gains; losses are valued relatively to the *a priori* income position, that is, before subtracting the losses. Thus since losses loom larger than gains, overall well-being is reduced even in fair games. Compared to the logarithmic utility function proposed by Daniel Bernoulli (1700–1782) which was published in 1738, losses are, at the margin, similarly valued, however, the Buffonian value function values gains less than losses. Loss aversion is quite accepted among modern researchers. For instance, Selten and Chmura (2008) assign a double weight to a loss than to a gain in their application of the impulse balance theory to experimental data. Such a double weighting of losses relative to gains was suggested by experimental evidence in Tversky and Kahneman (1992). We should mention here, even without further discussion, that modern loss aversion decisively differs from the Buffonian version in several respects. However, Buffon appears to have been the first to propose loss aversion within a utility approach. This utility approach, however and unfortunately, was not embedded in a fully-fledged and fully-developed theory.
2. He nevertheless suggests the existence of individual utility based on the requirements of personal needs according to one's position in society. He defines *necessary income* (that necessary to sustain the social status of an individual) and *superfluous income* (that over and above that which is necessary). This classification suggests that utility is structured in a way that necessary income represents a safety-first element, similar to Lopez's aspiration theory (1987).

Because of this classification of income, losses lead to a greater loss of overall well-being (caused by loss aversion), if the loss in the game is out of the necessary income and if the gain increases only the superfluous income, because necessary income must be valued higher than superfluous income. Finally, he assumes bounded utility in the sense of Cramer, that is, beyond a certain threshold superfluous income gives no extra utility.

Buffon dedicates a large part of the *Essays* to a presentation and discussion of the Petersburg gamble (articles XV–XXII). His discussion is so broad that it includes almost all currently-known 'solutions' to the Petersburg paradox.[3] In a footnote he quotes at length the letter he wrote to Gabriel Cramer in 1730. Thus he proves that some of his ideas preceded those of Daniel Bernoulli (1738). He concludes that the Petersburg paradox – that is, the discrepancy between the intuitive value and the mathematical expectation – arises from two causes; first, the small probabilities of the exorbitantly high payoffs are estimated as zero,[4] and second, because of the decreasing marginal utility of money the exorbitantly high payoffs lead to very low increased values.[5] Buffon next raises the solvency problem, according to which the payoff in the gamble can only be a finite amount, so that the value of the gamble must be based on a finite, rather short, gamble length. A remarkable contribution to the Petersburg paradox is his determination of the game value by repeated experiment.[6] A child played out 2^{11} Petersburg gambles to yield an average payoff of

[3]A concise survey of the literature and the solution concepts surrounding the Petersburg paradox accompanied by experimental-economics evidence is offered in Neugebauer (2010).

[4]The treatment of very small probabilities remains a very controversial issue today. As Selten (1998, p. 51) points out: "... In general, it is very difficult to judge how small a very small probability should be. Usually there will be no good theoretical reasons to specify a probability as 10^{-5} or rather than 10^{-10}. ... The value judgment ... that small differences between small probabilities should be taken very seriously and that wrongly describing something extremely improbable as having zero probability is an unforgivable sin... is unacceptable."

[5]One may criticise Buffon for not referring to the correspondences of Nicholas Bernoulli (1687–1759). Nicholas Bernoulli (1728) had referred to moral certainty in a letter written to Gabriel Cramer so that it is very likely that Buffon was aware of the fact that Nicholas Bernoulli had given this explanation earlier. Nevertheless, the value of presenting this explanation to a greater audience is an important contribution given that the statement of Nicholas Bernoulli had not been widely heard.

[6]This experiment seems to be the first conducted statistical experiment ever reported. It was replicated later by various researchers. Though the experimental approach is used to elicit the money value of the game, it is not an economic experiment, since the economic behaviour of subjects is not the purpose of study. According to the definition by Sauermann and Selten (1967, p. 8, our translation) this condition is crucial: "It is advisable to count only such experiments to experimental economic research in which the economic behavior of experimental subjects is observed. So-called simulation experiments, which only consist in the computation of numerical examples for theoretical models on electronic computers, do not belong, in this sense, to experimental economic research." An economic experiment on valuation of lotteries would have to elicit the willingness to pay from experimental subjects. A frequently used elicitation procedure in this context is the BDM approach as used in Selten et al. (1999). For a more general introduction to economic experiments see Hey (1991).

about 5 Ecu per game. From the outcomes of this experiment, Buffon motivated the geometric payoff distribution that results by application of the law of large numbers. Table 15.1 shows the outcomes of the repeated experiment and the statistically expected outcomes. By reflecting on the theoretical distribution, Buffon comes to the conclusion that the greater the number of repetitions the greater is the expected outcome. He, thus, anticipates partly the mathematical solution to the repeated Petersburg gamble as proposed by Feller (1945). In contrast to Bernoulli's utility function, his arguments stand the test of the Super-Petersburg paradox (Menger 1934), again by the cancelling of small probabilities and the zero marginal value of the superfluous income beyond a certain bound.

The contribution of the last two articles in the *Essays* preceding the concluding one (article XXV) is more to mathematics than to moral behaviour. Buffon presents his games of the *franc-carreau* and *Buffon's needle* (article XXIII); the latter is the reason why his name still remains in the mathematical sciences. The contribution is important since it introduces geometrical probability to the literature. According to available information on the internet, Buffon conducted experiments by throwing a stick over his shoulder into a tiled room. In similar random experiments on Buffon's needle, the number π was approximated in the nineteenth century.

Finally, Buffon discusses the meaning of infinity (article XXIV) by highlighting its value as a mathematical tool which allows the generalisation of results and by pointing out that it is not a 'real' number.[7]

Summing up, the *Essays* present a collection of articles of Buffon dedicated to judgment and behaviour, including experimental methods and mathematics. Regardless of a number of issues that one can take with his presentation from a modern-day perspective (as it lacks a fully-fledged and fully-developed approach, and is not always convincing), it appears remarkable to see how much thought,

Table 15.1 Buffon's results

Number of "tails"	Buffon's observations	Payoff[a]	Buffon's geometric approximation
0	1,060	1	$2^{10} = 1{,}024 + 1$
1	494	2	$2^9 = 512$
2	232	4	$2^8 = 256$
3	137	8	$2^7 = 128$
4	56	16	$2^6 = 64$
5	29	32	$2^5 = 32$
6	25	64	$2^4 = 16$
7	8	128	$2^3 = 8$
8	6	256	$2^2 = 4$
9	–	512	$2^1 = 2$
10	–	1,024	$2^0 = 1$

[a]The total payoff was 10,057 Ecu, an average of 4.91 Ecu

[7]Following the 25 described articles, Buffon dedicates another ten articles to "arithmetic and geometric measures". These articles look to us less relevant to the study of human sciences and are therefore omitted in the translation.

discussed today in the human sciences literature, was already expressed during the eighteenth century. Buffon's *Essays* is outstanding from the standpoint of today when compared to many other works of that age because of its many and remarkable contributions: the introduction of the valuation of a game by experimental method, thus highlighting also the importance of empirical evidence; the discussion based on philosophical arguments of levels of significant and negligible probabilities; the valuation of losses and its distinction to the valuation of gains; the distinguishing of necessary income from superfluous income; and, finally, the introduction of geometric probability.

Before concluding, one remaining question must evidently be addressed. Since this paper has been prepared for publication in honour of Reinhard Selten, the reader may ask what is the relationship between Buffon, his *Essays*, and Reinhard Selten and his work?-Thereto we reply that although we do not know of any direct link, we see relationships and analogies in (1) spirit, (2) presentation, (3) methodology, (4) thought, and (5) the grandeur of scientific contribution.

1. Although Reinhard Selten is evidently a modern researcher, in our view he shares the mind-set of the great savants of the age of the enlightenment, that is, the dedication to research driven only by the desire to find and communicate the truth about the nature of things. Owed to the conservativeness of academic reviewers who were not always open to Selten's unorthodox theories, and he being reluctant to making any changes to his work that could bias his vision of the truth, he would publish his famous papers unchanged in unknown academic journals rather than having the changed version published in a prestigious journal. Even though the prestige of publishing in such journals would have helped his career at that time, he accepted the facts in a humorous way. We remember him saying that by publishing in an unknown journal "you can make a journal famous." He surely made several journals famous as he was frequently quoted for his publications in journals that were quite unknown or nonexistent before his publication. This achievement is evidently more exceptional than publishing in a famous journal, but it is a rocky road to making any impact in the human sciences, even for an exceptionally brilliant mind. He also pointed to the fact that in earlier times, as in the Buffon's, researchers had to start writing an essay on scientific questions by first apologizing for daring to raise these questions at all (see article I).
2. *Selten's archer* (Selten and Buchta 1994) who represents direction learning theory is as original and as illustrative as the great Buffonian metaphor of the blind man who learns by experience about the return of the sunrise. Nevertheless both presentations share a similar description about the learning of information about unknown things and the adjustment in hindsight. The difference, indeed, is in the adjustment itself; while the blind man adjusts his belief, the archer adjusts his strategy. The archer learns hitting a target with an arrow, similarly to a blind man, only by repeated experience through feedback information. If he is informed that the arrow missed the target to the left, on the next trial, he is going to adjust the direction of the arrow to the right rather than to the left, and

vice versa. This direction learning theory has shown to explain the behaviour of most experimental subjects in economics laboratory experiments in very different scenarios (see Selten 2004 for a review). Impulse balance theory is based on this direction learning story; it allows point-predictions of behaviour by balancing potential positive and negative impulses in games of strategy (for a simple application of this concept, see Selten and Neugebauer 2006).

3. Selten has used mainly mathematical and experimental approaches in his research. Similarly to Buffon, he was fascinated by mathematics from his youth, received a degree in mathematics, before he turned to applied mathematics and experimental research. During his career, his main interests have involved the measurement and valuation of games, both theoretically and experimentally, and in the construction of solution concepts based on (levels of) rationality.
4. Selten sees limits to the benefit of mathematics in the understanding of human behaviour. In the language of Buffon (article II), we can imagine although we did not witness, Selten could have said: "maths involves the truth of definitions. This truth is only helpful if one understands the problem well, that is, if the analysis of the problem is based on the right definitions." The Reinhard Selten School believes in the merit of experimental research to uncover the fundamental definitions of human behaviour, and to support the building of a boundedly rational system in form of a toolbox if a fully-fledged theory and fully-developed approach is not available. Buffon also seems to favour a boundedly rational system to human beings over pure rationality, as he states that rationality is cold and does not make man really happy. In particular, he accepts that less rare people gamble from time to time because hope makes them happy. In the language of Selten, many people tend to avoid risk taking as much as they can, but on some occasions they enjoy and permit themselves to take risks; "today I take a chance."

   In relation to Buffon's discussion of the Petersburg gamble, Selten accepts, at least from the behavioural viewpoint, a treatment of very small probabilities as zero. Discussing the outcomes of valuations of the Petersburg gamble by experimental subjects, Reinhard Selten was not at all surprised that according to the data people behave as if they neglect very small probabilities in the Petersburg gamble; "of cause they do that."

5. Originality and the greatness of ideas which lead to path-breaking approaches must be used as a description of the work of both Buffon and Selten.[8] While we do not know about Buffon, Selten had a reason for being so original. He argued that he was slower than other researchers and therefore had to take greater steps.

[8]In fact, we do not intend to give here a fair and thorough appraisal of Reinhard Selten's contributions. Many important articles have been republished in Selten (1988, 1999).

Both had a great impact on the development of research in games and experiments; Buffon as the founding father of experimental statistics; Reinhard Selten as the founding father of experimental economics in Europe. Finally, both have influenced thinking, being thought provoking and created great interests in the subject of their research.

Reinhard Selten is very active with his current research. Recently, at the celebration of the twenty-fifth anniversary of the Bonn Laboratory of Experimental Economics, he said that he was the "money most-intensive researcher" at the economics department at Bonn. So looking forward, we expect many new researches, advancements of the field bounded rationality and enlightening ideas of and by Reinhard Selten.

Essays on Moral Arithmetic

I. I do not attempt to present general essays on morality here; that would demand more enlightenment than I can presume, and more art than I recognize. The first and most sensible part of morality is rather an application of the maxims of our divine religion than of a human science; and I cannot even dare to try matters where the law of God is our principle, and Faith our rationale. The respectful gratitude or rather the adoration, man has to his Creator; brotherly charity, or rather the love he has to his fellow man, are natural feelings and virtues written on an ingenuous mind; all that stems from this pure source bears the character of the truth; the light is so bright that the existence of error cannot obscure it, the evidence so great that it admits no argument or discussion, or doubt, and no other measures than conviction.

The measurement of uncertain things is my object here. I will try to give some rules to estimate likelihood ratios, degrees of probability, weights of testimonies, influence of risks, inconvenience of perils; and judge at the same time the real value of our fears and of our hopes.

II. There are truths of different kinds, certainties of different orders, probabilities of different degrees. The purely intellectual truths like those of Geometry all reduce themselves to truths of definition; it is a matter of resolving the most difficult problem only to understand it well, and there are no other difficulties in calculation and in the other purely theoretical sciences than to untangle what we put in, and to untie the knots that the human mind has created in the study of the implications of the definitions and the assumptions that are used as the foundation and framework of these sciences. All their propositions always can be proven evidently, because you can always go back from each of these propositions to other preceding propositions which are identical to them, and from these to others back to the definitions. It is for this reason that evidence[9] itself belongs to the mathematical sciences and only to them; because one must distinguish the evidence of reasoning from the evidence that comes through the senses, that is, the intellectual evidence from the physical intuition; the latter is only a clear apprehension of objects or of images; while reasoning is a comparison of similar or identical ideas, or rather it is the immediate perception of their identity.

III. In the physical sciences, evidence is replaced by certainty; evidence is not susceptible to measure, because it has only one absolute property, that is the clear negation or the affirmation of the matter it shows; but certainty is never a positive absolute one, it requires several relationships that we must compare and from which we can estimate the measure. Physical certainty, that is, the most certain of all certainties, is nevertheless only the almost infinite probability that an effect, an event that never failed to happen, will happen again; for example, because the sun has always risen, it is thenceforth physically certain that it will rise tomorrow; a reason for being is to have been, but a reason for ceasing to be is to have come into being; and consequently one cannot say that it is equally certain that the sun will always rise, at least one must assume a preceding eternity, equal to the subsequent perpetuity, otherwise it will end as it has begun. For we must judge the future by the evidence from the past, whenever something has always been, or if it always behaved the same way, we must be assured that it will be or will behave always in the very same way: by *always*, I mean a very long time, and not an absolute eternity, the always of the future never being equal to the always of the past. The absolute of any kind, whatsoever, is not the responsibility of Nature or of the human mind. Men have considered as ordinary, natural effects all the events that have this kind of physical certainty; an always occurring effect ceases surprising us; in contrast a phenomenon that has never occurred, or one that has always occurred in the same way but ceases to occur or occurs in a different way would surprise us with reason and would be an event that appears to us so extra-ordinary that we would consider it as supernatural.

IV. These natural effects that do not surprise us, however, have everything necessary to surprise us; what circumstances of causes, what collection of principles is not necessary to produce a single insect, a single plant! What a prodigious combination of elements, motions and springs in the animal machine! The smallest works of Nature are subjects of the greatest admiration. The reason why we are not at all surprised by all these wonders is that we were born into this world of wonders, that we have always seen them, that our understanding and our eyes are equally accustomed to them; finally, all were before and will be still after us. If we had been born in a

[9][comment: Buffon uses the word *évidence*, which we translate throughout by *evidence*, also in the sense of something obvious or evident].

different world with a different physical shape and different senses, we would have had other relationships with exterior objects, we would have seen different wonders and we would not have been more surprised; the wonders of this world and of the other are based on the ignorance of the causes, and on the impossibility of knowing the reality of the things, of which we are only permitted to see are the relationships they have with ourselves.

There are therefore two ways of looking at the natural effects; the first is to see them as they present themselves to us without paying attention to their causes, or rather without looking for their causes; the second is to examine the effects in the view to relating them to the principles and causes; and these two points of view are strongly different and produce different reasons for surprise, the one causes the sensation of surprise, and the other creates the feeling of admiration.

V. We will speak here only of this first way to look at the effects of Nature; as incomprehensible, as complicated as they present themselves to us, we will judge them as the most evident and the simplest ones, and judge them only by their results; for example, if we cannot conceive or even imagine why matter attracts each other, we will certainly be satisfied that it actually attracts each other, and we will judge from that time that it always has attracted each other and that it always will continue to attract each other: it is the same with other phenomena of all kinds, as unbelievable as they may appear to us, we will believe them if we are sure that they have occurred very often, we will doubt them if they have failed to happen as often as they occurred, finally, we will deny them if we believe for sure that they never occurred; in a word, according to having seen and recognized them, or to having seen and recognized the opposite.

But if the experience is the basis of our physical and moral knowledge, then analogy is the first instrument: when we observe that a thing occurs constantly in a particular way, we are assured by our experience that it will occur again in the same way; and when someone reports that a thing occurred in this or that way, if these facts have an analogy with the other facts that we know by ourselves, then we believe them; on the contrary, if the fact has no analogy with ordinary effects, that is, with the things that are known to us, we must be in doubt; and if it is directly opposed to what we know, we do not hesitate to deny it.

VI. Experience and analogy may give us different certainties, sometimes almost equal and sometimes of the same kind; for example, I am almost as certain of the existence of the city of Constantinople, which I never have seen, as I am of the existence of the Moon, that I have seen so often, and that because the testimonies in large number can produce a certainty almost equal to the physical certainty, when they concern things that have a full analogy with those that we know. The physical certainty must be measured by an immense number of probabilities, because such certainty is produced by a constant sequence of observations, which are what we call the *experience of all the times*. The moral certainty must be measured by a smaller number of probabilities, since it presupposes only a number of analogies with what is known to us.

Assuming a man that had never seen anything, heard anything, we investigate how the belief and the doubt generate itself in his mind; assume him struck for the first time by the appearance of the sun; he sees it shine from the top of the skies, then decline and finally disappear; what can he conclude? Nothing, except that he saw the sun, that he saw it follow a certain route, and that he no longer sees it; but this star reappears and disappears again on the next day; this second sight is a first experience, that must produce in him the hope to see the sun again, and he begins to believe that it could return, nevertheless he is very much in doubt; the sun reappears again; this third sight is a second experience which diminishes the doubt as much as it increases the probability of a third return; a third experience increases it to the point that he no longer doubts that the sun returns a fourth time; and finally when he will have seen this light-star appear and disappear regularly ten, twenty, hundred times again, he will believe to be certain that he will see it always appear, disappear and to move the same way; the more similar observations he will have, the greater will be the certainty to see the sun rise the next day; every observation, that is, every day, produces a probability, and the sum of these probabilities together, as it is very great, gives the physical certainty; one will therefore always be able to express this certainty by the numbers, dating back to the origin of the time of our experience and it will be the same for all the others effects of Nature; for example, if one wants to reduce here the seniority of the world and of our experience to 6,000 years, the sun has risen for us[10] only 2 million 190 thousand times, and as to date back to the second day that it rose, the probabilities to rise the next day increase, as the sequence 1, $2, 4, 8, 16, 32, 64. \dots$ or 2^{n-1}. One will have (where the natural sequence of the numbers, n is equal to 2,190,000), one will have, I say, $2^{n-1} = 2^{2,189,999}$; this already is such a prodigious number that we ourselves cannot form an idea, and it is by this reason that one must look at the physical certainty as composed from an immensity of probabilities; since by moving back the creation date by only 2,000 years, this immensity of probabilities becomes $2^{2,000}$ times more than $2^{2,189,999}$.[11]

VII. But it is not so easy to do the estimation of the value of analogy, nor consequently, to find the measure of moral certainty; it is in truth the degree

[10] I say for us, or rather for our climate, because it would not be exactly true for the climate of the poles.

[11] [comment: Buffon obviously means days rather than years].

of probability that gives analogical reasoning its power; and in itself the analogy is only the aggregate of the relationships with known things; nevertheless according to that aggregate or that relationship in general being more or less great, the consequence of the analogical reasoning will be more or less certain, but without ever being absolutely certain; for example, if a witness whom I suppose being of common sense, tells me that a child has just been born in this city, I will believe him without hesitating, as the fact of a child's birth is nothing other than very ordinary, but having on the contrary an infinity of relationships with the known things, that is, with the births of all other children, I will believe this fact therefore, however, without being absolutely certain about it; if the same man tells me that this child was born with two heads, I would believe it again, but more weakly, as a child with two heads has less relationships with known things; if he added that the newborn has not only two heads, but that it has also six arms and eight legs, I should have good reason to hardly believe it, but however weak my belief was, I could not refuse it in entirety; this monster, although very special, nevertheless being composed only of parts that all have some relationships with known things, and only having their assembly and their very extraordinary number. The power of the analogical reasoning will therefore be always proportional to the analogy itself, that is, to the number of the relationships with known things, and it is not a matter of making a good analogical reasoning, but to adjust oneself well to the fact of all the circumstances, to compare them with the analogous circumstances, to aggregate the number of these, to take next a model of comparison to which one will relate this found value, and one will have exactly the probability, that is, the degree of power of the analogical reasoning.

VIII. There is therefore a prodigious distance between the physical certainty and the certainty of the kind that one can deduce from most of the analogies; the first one is an immense sum of probabilities that forces us to believe; the other is only a smaller or greater probability, and often so small that it leaves us in the perplexity. The doubt is always inversely proportional to the probability, that is, it is always the greater the smaller the probability. In the order of the certainties produced by the analogy, one must place the moral certainty, which seems to even take the centre between doubt and physical certainty; and this centre is not a point, but a very extensive line, for which it is quite difficult to determine the limits: one can feel that it is a certain number of probabilities that equals the moral certainty, but what number is it? And can we hope to determine it as precisely as the one by which we have just represented the physical certainty?

After having reflected on it, I have thought that of all the possible moral probabilities, the one that most affects man in general is the fear of death, and I felt from that time that any fear or any hope, whose probability would be equal to the one that produces the fear of death, can morally be taken as the unit to which one must relate the measure of the other fears; and I relate

to the same even the one of hopes, since there is no difference between hope and fear, other than from positive to negative; and the probabilities of both must be measured in the same way. I seek therefore for what is actually the probability that a man who is doing well, and consequently has no fear of death, dies nevertheless in the 24 h: consulting the Mortality Tables, I see one can deduce that there are only ten thousand one hundred eighty-nine to bet against one, that a 56 year old man will live more than a day.[12] Now as any man of that age, when reason has attained its full maturity and the experience all its force, nevertheless has no fear of death in the 24 h, although there is only ten thousand one hundred eighty-nine to bet against one that he will die in this short interval of time; from this I conclude that any equal or smaller probability must be regarded as zero, since any fear or any hope below ten thousand must not affect us or even occupy for a single moment the heart or the mind.[13]
To make me better understood, suppose that in a lottery where there is only a single prize and ten thousand blanks, a man takes only one ticket, I say that the probability to obtain the prize is only as one against ten thousand, his hope is zero, since there is no more probability, that is, reason to hope for the prize, than there are fears of death within 24 h; and that this fear not affecting him in any way, the hope for the prize must not affect him more, and even again much less, since the intensity of the fear of death is much greater than the intensity of any other fear or of any other hope. If, despite the evidence of this demonstration, this man insists on wanting to hope, and if a similar lottery is played every day, and he persists in buying a new ticket every day, always hoping to win the prize, one could, to disabuse him, bet with him at equal odds that he was dead before he won the prize.

---

[12]See hereafter the result of the mortality Tables [comment: the tables are not included in the translation].

[13]Having communicated this idea to Mr. Daniel Bernoulli, one of the greatest geometricians of our century, and most experienced in all of the science of probabilities, here is the response that he gave me by his letter, dated from Bâle March 19, 1762.

"I strongly approve, Sir, your way to estimate the limits of moral probabilities; you consult the nature of man by his actions, and you suppose indeed, that no one worries in the morning to die that day; hence, he will die according to you, with probability one in ten thousand; you conclude that one ten-thousandth of probability should not make any impression in the mind of man, and consequently this one ten-thousandth has to be regarded as an absolute nothing. This is doubtless the reasoning of a Mathematician-Philosopher; but this ingenious principle seems to lead to a smaller quantity, because the absence of fear is certainly not in those who are already ill. I do not fight your principle, but it seems rather to lead to $1/100,000$ rather than to $1/10,000$."

I confess to Mr. Bernoulli that as the one ten-thousandth is taken according to the mortality Tables that never represent the *average man*, that is, men in general, well or sick, healthy or ill, strong or weak, there is maybe a little more than ten thousand to bet against one, that a man, healthy and strong will not die in the 24 h; but it is hardly necessary to increase this probability to one hundred thousand. Moreover, this difference, although very large, changes nothing of the main implications that I draw from my principle.

It is the same in all the games, the bets, the perils, the risks; in all cases, in a word, where the probability is smaller than $1/10,000$, it must be, and it is in fact for us, absolutely zero; and by the same reason in all cases where this probability is greater than 10,000, it makes for us the moral certainty most complete.

IX. From it we can conclude that the physical certainty relates to the moral certainty as $2^{2,189,999} : 10,000$; and that whenever an event, of which we absolutely ignore the cause, occurs in the same way thirteen or fourteen times in a row, we are morally certain that it will occur again even a fifteenth time, for $2^{13} = 8,192$, and $2^{14} = 16,384$, and consequently when this event has occurred thirteen times, there is 8,192 to bet against 1 that it will occur a fourteenth time; and when it has occurred fourteen times, there is 16,384 to bet against 1 that it will occur even a fifteenth time, which is a greater probability than the one of 10,000 against 1, that is, greater than the probability that makes moral certainty.

One will perhaps be able to tell me, that although we do not have the fear or the worry of sudden death, it must be that the probability of sudden death is zero, and that its influence on our conduct is morally zero. A man whose mind is beautiful, when he loves someone, would he not reproach himself to delay a day the measures that must assure the happiness of the loved person? If a friend entrusts us with a considerable deposit, would we not put the same day a written comment on this deposit? We act therefore in these cases, as if the probability of sudden death was something, and we have reason to act like this. Therefore one must not regard the probability of sudden death as zero in general.
This kind of objection will vanish if one considers that one often does more for others than one does for oneself! When one puts a written comment the very moment when one receives a deposit, it is only honesty to the owner of the deposit, for his tranquillity, and not at all the fear of our death in the next 24 h; it is the same attentiveness that makes the happiness of someone or of us, it is not the feeling of the fear of an approaching death that guides us, it is our own satisfaction that drives us, we seek to enjoy all at the earliest that is possible to us.
A reasoning that might appear more justified is that all men are inclined to flatter; that hope seems to arise from a lower degree of probability than fear; and that consequently one is not entitled to substitute the measure of the one with the measure of the other: fear and hope are feelings not concrete things; it is possible, it is even more likely, that these feelings do not measure the precise degree of probability; and in that case, must one give them an equal measure or even assign them no measure?
Thereto I reply that the measure in question is not on feelings, but on the reasons that must originate them, and that any wise man must estimate the value of these feelings of fear or of hope only by the degree of probability; because even when Nature, for the happiness of man, had given him more

inclination towards hope than towards fear, it is not less true that probability is the true measure of the one and of the other. It is not even by the application of this measure that one can figure out one's false hopes, or to reassure on its unfounded fears.

Before finishing this article, I must observe that one must beware of mistaking what I called effects of unknown cause; because I understand only the effects of which the causes, although unknown, must be supposed constant, such as the natural effects; any new discovery in physics noted by thirteen or fourteen experiences, all of which are confirmed, already has a degree of equal certainty to the one of moral certainty, and this degree of certainty doubles with each new experience; so that by multiplying, one approaches more and more the physical certainty. But we must not conclude from this reasoning that the effects of chance follow the same law; it is true that in a sense these effects are among those for which we ignore the immediate causes; but we know that in general these causes are quite far from being able to be supposed constant; on the contrary, they are necessarily variable and volatile as much as it is possible. Thus even by the notion of chance itself, it is evident that there is no connection, no dependence, between its effects; that consequently the past cannot have any influence on the future, and one would be very much and even completely mistaken, if one wanted to infer from previous events, any reason for or against the posterior events. If a card, for example, has won three times in a row, it is not less probable that it will win a fourth time, and one can bet equally that it will win or that it will lose, no matter how many times it has won or lost, as long as the laws of the game are such that the chances are equal. Presuming or believing the opposite, as some players do, is to go against the principle of chance itself, or not to remember that by the conventions of the game it is always equally distributed.

X. In the effects for which we see the causes, a single test is sufficient to cause the physical certainty; for example, I see that in a clock the weight makes the wheels turn, and that the wheels make the pendulum go; I am certain from that time on without the need of reiterated experiences, that the pendulum will always go the same as long as the weight makes the wheels turn; this is a necessary consequence of an arrangement that we made ourselves while constructing the machine; but when we see a new phenomenon, an effect in the still unknown Nature, as we are ignorant about its causes, and they can be constant or variable, permanent or intermittent, natural or accidental, we do not have other means to obtain the certainty than by repeating the experience as often as necessary; nothing here depends on us and we only know as much as we experiment; we are only assured by the effect itself and by the repetition of the effect. As soon as it has occurred thirteen or fourteen times in the same way, we already have a degree of probability equal to the moral certainty that it will occur for even a fifteenth time, and from this point we soon can cover an immense interval, and conclude by analogy that this effect depends on the general laws of Nature; that it is consequently as old as all the other effects and that there is physical certainty that it always will occur as it always has occurred and it only needed having been observed.

In the risks that we arranged, balanced and calculated ourselves, one must not say that we are ignorant of the causes of the effects: we are ignorant of the true immediate cause of each effect in particular; but we clearly see the first and general cause of all the effects. I do not know, for example, and I even cannot imagine in any way, what is the difference of the movements of the hand, to pass or not to pass ten with three dice, which nevertheless is the immediate cause of the event, but I see evidently by the number and the make of the dice, which are here the main and general causes the chances are absolutely equal, you are indifferent to betting on passing or not passing ten; I see moreover that these same events, if they happen, have no connection, since to every throw of the dice the risk always is the same, and nevertheless always new; that the past throw cannot have any influence on the throw to come; that one always can equally bet for or against; finally that the longer one plays, the greater the number of effects for and the number of effects against, they will approach equality. Making sure that every experience here gives a outcome exactly opposite to that of the experience on the natural effects, I want to say, the certainty of the inconstancy instead of the one of the constancy of the causes; in this one, each test leads to the doubling of the probability of the replication of the effect, that is, the certainty of the constancy of the cause; in the effects of risk, on the contrary, each test increases the certainty of the inconstancy of the cause by showing us always more and more that it is absolutely volatile and totally indifferent to produce the one or the other of these effects. When a gamble is in its nature perfectly fair, the player has no reason to choose this or that side; since finally, from the supposed fairness of this game, it necessarily follows that there are no good reasons to prefer the one or the other side; and consequently if one deliberated, one could only be determined by the wrong reasons; also the logic of the players has seemed to me completely wrong, and even the good minds that allow themselves to play fall as players into absurdities at which they soon blush as reasonable men.

XI. Besides, all this supposes that after having balanced the risks and having them rendered equal, as in the game *passing ten* with three dice, these same dice which are the devices of the risk are as perfect as possible, that is, they are exactly cubic, the material is homogenous, the numbers are painted and not marked hollow, so that the weight on one face is not more than on any other; but as it is not given to man to make anything perfect, and there is no such dice made with this strict precision, it is often possible to recognize by observation on which side the imperfection of the devices of chance tips the risk. Thus, it is only necessary to attentively observe for a long time the sequence of events, to count them exactly, to compare their relative numbers; and if one of these two numbers exceeds by far the other, one will be able to conclude, with great reason, that the imperfection of the devices of chance destroys the perfect equality of the risk, and gives it actually a stronger inclination to one side than to the other. For example, I suppose that before playing *passing ten*, one of the players was subtle enough, or rather, rogue enough to having thrown in advance thousand times the three dice to be used, and to having recognized that in these thousand trials there were six hundred that passed ten, he will have thenceforth a great advantage over his opponent by betting on passing, since by experience the probability of passing ten with these same dice relates to the probability of not passing-ten as 600:400 or 3:2. This difference that stems from the imperfection of the devices can therefore be recognized by observation, and it is for this reason that the players often change dice and cards, when their luck is against them.

Thus, however obscure the destinies may be, however opaque the future may appear to us, we may nevertheless by reiterated experiences, become, in some cases, also enlightened about future events, as if we were beings or rather superior natures who immediately deduce the effects from their causes. And among the very things that seem to be pure risk, as games and lotteries, one can again recognize the inclination of the risk. For example, in a lottery drawn every fortnight, and in which one publishes the winning numbers, if one observes those that most often won during one year, two years, three consecutive years, one can deduct, with reason, that the same numbers will win again more often than the others; because in any way one can vary the motion and the position of the device of chance, it is impossible to render them perfect enough to maintain the absolute equality of the chance; there is a certain routine to do, to place, to mix the tickets, which even in the midst of confusion produces a certain order, and makes that certain tickets must come out more often than the others; it is the same with the arrangement of cards to play; they have a kind of sequence of which one can grasp some terms by force of observations; because while assembling them with the hand one follows a certain routine, the player himself by mixing them has his routine; the whole is done in a certain way more often than in another, and thenceforth the attentive observer to results collected in large number will always bet with great advantage that such card, for example, will follow such other card. I say that this observer will have a great advantage, because the *a priori* risks must be absolutely equal, the least inequality, that is, the least degree of higher probability has very great influences on the game, which is in itself only a multiplied and always repeated bet. If this difference recognized by the experience of the inclination of the risk was only one hundredth, it is evident that in a hundred throws the observer would gain his stake, that is, the sum that he risks every time; so that a player equipped with these dishonest observations, cannot fail to ruin in the long run all his opponents. But we will give a powerful antidote against the epidemic evil of the passion of play, and at the same time some preventives against the illusion of this dangerous art.

XII. It is generally known that the game is a greedy passion, where the habit is ruinous, but this truth has perhaps never been shown unless through a sad experience on which one did not reflect enough to correct it by conviction. A player whose wealth is exposed every day to the draws of chance, exhausts himself little by little and finally finds himself necessarily ruined; he attributes his losses only to the same risk that he blames for unfairness; he equally regrets what he lost as what he did not win; the greed and the false hope gave him claims on the goods of others; so humbled to find himself in necessity as afflicted to no longer having the means to satisfy his cupidity; in his despair he blames his ill-fated star, he does not imagine that this blind power, the fortune of the game, marches in truth by an indifferent and uncertain pace, but that to every step it nevertheless tends to a goal, and pulls to a certain end that is the ruin of those who try it; he does not see that the apparent indifference it has for the good or for the evil produces with time the necessity of evil, that a long random sequence is a fatal chain whose extension causes the misfortune; he does not feel that regardless of the hard tax of the cards and of the even harder tribute paid to the roguery of some opponents, he has passed his life making ruinous conventions; that finally the game by its very nature is a vicious contract in its principle, a harmful contract to each contractor in particular, and contrary to the good of any society.

This is not at all a speech of vague morals, these are precise truths of metaphysics that I subject to calculation or rather to the strength of reason; truths that I pretend to mathematically show to all those who have their minds clear enough and the imagination strong enough to combine without geometry and to calculate without algebra.

I will not speak about these games invented by the artifice and worked out by the avarice, where the chance loses a part of its rights, where the fortune can never balance, because it is invincibly entailed and always obligated to lean to one side, I want to say all these games where the chances unequally divide up offer at once an assured as dishonest gain, and leave the other only with a sure and shameful loss, as in *Pharaoh*, where the banker is only a roguish solicitor and the punter a fool whom one agreed not to mock.

It is in games in general, in the most equal game, and consequently the most honest one that I find a vicious essence, I include even in the name of game all the conventions, all the bets where one puts at risk a part of his goods to obtain a similar part of the goods of others; and I say that in general the game is a misunderstood pact, a disadvantageous contract to both parties, whose effect is to make the loss always greater than the gain; and to remove the good to add to the evil. The demonstration of it is as easy as evident.

XIII. Take two men of equal fortune, who, for example, each have one hundred thousand pounds of goods, and suppose that these two men play in one or more throws of the dice fifty thousand pounds, that is, half of their goods; it is certain that whoever wins increases his goods only by a third, and that whoever loses diminishes his by half; each of them had one hundred thousand pounds before the game, but after the event of the game, one will have one hundred fifty thousand pounds, that is, a third more than he had, and the other has only fifty thousand pounds, that is, half less of what he had; so the loss is a sixth part bigger than the gain; since there is this difference between the third and the half, the agreement to play the game is detrimental to both, and consequently essentially vicious.

This reasoning is not fallacious, it is true and exact, although one player only lost precisely what the other won; this numerical equality of the sum does not prevent the true inequality of the loss and gain; the equality is only apparent, and the inequality very real. The agreement both men make when betting half of their goods is equal to the effect of another agreement that nobody has ever decided to make, that would be the agreement to throw each the twelfth part of his goods into the sea. For one can show them, before they risk half of their goods, that the loss necessarily being a sixth greater than the gain, this sixth must be considered as a real loss, that can indifferently fall on the one or on the other, consequently it has to be divided equally.

If two men decided to play all their goods, what would be the effect of this agreement? One would only double his fortune, and the other would reduce his to zero; now which proportion is here between the loss and the gain? the same as between all and nothing; the gain of the one is only equal to a rather modest sum, and the loss of the other is numerically infinite, and morally so great that the work of a lifetime would maybe not suffice to regain his goods.

The loss is therefore infinitely greater than any gain when one plays all his goods; it is greater by a sixth part when one plays half of his good, it is greater by a twentieth part when one plays the quarter of his good; in one word, however small the portion of his fortune that one risks in the game, there is always more to lose than to gain; the agreement to play the game is thus a vicious contract that tends to ruin the two contract parties. This is a new truth, but a very useful one, that I wish to be known to all of those who, for greed or for laziness, spend their life gambling.

One has often asked why one is more sensitive to loss than to gain; one could not give to this question a fully satisfactory answer, unless one does not doubt the truth I have just presented; now the response is easy: one is more sensitive to loss than to gain, because in fact, while supposing them numerically equal, the loss is nevertheless always and necessarily greater than the gain; the feeling is generally an implicit reasoning only less clear, but often brighter, and always surer than the direct product of the reason. One feels that the gain does not give us as much pleasure as the loss causes us pain; this feeling is only the implicit result of the reasoning that I have just presented.

XIV. Money must not be estimated by its numerical quantity: if the metal, that is merely the sign of wealth, was wealth itself, that is, if the happiness or the benefits that result from wealth were proportional to the quantity of money, men would have reason to estimate it numerically and by its quantity, but it is barely necessary that the benefits that one derives from money are in just proportion with its quantity; a rich man of one hundred thousand Ecus income is not ten times happier than the man of only ten thousand Ecus; there is more than that what money is, as soon as one passes certain limits it has almost no real value, and cannot increase the well-being of its possessor; a man that discovered a mountain of gold would not be richer than the one that found only one cubic fathom.[14]

Money has two values both arbitrary, both conventions, where one is the measure of the benefits to the individual, and where the other determines the rate of well-being of the society; the first of these values has never been estimated other than in a very vague way; the second is suitable for a just estimation by the comparison of the quantity of money with the proceeds of the land and the labour of men.

To succeed in giving some precise rules on the value of money, I will examine special cases of which the mind grasps easily the combinations, and that, as examples, drive us by induction to the general estimation of the value of the money for the poor, for the rich, and even for the more or less wise.

For the man who in his budget, whatever it is, has only what is necessary, money has an infinite value; for the man who in his budget abounds in superfluous, money has almost no value anymore. But what is necessary, what is superfluous? I understand by necessary the *income which one is obliged to spend to live as one has always lived*; with this necessary income, one can have its comforts and even pleasures; but soon the habit creates needs; so, in the definition of the superfluous I will not account for any of the pleasures we are used to, and I say that the superfluous is the *income that can bring us new pleasures*; the loss of the necessary income is a loss that one feels infinitely, and when one risks a considerable part of this necessary income, the risk cannot be offset by any hope, however great one may suppose; on the contrary the loss of the superfluous income has limited

[14] [comment: old measure; one cubic fathom ≈6.12 m^3].

effects; and if even superfluous income is still more sensitive to the loss than to the gain, it is because in fact the loss is generally always greater than the gain, this feeling is based on the principle that reasoning had not developed, because the ordinary feelings are based on common concepts or on simple inductions; but the delicate feelings depend on exquisite and elevating ideas, and are in fact only the results of several combinations often too subtle to be clearly noticed, and almost always too complicated to be reduced to a reasoning that can prove them.

XV. The mathematicians who have evaluated games of chance, and whose research in this field deserves praise, have considered money only as a quantity susceptible to growth and diminution, without other value than the number; they have estimated by the numerical quantity of money the relations of the gain and loss; they have calculated the risk and hope relatively to this very numerical quantity. We consider here the value of the money from a different point of view, and through our principles we will give the solution to some embarrassing cases for ordinary calculation. The problem, for example, of the game of heads and tails, where one supposes that two men (Peter and Paul) play against each other at these conditions that Peter will throw a coin in the air as many times as it will be necessary till tails shows up, and if that occurs at the first throw, Paul will give him one Ecu; if that occurs only on the second throw, Paul will give him two Ecus; if that occurs only on the third throw, he will give him four Ecus; if that occurs only on the fourth throw, Paul will give eight Ecus; if that occurs only on the fifth throw, he will give sixteen Ecus, and so on always doubling the number of the Ecus: it is obvious that in this condition Peter can only win, and that his gain will be at least an Ecu, maybe two Ecus, maybe four Ecus, maybe eight Ecus, maybe sixteen Ecus, maybe thirty-two Ecus, etc. maybe five hundred twelve Ecus, etc. maybe sixteen thousand three hundred eighty-four Ecus, etc. maybe five hundred twenty-four thousand four hundred forty-eight Ecus, etc. maybe even ten million, hundred million, hundred thousand millions of Ecus, maybe finally an infinity of Ecus. Because it is not impossible to throw five times, ten times, fifteen times, twenty times, thousand times, hundred thousand times the coin without tails showing up. One asks therefore how much Peter must give to Paul to compensate him, or what amounts to the same, what sum is equivalent to the hope of Peter who can only win.

This problem was proposed to me for the first time by the blessed Mr. Cramer, the famous professor of mathematics at Geneva, on a trip that I made to this city in the year 1730; he told me that it was previously proposed by Mr. Nicolas Bernoulli to Mr. de Montmort, as in fact it is found on the pages 402 and 407 of the *Analysis of the games of chance* by this author: I dreamed this problem some time without finding the knot; I could not see that it was possible to make agreed mathematical calculations with common sense without introducing some moral considerations; and having expressed my ideas to Mr. Cramer,[15] he told me that I was right, and that he had also resolved this question by a similar approach; he showed me then his solution almost identical to the one printed later in 1738 in the

[15]It follows what I left then in writing to Mr. Cramer, and of what I kept the original copy. "Mr. de Montmort is satisfied to reply to Mr. Nic. Bernoulli that the equivalent is equal to the sum of the sequence 1/2, 1/2, 1/2, 1/2, etc. Ecus continued to infinity, i.e., $\infty/2$, and I do not believe in fact that one can dispute his mathematical calculation; but far from giving an infinite equivalent, there is no man of common sense who would want to give twenty pounds, or even ten.

The reason of this contradiction between the mathematical calculation and the common sense seems to me to consist in the relation between money and its resulting advantage. A mathematician in his calculation estimates the money only by its quantity, i.e., by its numerical value; but the moral man must estimate it otherwise and only by the advantages or the pleasure that it can obtain; it is certain that he must behave in this view, and to estimate the money only in proportion to the resulting advantages, and not relatively to the quantity that, beyond certain limits, could not at all increase his happiness; he would, for example, hardly be happier with thousand million than he would be with hundred, or with hundred thousand million more than with thousand million; thus beyond certain limits, he would make a very big mistake to risk his money. If, for example, ten thousand Ecus were all his goods, he would make an infinite mistake to risk them, and the more these ten thousand pounds will be an object in relation to it, the bigger will be the mistake; I believe therefore that his mistake would be infinite, when these ten thousand pounds were a part of his necessary income, that is, when these ten thousand pounds will be absolutely necessary for him to live, as he was raised and as he has always lived; if these ten thousand pounds are of his superfluous income, his mistake diminishes, and the more they will be a small part of his superfluous income the more will diminish his mistake; but it will never be zero, unless he can consider this part of his superfluous income with indifference, or that he only considers the expected sum necessary to succeed in a purpose that will give him in proportion as much of pleasure as this very sum is larger than the one that he risks, and it is by this way of foreseeing a future happiness, that one can not at all give rules, there are people for whom hope itself is a greater pleasure than they could obtain by the enjoyment of their stake; to reason therefore more certainly on all these things, one must establish some principles; I would say, for example, that necessary income is equal to the sum that one is obliged to spend to continue to live as one always lived; the necessary income of a King will be, for example, ten millions of revenues (because a king that had less would be a poor king); the necessary income of a nobleman, will be ten thousand pounds of revenue (because a decent man who has less would be a poor nobleman); the necessary income of a peasant will be five hundred pounds, because with less he would be in misery, he cannot spend less to live and to nourish his family. I would suppose that the necessary income cannot give us new pleasures, or to speak more exactly, I would not account for any of the pleasures or advantages that we always had, and according to that, I would define the superfluous income as what could bring us other pleasures or new advantages; I would further say, that the loss of the necessary income makes itself feel infinitely; thus that it cannot be compensated by any hope, that on the contrary the feeling of the loss of the superfluous income is limited, and that consequently it can be compensated; I believe that oneself feels this truth when one plays, because the loss, even if small, gives us always more pain than an equal gain gives us pleasure, and is that which leads to wounded pride, since I suppose the game is entirely one of pure chance. I would also say that the quantity of the money within the necessary income is proportional to what it returns to us, but that within the superfluous income this proportion begins diminishing, and diminishes the more the greater the superfluous income becomes.

I leave you, Mister, to judge these ideas, etc. Geneva, October 3 1730. *Signed*, Le Clerc de Buffon.

Memories of the Academy of Petersburg following an excellent work of Mr. Daniel Bernoulli on the measure of chances where I saw that most of the ideas of Mr. Dan. Bernoulli agreed with mine, which gave me great pleasure since I always have, in addition to his great talents in geometry, considered and acknowledged Mr. Dan. Bernoulli as one of the best minds of this century. I found also the idea of Mr. Cramer very justified, and worthy of a man who has given us proofs of his skill in all the mathematical sciences, and to whose memory I do justice with so much more pleasure than there is to the company and to the friendship of this scholar whom I owe part of the first knowledge that I acquired in this field. Mr. de Montmort gives the solution to this problem by the ordinary rules, and he says, that the equivalent sum to the hope of the one who can only win, is equal to the sum of the sequence 1/2, 1/2, 1/2, 1/2, 1/2, 1/2, 1/2 Ecu, etc. continued to infinity, and that consequently this equivalent sum is an infinite sum of money. The reason on which this calculation is based is that there is one half of probability that Peter who can only win, will have one Ecu; one quarter of probability that he will have two; one eight of probability that he will have four; one sixteenth of probability that he will have eight; one thirty-second of probability that he will have sixteen, etc. to infinity; and that consequently his hope for the first case is one half Ecu, because the hope measures itself by the probability multiplied by the amount to be obtained; now the probability is one half, and the sum to be obtained for the first throw is one Ecu; therefore hope is one half Ecu: equally his hope for the second case is again one half Ecu, because the probability is one quarter, and the sum to be obtained is two Ecus; now a quarter multiplied by two Ecus gives again one half Ecu. One will find likewise that his hope for the third case is again one half Ecu; for the fourth case one half Ecu, in a word for all the cases to infinity always one half Ecu each, since the number of Ecus grows in the same proportion as the number of the probabilities diminishes; therefore the sum of all these hopes is an infinite sum of money, and consequently it is necessary that Peter gives to Paul the equivalent half of an infinity of Ecus.

That is mathematically true and one cannot dispute this calculation; also Mr. de Montmort and the other geometricians considered this problem as well resolved; but this solution is so far from being the true one that instead of giving an infinite sum, or even a very large sum, which already is quite different, there is no man of common sense who would give twenty Ecus or even ten to buy this hope of putting himself in the place of the one who can only win.

XVI. The reason for this extraordinary contradiction to common sense and calculation comes from two causes, the first one is that the probability must be considered as zero as soon as it is very small, that is, below 1/ 10,000; the second cause is the relation between the quantity of money and its resulting benefits; the mathematician in his calculation estimates the money by its quantity, but the moral man must estimate it otherwise; for example, if one proposes to a man with a mediocre fortune to put one hundred thousand pounds in a lottery, because there is only one hundred thousand to bet against one that he will win hundred thousand times one hundred thousand pounds; it is certain that the probability to obtain hundred thousand times one hundred thousand pounds, being one against hundred thousand, it is certain, I say, mathematically speaking, that his hope will be worth his stake of one hundred thousand pounds; but this man would make a very big mistake to risk this sum, and an even bigger mistake, since the probability of winning is very small, although the money to win increases in proportion, and that because with hundred thousand times one hundred thousand pounds he will not have double the benefits that he will have with fifty thousand times one hundred thousand pounds, or ten times as much benefit as he will have with ten thousand times hundred thousand pounds; and as the value of money, for the moral man, is not proportional to its quantity, but rather to the benefits that money can buy, it is obvious that this man must risk only in proportion to the hope of these benefits, which he must not calculate by the numerical quantity of the sums that he could obtain, since the quantity of money, beyond certain limits, could not further increase his happiness, and he would not be happier with hundred thousand millions of income than with one thousand million.

XVII. To understand the connection between and the truth of all that I have just advanced, we examine more closely than the geometricians did, the question that has just been proposed; since the ordinary calculation cannot resolve it, because of the morale which is causing difficulties with the mathematics, let us see if other rules enable us to reach a solution which does not violate common sense, and which is at the same time in accordance with experience; this research will not be useless, and we will furnish the means to estimate exactly the price of money and the value of hope in all cases. The first thing I remark is that in the mathematical calculation that gives as equivalent to the hope of Peter an infinite sum of money, this infinite sum of money is the sum of a sequence composed of an infinite number of terms each worth one half Ecu, and I notice that this sequence that mathematically must have an infinity of terms, cannot morally have more than thirty, since if the game takes until this thirtieth term, that is, if tails shows up only after twenty nine throws, Peter will be due a sum of 520 million 870 thousand 912 Ecus, that is, as much money as exists perhaps in the whole kingdom of France. An infinite sum of money is a thing that does not exist, and all hopes based on the terms to infinity beyond thirty do not exist either. There is here a moral impossibility that destroys the mathematical possibility; because mathematically it is possible and even physically to throw thirty times, fifty, hundred times in a row, etc. the coin without tails showing up; but it is impossible to satisfy the condition of the problem,[16] is to be said, to pay the number of Ecus that are due, if that occurs; because any money on the earth would not suffice to give the due sum, only up to the fortieth throw, since that would require thousand twenty-four times more money than exists only in the whole kingdom of France, and that it would be necessary that on the whole earth there are one thousand and twenty four kingdoms as rich as France.

Now the mathematician has only found this infinite sum of money equivalent to the hope of Peter, because the first case gives him one half Ecu, the second case one half Ecu, and every case to the infinite one always one half Ecu; therefore the moral man, also counting in the same way at the beginning, will find twenty Ecus instead of the infinite sum, since all the terms that are beyond the fortieth one yield such great amounts of money that do not exist; so that it is necessary to count only one half Ecu for the first case, one half Ecu for the second, one half Ecu for the third one, etc. until forty, which gives in total twenty Ecus as equivalent for Peter's hope, a sum already well reduced and well different from the infinite sum. This sum of twenty Ecus will be reduced again a lot considering that the thirty-first term would give more than one thousand million Ecus, that is, it would suppose that Peter had a lot more money than the richest kingdom of Europe, an impossible thing to suppose, and thenceforth the terms between thirty and forty are again imaginary, and hopes based on these terms must be looked at as zero, thus the equivalent of Peter's hope, already is reduced to fifteen Ecus.

One will reduce it again considering that the value of money should not be estimated by its quantity, Peter must not count thousand millions of Ecus as if they served him to the double of five hundred millions Ecus, or to the quadruple of two hundred fifty millions of Ecus, etc. and that consequently the hope of the thirtieth term is not one half Ecu, nor the hope of the twenty ninth, of the twenty eighth, etc. the value of this hope, that mathematically finds itself to be one half Ecu for every term, must be diminished from the second term, and always diminished until the last term of the sequence; because one must not estimate the value of the money by its numerical quantity.

XVIII. But how to estimate it, how to find this value case by case for different quantities? What are accordingly two millions of money, if this is not the double of one million of the same metal? Can we give precise and general rules for this estimation? It seems that everyone must judge his state, and next estimate his utility and the quantity of money proportional to this state and to the usage he can make from it; but this approach is still vague and too

[16] For this reason, one of our most skilful geometricians, the blessed M. Fontaine, introduced in the solution he gave us to this problem the declaration of Peter's goods, because indeed he can give for equivalent only the totality of goods he owns. See this solution in the mathematical works of M. Fontaine, in-4.° Paris, 1764.

special for it to serve as a principle, and I believe that one can find more general methods and safer ones to do this estimation; the first method that presents itself is to compare the mathematical calculation with experience; because in many cases, we can by repeated experiences, reach, as I have said, knowing the effect of chance as surely as if we deduced it immediately from the causes.

I have therefore made two thousand and forty eight experiences on this question, that is, I played two thousand and forty eight times this game by letting a child throw the coin in the air; the two thousand and forty eight trials produced ten thousand and fifty seven Ecus in all, thus the equivalent sum to the hope of the one who can only win is almost five Ecus for every trial. In this experiment, there were one thousand and sixty trials that produced only one Ecu, four hundred and ninety four trials that produced two Ecus, two hundred and thirty two trials that produced four Ecus, one hundred and thirty seven trials that produced eight Ecus, fifty six trials that produced sixteen Ecus, twenty nine that produced thirty-two Ecus, twenty five trials that produced sixty four Ecus, eight trials that produced some one hundred and twenty eight, and finally six trials that produced two hundred and fifty six. I regard this result generally as good, because it is based on a large number of experiences, and besides it agrees with another mathematical and indisputable reasoning by which one finds almost the same equivalent of five Ecus. Here is the reasoning. If one plays two thousand and forty eight trials, there must be naturally one thousand and twenty-four trials that will produce only one Ecu each, five hundred and twelve trials that will produce two, two hundred and fifty six trials that will produce four, one hundred and twenty eight trials that will produce eight, sixty-four trials that will produce sixteen, thirty-two trials that will produce thirty-two, sixteen trials that will produce sixty-four, eight trials that will produce one hundred and twenty eight, four trials that will produce two hundred and fifty six, two trials that will produce five hundred and twelve, one trial that will produce one thousand and twenty four; and finally one trial that one cannot estimate, but that one can neglect without appreciable error, because I can suppose without violating more than slightly the equality of chance, that there were one thousand and twenty five instead of one thousand and twenty four trials that produced only one Ecu, besides the equivalent on this trial being placed at the most extreme cannot be more than fifteen Ecus, since one has seen that for a trial of this game all the terms beyond the thirtieth term of the sequence give sums of such great money that they do not exist, and that consequently the greatest equivalent one can suppose is fifteen Ecus. Adding together all these Ecus which I naturally must expect by the indifference of risk, I have eleven thousand two hundred and sixty five Ecus for two thousand and forty eight trials. Thus this reasoning gives very roughly five Ecus and one half as the equivalent which agrees with the experience to within 1/11. I feel, although one will be able to criticise me, that this type of calculation that gives five Ecus and one half for equivalent when one plays two thousand and forty eight trials, would give a greater equivalent, if one added a much larger number of trials; because, for example, it turns out that if instead of playing two thousand and forty eight trials, one only plays one thousand and twenty four, the equivalent is very roughly five Ecus; that if one plays only five hundred and twelve trials, the equivalent is no more than very roughly four Ecus and one half; that if one plays only two hundred and fifty six, it is no more than four Ecus, and thus always diminishing; but the reason for concern is in the trial outcome that one cannot estimate which makes a considerable part of the totality, and it is even much more considerable, if one plays less than that, and consequently a large number of trials is necessary, like one thousand and twenty four or two thousand and forty eight which are so large that this trial outcome can be looked at as negligible value, or even as zero. While following the same argument, one will find that if one plays one million and forty eight thousand five hundred and seventy six trials, the equivalent by this reasoning itself would be found to be almost ten Ecus; but one must consider all in the morale, and from there one will see that it is not possible to play one million and forty eight thousand five hundred and seventy six trials of this game, because even if one suppose only 2 min of time for every trial, including the time that it is necessary to pay, etc. one would find that it was necessary to play during two million ninety seven thousand one hundred and fifty two minutes, that is, more than 13 years of sequence, 6 h a day, which is morally impossible. And if one pays attention, one will find that between playing only one trial and playing the largest morally possible number of trials, this reasoning that gives different equivalents for all the different numbers of trials, yields an average equivalent of five Ecus. Thus I still say that the equivalent sum to the hope of the one who can only win is five Ecus instead of the half of an infinite sum of Ecus, as said by the mathematicians, and as their calculation seems to demand.

XIX. Let us see if according to this estimation it would not be possible to derive the ratio of the money value that corresponds to the benefits resulting from it.

The progression of the probabilities is	$\frac{1}{2}$	$\frac{1}{4}$	$\frac{1}{8}$	$\frac{1}{16}$	$\frac{1}{32}$	$\frac{1}{64}$	$\frac{1}{128}$	$\frac{1}{256}$	$\frac{1}{512}$	$\frac{1}{2^\infty}$
The progression of the money amounts to win is	1	2	4	8	16	32	64	128	256	2^∞.

The sum of all these probabilities, multiplied by the one of the money at stake is $\frac{\infty}{2}$, that is the equivalent data to the mathematical calculation for the hope of the one who can only win. But we saw that this sum $\frac{\infty}{2}$ can, in reality, only be five Ecus; it is necessary therefore to look for a sequence, such that the sum multiplied by the sequence of probabilities equals five Ecus, and this sequence being geometric as that of the probabilities, one will find that it is 1, $\frac{9}{5}$, $\frac{81}{25}$, $\frac{729}{125}$, $\frac{6{,}561}{625}$, $\frac{59{,}049}{3{,}125}$, instead of 1, 2, 4, 8, 16, 32.

Now this sequence 1, 2, 4, 8, 16, 32, etc. represents the quantity of money, and consequently its numerical and mathematical value.
And the other sequence 1, $\frac{9}{5}$, $\frac{81}{25}$, $\frac{729}{125}$, $\frac{6{,}561}{625}$, $\frac{59{,}049}{3{,}125}$, represents the geometric quantity of money given by the experience, and consequently its moral and real value.
Here is therefore a general estimation sufficiently close to the value of money for all possible cases, and independent of any assumption. For example, one sees, comparing the two sequences, that two thousand pounds do not produce double the benefit of one thousand pounds, that $\frac{1}{5}$ is being discounted, and that two thousand pounds are in the morale and in reality only $\frac{9}{5}$ of two thousand pounds, that is, eighteen hundred pounds. A man who has twenty thousand pounds of goods must not estimate it as the double of the goods of another one who has ten thousand pounds, because he actually does only have eighteen thousand pounds of money of this same currency whose value is calculated by the resulting benefits; and a man who even has forty thousand pounds is not four times richer than the one who has ten thousand pounds, because in this real currency he has only 32 thousand and 400 pounds; a man who has 80 thousand pounds, has, by the same currency, only 58 thousand 300 pounds; the one who has 160 thousand pounds, must only count 104 thousand 900 pounds, that is, that although he has sixteen times more than the first one, he hardly has only ten times as much of our real currency. Or even again a man who has thirty-two times as much of money as another, for example 320 thousand pounds in comparison to a man who has 10 thousand pounds, his wealth in the real currency is only 188 thousand pounds, that is, he is eighteen or nineteen times richer, instead of thirty-two times, etc. The miser is like the mathematician; both of them esteem the money by its numerical quantity, the sensible man considers neither the mass nor the number, he sees only the benefits that he can derive from it; he reasons better than the miser, and discerns better than the mathematician. The Ecu that the poor has set aside to pay a necessary tax and the Ecu that completes the bags of the financier have for the miser and for the mathematician just the same value, the latter will count them by two equal quantities, the other will take it from them with equal pleasure, the sensible man instead will count the Ecu of the poor for a Louis, and the Ecu of the financier for a Liard.

XX. Another consideration that comes to the support of this estimation of the moral value of money is that a probability must be regarded as zero as soon as it is only $\frac{1}{10{,}000}$, that is, as soon as it is as small as the unfelt fear of death in the 24 h. One can even say that given the intensity of this fear of death which is considerably larger than the intensity of all the other feelings of fear or of hope, one must consider as almost zero, a fear or a hope that has only $\frac{1}{1{,}000}$ of probability. The weakest man could draw the lots without any emotion, if the ticket of death was mixed with ten thousand tickets of life; and the strong man must draw without fear, if this ticket is mixed with a thousand; thus in any case where the probability is under a thousandth, one must look at it as almost zero. Now, in our question, the probability is found to be $\frac{1}{1,024}$ as early as the tenth term of the sequence $\frac{1}{2}, \frac{1}{4}, \frac{1}{8}, \frac{1}{16}, \frac{1}{32}, \frac{1}{64}, \frac{1}{128}, \frac{1}{256}, \frac{1}{512}, \frac{1}{1,024}$, it follows that morally thinking, we must neglect all the following terms, and limit all our hopes at this tenth term; which again produces five Ecu as the desired equivalent, and consequently confirms the accuracy of our estimation. By rearranging and cancelling all the calculations where the probability becomes smaller than one thousandth, there will no longer remain any contradiction between the mathematical calculation and common sense. All difficulties of this kind disappear. The man pervaded by this truth will not dedicate more to vain hopes or to false fears; he will not gladly give his Ecu to obtain a thousand, unless he does clearly see that the probability is greater than one thousandth. Finally, he will correct his frivolous hope of making a great fortune with small means.

XXI. So far I have only reasoned and calculated for the truly wise man who is determined only by the weight of reason; but must we not give some attention to the great number of men whom illusion or passion deceive, and who are often quite comfortable with being deceived? Is there yet nothing to lose if things are always presented just as they are? Is hope, however small the probability, not a good for all men, and the only good for the unfortunate ones? After having calculated for the wise man, we therefore calculate also for the much less rare man who enjoys his mistakes often more than his reason. Regardless of cases where for lack of all means, a glimmer of hope is a supreme good; regardless of these circumstances where the restless heart can only rest on the objects of its illusion, and only enjoys its desires; are there not thousands and thousands of occasions where even wisdom must throw forward a volume of hope in the absence of real evidence? For example, the desire to do good, recognized in those who hold the reins of government, albeit without exercise, spreads on all the people a sum of happiness one cannot estimate; hope, albeit vain, is therefore a real good, the enjoyment is taken by anticipation of all other goods. I am forced to admit that full wisdom does not imply full happiness of man, that unfortunately the mere reason had at any time only a small number of cold listeners, and never created enthusiasts; that man stuffed with goods would not yet be happy if he did not hope for new ones; that the superfluous income becomes with time a very necessary thing, and that the mere difference between the wise and the non-wise is that the latter, at the very moment he reaches an overabundance of goods, converts this lovely superfluous income into the sad necessary income, and raises his state equal to that of his new fortune; while the wise man would use this overabundance only to spread the benefits and to obtain himself some new pleasures, sparing the consumption of the superfluous income at the same time as multiplying the enjoyment of it.

XXII. The display of hope is the business of all money swindlers. The big art of the lottery maker is to present large sums with very small probabilities, soon swollen by the spring of greed. These swindlers still enlarge this ideal product, dividing it and giving it for very little money, which everyone can lose, a hope that, though much smaller, seems to be a part of the grandeur of the total sum. One does not know that when the probability is below one thousandth, hope becomes zero however large the promised sum, since anything, however great it may be, is necessarily reduced to nothing as soon as it is multiplied by nothing, as is here the large sum of money multiplied by the probability of zero, as in general any number, multiplied by zero, is always zero. One is unaware again that independently of this reduction of the probabilities to nothing, as soon as they are below one thousandth, hope suffers a successive and proportional decline of the moral value of money, always less than its numerical value, so that hope which numerically appears to double from one to another, has nevertheless only $\frac{9}{5}$ of the real hope instead of 2; and correspondingly the one whose numerical hope is 4 does only have $3\frac{6}{25}$ of this moral hope, this outcome being the mere reality. Instead of 8, this product is only $5\frac{104}{125}$; instead of 16, it is only $10\frac{311}{625}$; instead of 32, $18\frac{2,799}{3,125}$; instead of 64,$34\frac{191}{15,625}$; instead of 128, $61\frac{17,342}{78,625}$; instead of 256, $110\frac{77,971}{390,625}$; instead of 512, $198\frac{701,739}{1,953,125}$; instead of $1,024,357\frac{456,276}{9,765,625}$, etc. thus one sees how much the moral hope differs in each case from the numerical hope for the real outcome that results from it; the wise man must therefore reject as false all the propositions, though proven by calculation, where the very large quantity of money seems to compensate the very small probability; and if he wants to risk with less disadvantage, he must never allocate his funds to the large venture, it is necessary to divide them. To risk one hundred thousand francs on a single vessel or twenty-five thousand francs on four vessels is not the same thing, because one will have one hundred for the outcome of moral hope in the latter case, while one will have only eighty-one for the same outcome in the first case. It is by this same reason that the most surely profitable businesses are those where the mass of the debt is divided between a large number of creditors. The owner of the mass must suffer only light bankruptcies instead of the one that ruins him when this mass of his business can be placed only with a single hand or similarly only be divided between a small number of debtors. Playing for high stakes in the moral sense is to play a bad game; a punter in the game of pharaoh who gets into his head to push all his cards until fifteen would lose close to a quarter on the moral product of his hope, because while his numerical hope is to pull 16, his moral hope is only $13\frac{104}{125}$. It is the same for countless other examples that one could give; and in all it will always result that the wise man must put at risk the least possible, and that the prudent man who, through his position or his business, is forced to risk large funds, must divide them, and subtract from his speculations the hopes for which the probability is very small, albeit the sum to obtain is proportionally also large.

XXIII. Analysis is the only instrument that has been used up-to-now in the science of probabilities to determine and to fix the ratios of chance; Geometry appeared hardly appropriate for such a delicate matter; nevertheless if one looks at it closely, it will be easily recognized that this advantage of Analysis over Geometry is quite accidental, and that chance according to whether it is modified and conditioned is in the domain of geometry as well as in that of analysis; to be assured of this, it is enough to see that games and problems of conjecture ordinarily revolve only around the ratios of discrete quantities; the human mind, rather familiar with numbers than with measurements of size, has always preferred them; games are a proof of it because their laws are a continual arithmetic; to put therefore Geometry in possession of its rights on the science of chance is only a matter of inventing some games that revolve on size and on its ratios or to analyse the small number of those of this nature that are already found; the clean-tile [franc-carreau] game can serve us as example: here are its very simple terms.

In a room parquetted or paved with equal tiles of an unspecified shape one throws an Ecu in the air; one player bets that after its fall this Ecu will be located clean-tile, that is, on a single tile; the second bets that this Ecu will be located on two tiles, that is, it will cover one of the joints which separates them; a third player bets that the Ecu will be located on two joints; a fourth one bets that the Ecu will be located on three, four or six joints: one asks for the chances of each of these players.

To begin with, I seek the chances of the first and second player; to find them, I inscribe in one of the tiles a similar figure, distant from the tile borders, by the length of half the diameter of the Ecu; the chances of the first player will relate to the one of the second as the area of the circumscribed annulus relates to the area of the inscribed figure; that can be easily shown, because as long as the centre of the Ecu is in the inscribed figure, this Ecu can be located only on a single tile, since by construction this inscribed figure is everywhere distant from the contour of the tile, of a distance equal to the radius of the Ecu; and in contrast, as soon as the centre of the Ecu falls outside of the inscribed figure, the Ecu is necessarily located on two or several tiles, since then its radius is greater than the distance of the contour of this inscribed figure from the contour of the tile; and yet, all points where this centre of the Ecu may fall are represented in the first case by the area of the annulus which is the remainder of the tile; therefore the chances of the first player relate to the chances of the second, as this first area relates to the second; thus to render equal the chances of these two players, it is necessary that the area of the inscribed figure is equal to the one of the annulus, or equivalently, that it is the half of the total surface of the tile.

I enjoyed to do the calculation, and I found that to play a fair game on two square tiles, the border length of the tile must relate to the diameter of the Ecu as $1 : 1 - \sqrt{1/2}$; that is, almost three and half times greater than the diameter of the coin with which one plays.
To play on triangular equilateral tiles, the border length of the tile must relate to the diameter of the coin as $1 : \frac{(1/2)\sqrt{3}}{3+3\sqrt{1/2}}$, that is, almost six times greater than the diameter of the coin.
On lozenge tiles, the border length of the tile must relate to the diameter of the coin as $1 : \frac{(1/2)\sqrt{3}}{2+\sqrt{2}}$, that is, almost four times greater.
Finally, on hexagon tiles, the border length of the tile must relate to the diameter of the coin as $1 : \frac{(1/2)\sqrt{3}}{1+\sqrt{1/2}}$, that is, almost double.
I have not done the calculation for other figures, because these are the only ones by which one can fill a space without leaving some intervals for other figures; and I do not think it is necessary to tell that if the joints of the tiles have some width, they give advantage to the player who bets on the joint, and that consequently one will do well to render the game even more equal by giving to the square tile a little more than three and half times, to the triangular six times, to the lozenges four times, and to the hexagons two times the length of the diameter of the coin with which one plays.
Now I seek the chances of the third player who bets that the Ecu will be located on two joints; and to find it, I inscribe in one of the tiles a similar figure as I have already done, next I extend the inscribed borders of this figure until they meet those of the tile, the chances of the third player will relate to the one of his opponent as the sum of the spaces enclosed between the extension of these lines and the borders of the tile relates to the remainder of the surface of the tile. To be fully shown this requires only to be well understood.
I have also done the calculation for this case, and I found that to play a fair game on square tiles, the border length of the tile must relate to the diameter of the coin as $1 : 1/\sqrt{2}$, that is, greater than by a little less than a third.
On triangular equilateral tiles, the border length of the tile must relate to the diameter of the coin as $1 : 1/2$, that is, double.
On lozenge tiles, the border length of the tile must relate to the diameter of the coin as $1 : \frac{(1/2)\sqrt{3}}{\sqrt{2}}$, that is, greater by about two-fifth.
On hexagon tiles, the border length of the tile must relate to the diameter of the coin as $1 : (1/2)\sqrt{3}$, that is, greater by a half-quarter.
Now, the fourth player bets that on triangular equilateral tiles the Ecu will be located on six joints, that on square or on lozenge tiles it will be located on four joints, and on hexagon tiles it will be located on three joints; to determine his chances, I describe around the vertex of the tile a circle equal to the Ecu, and I say that on triangular equilateral tiles, his chances will relate to the one of his opponent as the half of the area of this circle relates to the one of the rest of the tile; that on square or on lozenge tiles, his chances will relate to the one of the other as the entire area of the circle

relates to the one of the rest of the tile; and that on hexagon tiles, his chances will relate to the one of his opponent as the double of this area of the circle relates to the remainder of the tile. While supposing therefore that the circumference of the circle relates to the diameter as 22 relates to 7; one will find that to play a fair game on triangular equilateral tiles, the border length of the tile must relate to the diameter of the coin as $1 : \frac{\sqrt{7}\sqrt{3}}{22}$, that is, greater by a little more than a quarter.

On lozenge tiles, the chances will be the same as on triangular equilateral tiles.

On square tiles, the side of the tile must relate to the diameter of the coin, as $1 : \frac{\sqrt{11}}{7}$, that is, greater by about one fifth. On hexagon tiles, the border length of the tile must relate to the diameter of the coin as $1 : \frac{\sqrt{21}\sqrt{3}}{44}$, that is, greater by about one thirteenth.

I omit here the solution of several other cases, like when one of the players bets that the Ecu will fall only on one joint or on two, on three, etc., these are not more difficult than the preceding ones; and besides, one plays rarely this game under different conditions than the mentioned ones.

But if instead of throwing in the air a round piece, as an Ecu, one would throw a piece of another shape, as a squared Spanish pistole, or a needle, a stick, etc. the problem demands a little more geometry, although in general it is always possible to give its solution by space comparisons, as we will show.

I suppose that in a room where the floor is simply divided by parallel joints one throws a stick in the air, and that one of the players bets that the stick will not cross any of the parallels on the floor, and that the other in contrast bets that the stick will cross some of these parallels; one asks for the chances of these two players. One can play this game on a checkerboard with a sewing needle or a headless pin.

To find it, I first draw between the two parallel joints on the floor, *AB* and *CD*,[17] two other parallel lines *ab* and *cd*, at a distance from the primary ones of half the length of the stick *EF*, and I evidently see that as long as the middle of the stick is between these second two parallels, it never will be able to cross the primary ones in whatever position *EF*, *ef*, it may be located; and as everything that can occur above *ab* alike occurs below *cd*, it is only necessary to determine the one or the other; that is why I notice that all the positions of the stick can be represented by one quarter of the circumference of the circle of which the stick length is the diameter; letting therefore $2a$ denote the distance *CA* of the floor-joints, c the quarter of the circumference of the circle of which the length of the stick is the diameter, letting $2b$ denote the length of the stick, and f the length *AB* of the joints,

[17][comment: see Fig. 15.1].

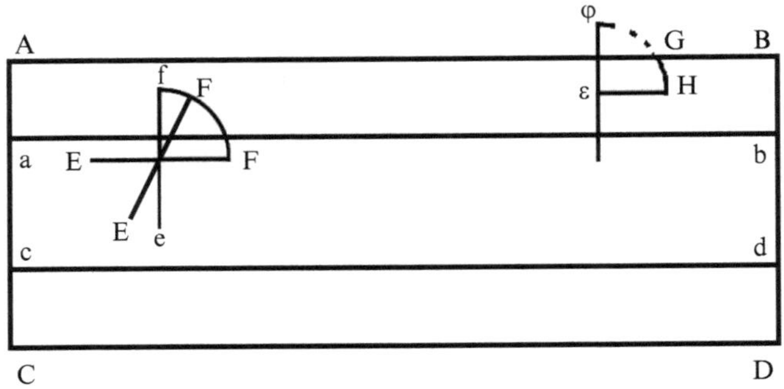

Fig. 15.1 Buffon's needle

I will have $f\overline{(a-b)c}$ as the expression that represents the probability of not crossing the floor-joint, or equivalently, as the expression of all cases where the middle of the stick falls below the line ab and above the line cd. But when the middle of the stick falls outside the space $abcd$, enclosed between the second parallels, it can, depending on its position, cross or not cross the joint; so that the middle of the stick being located, for example, in ε, the arch φG represents all the positions where it will cross the joint, and the arch GH all those where it will not cross, and as it will be the same for all the points on the line $\varepsilon\varphi$, I denote by dx the small parts of this line, and γ the arches of the circle φG, and I have $f(sin\gamma dx)$ as the expression of all the cases where the stick crosses, and $f\overline{(bc - sin\gamma dx}$ as the expression of the cases where it does not cross; I add this last expression to the one noticed above $f\overline{(a-b)c}$ in order to have the entirety of the cases where the stick will not cross, and from that time I see that the chances of the first player relates to the one of the second as $ac - sin\gamma dx : sin\gamma dx$.

If one wants therefore that the game is fair, one will have $ac = 2sin\gamma dx$ or $a = \frac{sin\gamma dx}{(1/2)c}$, that is, equal to the area of a part of the cycloid whose generating circle has as diameter the stick-length $2b$; now, one knows that this cycloid-area is equal to the square of the radius, therefore $a = \frac{b^2}{(1/2)c}$, that is, the length of the stick must be almost three quarters of the distance of the floor-joints.

The solution of this first case drives us easily to the one of another which before would have appeared more difficult, that is to determine the chances of these two players in a paved room of square tiles, because by inscribing in one of the square tiles, a square distant from all borders of the tile by the length b, first one has $c\overline{(a-b)^2}$ as the expression for one part of the cases where the stick does not cross the joint; next one finds $\overline{(2a-b)sin\gamma dx}$ for those of all the cases where it does cross, and finally $cb\overline{(2a-b)} - \overline{(2a-b)sin\gamma dx}$ for the rest of the cases where it does not

cross; thus the chances of the first player relates to the one of the second as $c\overline{(a-b)^2} + cb\overline{(2a-b)} - \overline{(ca-b)sin\gamma dx} : \overline{(2a-b)sin\gamma dx}$.[18]
If one wants therefore that the game be fair, one has $c\overline{(a-b)^2} + cb\overline{(2a-b)} = \overline{(2a-b)^2 sin\gamma dx}$. or $\frac{(1/2)ca^2}{2a-b} = sin\gamma dx$; but as we have seen above, $sin\gamma dx = b^2$; therefore $\frac{(1/2)ca^2}{2a-b} = b^2$; thus the border length of the tile must relate to the length of the stick, almost as $\frac{41}{22} : 1$, that is, not completely double. If one played therefore on a checkerboard with a needle of half the length of the side-length of the squares on the checkerboard, it is advantageous to bet that the needle will cross the joints.
One finds by a similar calculation that if one plays with a squared money coin, the sum of the chances will relate to the chances of the player that bets on the joint as $a^2c : 4ab^2\sqrt{1/2} - b^3 - (1/2)Ab$, where A denotes the excess of the area of the circle circumscribed around the square, and b the semi-diagonal of this square.
These examples suffice to give an idea of the games that one can imagine on the relationships of size; one could propose several other problems of this type, which do not cease to be interesting and even useful: if one asked, for example, how much one risks passing a river on a more or less narrow plank; what must be the fear one must have of lightning or of a bomb drop, and a number of other problems of conjecture where one must consider only the ratio of the size, and that consequently belong to geometry as much as to analysis.

XXIV. From the first steps that one takes in Geometry, one finds Infinity, and since the earliest times the Geometricians caught sight of it; the squaring of the parable and the treatise on *Numero Arenae* by Archimède prove that this great man gave thought to infinity, and even some of his thoughts we must share; we have extended his ideas, though handled in different ways, finally we have found the art of applying calculus to his ideas: but the basis for the metaphysics of the infinite has not changed at all, and it has only been recently that some Geometricians gave us views on infinity that are different from those Ancients, and so far from the nature of the things and truth, that they were even neglected in the Works of these great Mathematicians. Hence arose all the oppositions, all the contradictions that one suffered in calculus; hence arose the disputes between the Geometricians on how to make this calculation, and on the principles from which it derives; one was surprised by the kinds of miracles this calculation produced, this surprise was followed by confusion; it was believed that infinity produced all these wonders; it was imagined that the knowledge of infinity had been refused in

[18][comment: Buffon obviously means $c(a-b)^2 + cb(2a-b) - (2a-b)\sin\gamma dx : (2a-b)\sin\gamma dx$, and in consistence with his earlier notation, one should have the term $(2a-b)\sin\gamma\, dx$ without cap for the cases where it does cross].

all the centuries and has been reserved for ours; finally it was built on systems that have only served to obscure thought. Let us say therefore a few words on the nature of infinity, which while enlightening seems to have blinded men.

We have clear ideas about the magnitude, we see that things in general can be augmented or diminished, and the idea that a thing becomes larger or smaller is an idea as present and as familiar to us as the one about the thing itself; anything being thus presented to us or being only imagined, we see that it is possible to augment or diminish it; nothing stops, nothing destroys this possibility, one can always conceive the half of the smallest thing and the double of the largest thing; one even can conceive that it can become one hundred times, one thousand times, one hundred thousand times smaller or larger; and it is this possibility of growth without limits in what consists the true concept one must have on infinity; this concept is derived by us from the concept of the finite; a finite thing is a thing that has ends, limits; an infinite thing is only the same finite thing on which we remove these ends and limits; the concept of infinity is thus only a concept of deprivation, and has nothing of a real object. Here is not the place to show that space, time, duration, are no infinite realities; it will suffice to prove that there exists no number currently infinite or infinitely small, or larger or smaller than an infinite one, etc.
Numbers are only an assembly of units of the same kind; the unit is not at all a number, the unit designates a single thing in general; but the first number 2 denotes not only two things, but again two similar things, two things of same kind; it is the same for all the other numbers: yet these numbers are only representations, and never exist independently of the things that they represent; the characters that they designate do not give them any reality at all, they require a subject or rather an assembly of subjects to represent, to make their existence possible; I understand their intelligible existence, because they can only have real values; now an assembly of units or of subjects can never be other than finite, that is, that one always will be able to assign the parts of which it is composed; consequently numbers cannot be infinite whatever the growth one gives them.
But, one will say, the last term of the natural sequence 1, 2, 3, 4, etc. is it not infinite? Are there no last terms of other even more infinite sequences than the last term of the natural sequence? It seems that in general the numbers have to become infinite in the end, but can they still grow? Thereto I reply, that this growth to which they are susceptible proves evidently that they can not at all be infinite; I say further that in these sequences there is no last term; that only even to suppose them a last term, is to destroy the quintessence of the sequence that consists in the succession of the terms that can be followed by other terms, and these other terms again by others; but all are of the same nature as the preceding ones, this is to say, all finite, all

composed of units; thus when one supposes that a sequence has a last term, and that the latter term is an infinite number, one contradicts to the definition of the number and to the general law of sequences.

Most of our errors in metaphysics come from the reality that we give to ideas of deprivation; we know the finite, we recognize its real properties, we examine it, and by considering it after this examination we do not recognize it anymore, and we believe having created a new being, while we did only destroy some part of what we formerly had known.

One must therefore consider infinity, in small, in large, only as a deprivation, an entrenchment of the concept of the finite, which can be used like an assumption that, in some case, can help to simplify the concepts, and must generalize their results in the practice of the Sciences; thus all the art is reduced to capitalizing on this assumption, trying to apply it to the subjects that one considers. The whole merit is therefore in the application, in a word in the use one makes of it.

XXV. All our knowledge is based on relationships and comparisons, everything is therefore relation in the Universe; and hence everything is subject to measure, even our ideas being all relative have nothing absolute. There are, as we have explained, different degrees of probabilities and of certainty. And even the evidence is more or less clear, more or less intense, according to the different aspects, that is, according to the relationships under which it is presented; the truth transmitted and compared by different minds appears under more or less large relationships, since the result of the affirmation or negation of a proposition by all men in general seems to still give weight to the truths that are best shown and most independent of any convention.

The properties of matter that appear to us evidently distinct from each other have no relation between them; size cannot be compared to gravity, inscrutability not to time, movement not to surface, etc. These properties have in common only the underlying subject, and that gives them their being; each of these properties considered separately asks therefore for a measure of its kind, that is, a measure different from all the others.

References

Bernoulli N (1728) Letter #9 to Gabriel Cramer. In: van der Waerden BL (ed) Die Werke von Jakob Bernoulli 3, K 9. Birkhäuser Verlag, Basel, pp 562–563

Bernoulli D (1738) Specimen theoriae novae de mensura sortis. Comentarii Academiae Scientarium Imperialis Petropolitanae V, 175–192, 1738; English translation 1954 by L. Sommer with footnotes by Karl Menger, Exposition of a new theory on the measurement of risk. Econometrica 22(1):23–36

Buffon GLC (1777a) Probabilités de la Durée de la Vie. In: Oeuvres Complètes de Buffon, Tome III Nouveau Edition, 95–268, 1829

Buffon GLC (1777b) Essais d'Arithmetique Morale. In: Oeuvres Complètes de Buffon, Tome III Nouveau Edition, 338–405, 1829
Feller W (1945) Note on the law of large numbers and 'Fair' games. Ann Math Stat 16:301–304
Gigerenzer G, Selten R (2001) Bounded rationality: the adaptive toolbox. MIT Press, Cambridge
Hey JD (1991) Experiments in economics. Blackwell, Oxford
Lopez LL (1987) Between hope and fear: the psychology of risk. In: Berkowitz L (ed) Advances in experimental social psychology, vol 20. Academic, San Diego, CA, pp 255–295
Menger K (1934) Das Unsicherheitsmoment in der Wertlehre. Z Natl Okon 51:459–485
Neugebauer T (2010) Moral impossibility in the Peterburg gamble: a literature survey and experimental evidence. LSF working papers, University of Luxembourg
Sauermann H, Selten R (1967) Zur Entwicklung der Experimentellen Wirtschaftsforschung. In: Sauermann H (ed) Beiträge Zur Experimentellen Wirtschaftsforschung. Paul Siebeck, Tübingen, pp 1–8
Selten R (1988) Models of strategic rationality, Theory and decision library, Series C: Game theory, mathematical programming and operations research. Kluwer, Dordrecht
Selten R (1998) Axiomatic characterization of the quadratic scoring rule. Exp Econ 1(1):43–61
Selten R (1999) Game theory and economic behavior: selected essays, vol 2. Edward Elgar, Cheltenham-Northhampton
Selten R (2004) Learning direction theory and impulse balance equilibrium. In: Cassar A, Friedman D (eds) Economics lab: an intensive course in experimental economics. Routledge, NY, pp 133–140
Selten R, Buchta J (1994) Experimental sealed-bid first price auctions with directly observed bid functions. Discussion Paper B-270, University of Bonn. In: Budescu D, Erev I, Zwick R (eds) Games and human behavior: essays in the Honor of Amnon Rapoport. Lawrenz Associates, Mahwah, NJ, 1999.
Selten R, Chmura T (2008) Stationary concepts for experimental 2x2 games. Am Econ Rev 98(3):938–966
Selten R, Neugebauer T (2006) Individual behaviour of first-price auctions: the importance of feedback information in experimental markets. Games Econ Behav 54:183–204
Selten R, Sadrieh A, Abbink K (1999) Money does not induce risk neutral behavior, but binary lotteries do even worse. Theory Decis 46:211–249
Taleb N (2007) The Black Swan: the impact of the highly improbable. Random House, New York
Tversky A, Kahneman D (1992) Advances in prospect theory: cumulative representation of uncertainty. J Risk Uncertain 5:297–323
Weil F (1961) La correspondance Buffon-Cramer. Rev Hist Sci (Paris) 14(2):97–136

Chapter 16
Risky Choice and the Construction of Preferences

Abdolkarim Sadrieh

Introduction

Imagine your doorbell rings. You open to find a young professional salesperson offering you a new type of electronic data loss insurance. You may wish to send him away immediately, but he talks so fast that you cannot help hearing the details of the deal. The contract will pay you a certain amount, if your electronic data storages are lost due to power blackouts or irregularities, due to fluid damages (water, coffee, soft drinks, etc.), or due to other unforeseen accidents or natural disasters. The insurance agent has a certified table listing the historic frequencies of all these events, i.e. you have reliable estimates of the event probabilities. The agent also quotes a price (or perhaps a number of prices for variants of the contract, insuring more or less of the data loss events). All you have to do now is to assess the value of your electronic data, match it up with your risk preferences, compare cost to benefit, relate the net benefit of the insurance contract to all other prospects that life offers, and you can immediately tell the friendly salesperson, whether the suggested insurance contract is a deal or is no deal. This should not be a big deal, right?

This paper is an attempt to answer the question above from three different perspectives: from the perspective of a classical economist, from the perspective of a decision behavior researcher, and from Reinhard Selten's perspective. As we move down that line, we find that the answer goes from a light-hearted response by the classical economist (something like: "Making that decision is a simple task and not a big deal at all.") to a contemplative and finger-waving reply by Reinhard Selten (something like: "How such decisions are made is a fundamental mystery of economic behavior!"). The reason for this disparity in responses lies in a subtle, but important detail of the way, we see our mission as economists. While classical

A. Sadrieh (✉)
Fakultät für Wirtschaftswissenschaft, Otto-von-Guericke-Universität Magdeburg, Postfach 41 20 Magdeburg 39106, Germany
e-mail: sadrieh@ovgu.de

A. Ockenfels and A. Sadrieh (eds.), *The Selten School of Behavioral Economics*, DOI 10.1007/978-3-642-13983-3_16,

economists – even those who resolutely refer to themselves as "behavioral economist" – are merely interested in the way choices are made *given* a complete and stable preference ordering, a behavioral decision researcher (usually a scholar in marketing or economic psychology) is also interested in the *construction of preferences* from the parameters of the decision environment (Cubitt et al. 2001). Reinhard Selten's scientific curiosity goes even one step further, not simply seeking the (statistically) best mapping from the decision environment parameters to the observed choice behavior, but trying to uncover the mental processes that govern decision making. Reinhard Selten believes that these *decision emergence* processes (Selten 1990, 1991) are only partially non-routine choices that result from conscious deliberations. And even the deliberate choices, he believes, are not based on the optimization of an objective function, but on adaptation of aspirations to the perceived environment (Selten 1998a, b).

After having discussed the intricate interplay between ideas of the "formula believers" (classical economists), the "psychological phenomenonists" (decision behavior researchers), and the "full-fledged constructionist" (Reinhard Selten), we will turn to some of methods that may be helpful tools for uncovering the mental processes that precede a risky choice decision. Our preliminary results seem to point in a direction that has hardly been explored in the past. We find decision-making to be highly heterogeneous. While some individuals seem to actually use simple optimization procedures, most others use simple non-optimization procedures, and some of the remaining engage in highly complicated non-optimal decision-making. We conclude that any empirically relevant model of risky-choice must allow for heterogeneity in decision procedures and not just in preferences.

What If You Ask a Classical Economist?

If you ask a classical economist, whether the decision on buying the new insurance contract is a big deal, there is no doubt that the response will be negative. An insurance decision like the one above is a simple exercise in basic micro-economics (or decision science, if you prefer). In fact, the task is so simple that most economists would complicate it in several ways, before using it as an exercise in class or as a question in a mid-term exam. Since the students, who are trying to solve the exercise problem, cannot look into the decision-maker's mind to observe his preferences, the economist provides them with a *preference function* that (so the economist claims) represents the decision-maker's preferences in a beautifully simple way, fulfilling a number of consistency conditions.[1] Given the stable and

[1]Typically preference functions represent preference orderings that are *complete*, *monotonic* (i.e. not contradicting stochastic dominance criteria), *transitive*. This is true for expected utility theory (von Neumann and Morgenstern 1947) and for most of the non-expected utility (or "generalized expected utility") theories that provide a preference function (e.g. Machina 1982; Quiggin 1982; Tversky and Kahneman 1992). In expected utility theory preference orderings are also assumed to

consistent preferences that classical economists tend to postulate and given the simple rules of rational decision-making, solving this problem is in fact no big deal.

Note that the "modern" classical economists do not deny the evidence for decision behavior that does not conform to expected utility theory. They have accepted that – while being the most beautifully rational theory – expected utility theory may have deficiencies in dealing with observed behavior. Admitting to these deficiencies has led to three general lines of defense, but to no retreat. The first line of defense (and probably the crudest) has been to deny the external validity of the experimental findings on the grounds that the decision-makers in the artificial situation were either not given a chance to "discover" their "real" underlying preferences (Plott 1996) or were not given enough training to understand the game (Binmore 1999). The obvious question to ask here is: How much "training" a decision-maker in reality has when choosing a new insurance contract as the one in our example above?

It seems straightforward to believe that experience with a decision situation will increase the stability of choices in that situation. However, in a recent line of research that they have dubbed "coherent arbitrariness," Ariely et al. (2003) show that repeated choices (and presumably the underlying preferences) do stabilize in a coherent manner, but they stabilize relative to an (almost) arbitrary first choice that can be easily influenced by random cues or by adding dominated (hence irrelevant) alternatives to the decision task (Ariely et al. 2006). These results make the case even worse for the discovery of preferences argument, because they indicate that with experience people may not be discovering their true underlying preferences, but simply learning to stick to an arbitrary first choice (Hoeffler et al. 2006).

The second line of defense that is used by classical economists is based on a notion often referred to as *stochastic choice*. The idea is that individuals have well-defined preferences and do plan to maximize expected utility, but err when making decisions (Loomes and Sugden 1995, 1998; Loomes et al. 2002; Hey 2005). These models seem perfectly in line with the econometric tradition of economics, where "representative household" decisions are modeled stochastically to allow for idiosyncratic noise. In fact, if the goal is to measure the aggregate behavior (i.e. as in a macro-economic analysis) this approach may be a promising route to take. But, if the idea is to understand decision-making on a micro-level and to develop micro-founded economic models, this approach may be too imprecise, especially if the instable preference phenomena exhibit predictable biases in behavior (Tversky and Kahneman 1974; Selten 1998b; Ariely et al. 2006). Systematic and predictable biases in choice behavior cannot easily be explained using stochastic choice (or stochastic preference) models, because errors are usually thought to be non-systematic in their distribution, i.e. non-informative "white noise".

The third line of defense for "modern" classical economists is to integrate some of the findings of behavioral decision research into preference functions

---

fulfill the *independence* axiom, which basically states that if one prospect is preferred to another, combining both with a third common prospect should lead to the same preference relation between the combined prospects as between the original prospects.

("non-expected utility theory") has clearly been the most popular approach (Starmer 2000). In fact, non-expected utility was so fruitful as a new field of economic research that in 1988 the active researchers decided to establish a specialized field journal (*The Journal of Risk and Uncertainty*) that has been thriving ever since. As if to certify that the approach has "won over" the behavioral decision research, two of the staunchest adversaries of the classical economic approach, Amos Tversky and Daniel Kahneman, caved in and republished their highly successful behavioral model "Prospect Theory" (Kahneman and Tversky 1979) as pure generalization of expected utility theory in an early issue of the *Journal of Risk and Uncertainty* (Tversky and Kahneman 1992). The new version was deprived of the "editing phase," an originally central part of the model in which the decision-maker is believed to code, aggregate, or eliminate prospects (or consequences in prospects) in order to facilitate the task of choosing.

What If You Ask a Behavioral Decision Researcher?

If you ask a behavioral decision researcher (or a consumers' choice psychologist), whether the decision on buying the new insurance contract is a big deal, you will realize that the answer may be much more involved than the classical economist had indicated. Behavioral decision research (Payne et al. 1992) identifies two important issues in the decision situation above that the classical economist ignores. For one thing, every deliberation is costly in terms of cognitive effort. Hence, creating a complete preference ordering that includes all options, including those that are not in the immediate choice set, is much too costly for most individuals. If contemplating about the benefits of alternatives is costly, then it may "pay" to construct preferences on-the-fly, at least to some extent. Hence, when faced with a new type of insurance, an individual most probably will not have evaluated and ranked this new prospect within his preference ordering yet. Instead, the individual may start contemplating on the prospect ex-post, i.e. after the prospect has (involuntarily) entered his consciousness. Note, that this decision-maker is a different creature than the one the classical economist has in mind, because there is no prespecified preference ordering that can be simply "looked up" without incurring an effort cost. This creature exerts cognitive effort to *construct* his preferences (Bettman et al. 1998).

The second line of reasoning is based on the observation that perception generally is imperfect and prone to misperception biases (Tversky and Kahneman 1974). Obviously, if stochastic events, statistical measures, and even own future satisfaction are systematically misperceived, then preferences are no longer invariant to the problem frame, but are constructed depending on when, where, and how the choice arises and the information about the decision is accumulated. Note that the "when, where, and how" mentioned here are akin to the "place" and "promotion" in marketing's four P's paradigm (the other two P's are "product" and "price", which may correspond to the outcome and risk attributes of the prospect). If preferences were completely invariant to the decision frame, many marketing activities would be futile.

While behavioral decision researchers clearly complicate things by introducing a growing set of framing issues (Griffin et al. 2005) and by more-or-less invalidating all three rationality provisions of classical economics,[2] they do believe that their predictions of the observed choices are actually enhanced using the more elaborate models (Bettman et al. 1998). This assessment, however, is challenged by classical economists on the account of the loss of parsimony (e.g. Hey 2005). The argument is based on an idea that is similar to the econometric trade-off between the increasing the number of exogenous parameters in a regression model (i.e. decreasing parsimony) and increasing the degree of variance in the data that is explained by the regression (i.e. increasing R^2). Obviously, in both cases the model complication and ex-post explanatory power are positively correlated, justifying some degree of scientific suspicion to increased model complication. It, however, remains unclear, why model parsimony in itself is of value, if the predictive power of models can be increased by increasing their complexity. Just imagine an astrophysicist or a meteorologist saying: "We know that our forecast for some phenomena would be enhanced with a more complicated model, but we prefer parsimony, because it keeps things simple."

What If You Ask Reinhard Selten?

If you ask Reinhard Selten, whether the decision on buying the new insurance contract is a big deal, you may be surprised to find that he is even less inclined to believe in a simple, comprehensive catch-all rule than most behavioral decision researchers are. In a response to Kahneman's (1994) overview of "new challenges" to rationality (Kahneman 1994), Reinhard Selten points out that his "own view is more radically constructivist" than Kahneman's (Selten 1994). This statement perhaps conveys the disappointment that Reinhard Selten felt with the work of Kahneman and Tversky, after they had published the revised version of Prospect Theory (Tversky and Kahneman 1992), in which the constructive model elements were purged and the probability weighting function was specified as in Quiggin (1982), in order to guarantee monotonicity and transitivity of predicted choice behavior. While these features made the model more viable for classical economists, in Reinhard Selten's view, they also substantially reduced the behavioral validity of new Prospect Theory, reducing it to "just another" generalization of expected utility theory with all the same weaknesses.

[2]McFadden (1999, p. 75) sums up the classical economic model as follows: "The *standard model* in economics is that consumers behave *as if* information is processed to form perceptions and beliefs using strict Bayesian statistical principles (*perception-rationality*), preferences are primitive, consistent and immutable (*preference-rationality*), and the cognitive process is simply preference maximization, given market constraints (*process-rationality*)."

In contrast, the original version of Prospect Theory (Kahneman and Tversky 1979) is well in line with Reinhard Selten's (1990) and Herbert Simon (1957) picture of boundedly rational decision making.[3] The common element in the original Prospect Theory model and the work of the Reinhard Selten and Herbert Simon is that individuals facing a choice task first try to solve the problem using procedures that incur low cognitive cost and only move on to cognitively more costly procedures, if a simple solution cannot be found. This notion of stepwise increased cognitive effort has an intuitively "rational" appeal, because it avoids wasting cognitive effort on easy problems. But, note that these procedures are not rational, because instead of optimizing a net benefit function, categorical heuristics are employed.

Reinhard Selten's earliest model of boundedly rational decision making appears in the "legendary" section 3 of his paper on the Chain-Store-Paradox (Selten 1978).[4] In that section, he introduces a three-level model of decision making. In later publications this model is sometimes referred to as *decision emergence* (Selten 1990). The three levels of the process consist of (1) the *level of routine*, (2) the *level of imagination*, and (3) the *level of reasoning*. Cognitive effort increases going from one level to the next. In the level of routine decisions are based on past experience with similar situations and "are made without any conscious effort." (Selten 1978, p. 147) The level of imagination is creative in that the decision maker "imagines" the outcomes of several new scenarios, using the routine knowledge to make guesses. On the level of reasoning, the decision maker consciously uses all current and past information and draws conclusions using logical reasoning.

For any given decision situation the decision maker first uses a routine level decision process to decide on how to decide, i.e. to decide which level of reasoning to employ. This *predecision* is the most critical element of the model, because it is this meta-decision that specifies the degree of rationality and the level of cognitive cost that is incurred for each decision. It is crucial that the predecision in this model is set on the level of routine decisions. There is no doubt that this assumption is to some extent *ad hoc*, but it does help to avoid the problem of introducing higher degree complexities and cognitive cost that rational models specifying decision complexity and cost often encounter (e.g. Kalai and Stanford 1988; Wilcox 1993).

If the predecision has selected a higher level of decision making, the *final decision* requires a meta-decision to select from the suggested choices on the different levels. The crucial element here is that the model does not assume that

[3]Herbert Simon's work on boundedly rational decision making was an important early influence for Reinhard Selten, who says that in the late 1905s he was "converted to Simon's views." (Selten 1993, p.117).

[4]The "legend" of Sect. 3 is that the paper had be accepted for publication in *Econometrica* under the provision that some minor revisions are made and that the paper's Section 3 is removed. Instead of revising and resubmitting the paper, however, Reinhard Selten withdrew the paper from *Econometrica* and submitted it to *Theory & Decision* insisting that it is either rejected or taken as it is. The paper was published in *Theory & Decision* without major revisions and turned out to be one of the most frequently cited papers in game theory.

the solution offered by the highest level is generally followed. Hence, decision makers may actually know the rational (or optimal) decision, but still decide to follow the suggestion of a lower level. The possibility that rational reasoning may be overruled by a lower level decision is essential to Reinhard Selten's view of boundedly rational decision making, not only because there is abundant experimental evidence that subjects systematically deviate from rational choice, even after being trained explicitly (Grether and Plott 1979), but also because of the introspective assessment of the conflict between rational and intuitive choice in the Chain-Store-Paradox that he describes in the introduction of the paper (Selten 1978).

Both the predecision and the final decision are made on the routine level, i.e. intuitively without a conscious deliberation. To be successful these decisions must be based on past experience and adapted through a learning process. The idea that routine decision behavior is adapted to an ex-post assessment of short-run success and failure later enters into the widely cited learning direction theory (Selten et al. 2005) and the related impulse balance theory (Ockenfels and Selten 2005).

Summarizing, we find that Reinhard Selten's view on how risky choice decisions are made fundamentally differs from the formula optimization (be it expected utility, generalized expected utility, stochastic expected utility, or any other value function) approach of the classical economists. It also differs in many ways from the diverse and dispersed behavioral decision research models that describe the construction of preferences based on numerous (partially unrelated) parameters of the decision environment. Instead, Reinhard Selten suggests taking all elements of human decision making (including routine decisions, emotional and intuitive decisions, and rational decisions) together and creating a joint framework, in which individuals learn to use different decision types in different situations. Learning will certainly be guided by the cost of cognition and the payoff obtained through the decisions. However, the learning procedure is not an optimization process, but an adaptive and satisficing process as suggested by Herbert Simon (1957).

Using the Strategy Method to Uncover Decision Processes

One of the first methods that Reinhard Selten devised to gain a better understanding of decision behavior is the *strategy method* (Selten 1967). The strategy method – as Reinhard Selten defined it – is a rather complicated procedure including an initial *spontaneous play* phase, a *strategy elicitation* phase, and at least one round of *strategy evaluation and enhancement*. The only group of researchers, who have used the entire complicated procedure in all details, are Reinhard Selten and his students (e.g. Selten 1967; Selten, Mitzkewitz, Uhlich 1997; Selten, Abbink, Buchta, Sadrieh 2003; Keser and Gardner 1999; Rockenbach and Wolff 2009).

Most experimental economists, however, do not relate the term "strategy method" to the elaborate procedure Reinhard Selten has in mind, but only to the strategy elicitation phase.[5] Hundreds of studies now exist that make use of the strategy elicitation phase of the strategy method.[6] Furthermore, there is some research on tournaments amongst software agents (human strategies) that combine the strategy elicitation and the strategy evaluation and enhancement phase.[7] In any of these formats, the strategy method has the advantage of revealing the subjects' entire strategy and, thus, giving the researcher an insight into the counterfactuals of the game. There has been some debate on whether eliciting complete strategies, forces subjects to think differently – or more thoroughly – about the problem (Brandts and Charness 2009). The argument, however, seems peculiar, because hindering subjects to think about the game more deeply is almost like giving them instructions that are too complicated or misleading. Hence, if it is true that the strategy elicitation makes subjects think about the situation more thoroughly, then all economic experiments should use strategy elicitation.

While the strategy elicitation reveals a subject's complete strategy, it does not reveal the reasoning that led to the chosen strategy. However, to be able to create a behavioral model of (perhaps non-optimizing) decision-making, it would be helpful to gain insight into the mental deliberations of a decision-maker. To do so, psychologists sometimes use think-aloud studies. In these studies, the decision-maker is asked to communicate his deliberations out loud, so that a monitor can record the on-going thought process. The procedure obviously has the disadvantage that the subjects cannot be incentivized to reveal their thought process completely or truthfully. Early in his experimental research (Sauerman and Selten 1959), Reinhard Selten realized that the incentive problem may be solved, if not one subject decides, but a team. Since each team member has an incentive to convince the others of his most favored strategy, Reinhard Selten asserted that the protocol of the team discussion should contain most (if not all) relevant arguments. The team protocol method was later improved by videotaping the team discussion, instead of letting one of the subjects write a protocol of the team discussion. This then resulted in the video-taped team decision method (often abbreviated as "video experiment") that is the topic of one of the chapters of this book.

In mid 1990s, the team at the laboratory of experimental economics of the University of Bonn was heavily involved in video experimentation. The method proved to be very informative in bargaining (e.g. Hennig-Schmidt 1999) and in trust settings (Jacobsen and Sadrieh 1996). In lottery choice tasks, however, we encountered an unexpected problem with the method. In many cases, no significant discussion came up in the team. If, for example, the team was to choose between

[5]While the distorted usage of the term used to infuriate Reinhard Selten early on, he has now accepted it as the result of an evolutionary process (personal communication).

[6]See Brandts and Charness (2009) for a survey comparing the strategy elicitation method to spontaneous play.

[7]Amongst these the most notable are the *strategy tournaments* that ask subjects (often self-selected subjects) to send in a full strategy that can be run in a tournament against others. See e.g. Axelrod (1984), Friedman and Rust (1993), and Erev et al. (2010).

the lottery A and the lottery B, we frequently observed a discussion of the following type:

Team member 1: "Let's take A."
Team member 2: "Good choice."
Team member 3: "OK."

Obviously, there is little to learn about the thought processes in such a discussion. It seems that the limited number of alternatives (just two) often leads to an immediate and implicit agreement with no need for further discussions. But, increasing the number of choice alternatives is not an option in these experiments, because the behavioral regularities that we want to learn more about are generally parameterized in simple lottery choices of the type described. To overcome the problems with the standard think-aloud studies and video-taped team decisions, we devised a new experimental design.

Introducing the Verbal Strategy Elicitation Method

In Sadrieh et al. (2010), we introduce the *verbal strategy elicitation method*, a new incentive compatible method to uncover the heuristics that subjects use to make their risky choices. To be able to uncover more than just a simple choice for a specific task (i.e. more than just "In this pair, I choose A."), the verbal strategy elicitation uses a "veil of uncertainty" trick. The subjects know what "type" of lotteries will be coming up in the next lottery choice task, but they do not know exactly which payoffs and probabilities are assigned to the outcome states of the two lotteries A and B. Each lottery has two possible prizes, a_1 and a_2 for lottery A and b_1 and b_2 for lottery B. Each prize is drawn randomly from a uniform distribution of values, where each distribution has a potentially different support. The high prize a_1 of lottery A is drawn from [600, 650, 700, 750], while the low prize a_2 is draw from [0, 50, 100, 150]. Both of the prizes of lottery B, b_1 and b_2, are drawn from the interval [200, 250, 300, 350, 400, 450, 500, 550]. Hence, it is always the case that the lottery A is the lottery with the "extreme" outcomes, i.e. it has one prize that is greater than and one prize that is smaller than the prizes in lottery B.

Not only the lottery outcomes are randomly drawn, but also the probabilities of winning, where p denotes the probability of winning the high prize in lottery A and q denotes the probability of winning the high prize in B. Hence, the two lotteries are specified as follows: lottery A $= [(a_1, p),(a_2, 1 - p)]$ and lottery B $= [(b_1, q),(b_2, 1 - q)]$. Both p and q are randomly drawn from the interval [0.1, 0.2, . . ., 0.9] with equal probabilities.

After making 20 spontaneous decisions on randomly generated tasks, subjects are asked to report the decision rules that they intend to use in the next five choice tasks. The experimenter records the verbally reported decision rules and double checks with the subject that the rules are correctly understood. There is no limit to the number and the complexity of the rules, as long as the subject can precisely demonstrate that each rule can be executed in no more than a minute and using a simple pocket calculator at most. After the rules are fixed, the experimenter goes through a set of five decision tasks together with the subject, making choices according to the reported decision rules.[8] The subject monitors this process.[9] The subject is then given the opportunity to revise the rules. Once the revised rules are set, they are applied to the next five choice tasks. The process of rule revision and application is repeated three times, before the subjects are asked to report their final decision rules that are applied to the final set of 20 choice tasks. Subjects receive financial rewards based on the decisions made by their decision rules. Hence, unlike other self-reporting procedures, the verbal strategy elicitation method entails salient incentives for the truthful revelation of subjects' decision heuristics.

Our experiment includes two treatments. In the treatment *Stats*, the subjects are provided with the expected value (EV) and the mean absolute deviation from the expected value (MAD) of every lottery under consideration.[10] In the treatment *NoStats*, statistical measures are not provided. Obviously, any type of optimization-based choice theory (including expected utility theory, cumulative prospect theory, and all other non-expected utility theories that ignore framing) predicts no difference between the decision rules that subjects report in the two treatments, since the set of lotteries to be chosen from and the information set are identical across the two treatments. Adding the measures simply provides a predefined aggregation of information that was already available. An earlier study, however, shows that providing the measures may significantly affect lottery choice behavior (Selten, Sadrieh, and Abbink 1999). Given that evidence, we conjectured that most of the subjects in the *Stats* treatment will not simply ignore the statistical measures, just as most of the subjects in the *NoStats* treatment will not readily specify rules that contain the measures.

The main results of our verbal strategy elicitation experiment are summarized in Table 16.1. About two-thirds of all subjects use production systems instead of optimization procedures in their final decision rules. A production system consists

Table 16.1 Frequency of final decision rules

Treatment	Production system	Not production system	Use EV(A)	Use EV(B)	Use MAD(A)	Use MAD(B)
NoStats	19(76%)	6(24%)	7(28%)	7(28%)	0(0%)	0(0%)
Stats	12(55%)	10(45%)	11(50%)	11(50%)	9(40%)	8(36%)

[8] If the decision rules do not specify a unique choice, the task is skipped and the subject receives no payment for the specific choice task. Note, however, that subjects are allowed to include a random draw rule (e.g. "flip a coin") to make a decision when indifferent.

[9] Many subjects take notes that are later used to update their set of decision rules.

[10] We use MAD instead of SD or VAR, because MAD is much simpler to explain than the other measures of dispersion. It is also easier to calculate MAD without a calculator, because there are no exponents involved.

of a series of conditional clauses (e.g. "If $a_1 > 650$ and $p > 0.3$, then choose A.") Moreover, most of these production systems begin with a clause conditioned on the highest possible prize a_1. The only statistical measure that is used by subjects in the *NoStats* treatment is the EV, while subjects in the *Stats* treatment also included the MAD in their final decision rules. Quite a number of subjects, who use EV and/ or MAD in their final decision rules, include these measures in a complex production system. A few subjects use unusual functions to combine the measures. Finally, we observe an enormous amount of heterogeneity in the final decision rules both amongst the production systems and amongst those that use a numerical decision function. The only decision function that we observe more than once is the EV maximization rule. Amongst the observed production systems, basically no two systems are identical.

Conclusions

While we are still far away from fully understanding choice processes and the emergence of preferences, some of the ideas that Reinhard Selten has promoted and – perhaps even more so – some of the research methods that he has introduced have gradually proliferated. Especially in consumer research, the notion that preferences are constructed and not merely discovered is a widely spread paradigm that is mainly based on the observation that many decision are not framing invariant.[11] But, significant attempts to compile the boundedly rational phenomena into a model of non-optimizing decision-making have not been made in consumer research either. It seems that Selten's aspiration adaptation model (Selten 1998a), which is inspired by Simon's approach (Simon 1957), remains unrivaled as a non-optimizing model of decision-making. The model, however, must first be supplemented to provide a satisfactory representation of the emergence or the construction of preferences.

To achieve progress in this area, we have suggested a novel experimental method, the *verbal strategy elicitation method* (Sadrieh et al. 2010). The method provides information on the process of decision-making and, thus, allows us to enhance the procedural models of risky choice. The new method certainly will need some time before finding a general acceptance in academics. But, it seems worthwhile to go down that dusty road. Paraphrasing one of Reinhard Selten's favorite lines: You cannot discover new plants without going out to hike. You cannot discover new chemicals without experimenting in the lab. Why should you expect to discover anything about human behavior, while sitting comfortably and reasoning in your armchair?

[11]See e.g. Ariely et al. (2006) and the references therein. Amir and Levav (2008) study the stability of preferences that have been constructed in different framing tasks.

References

Amir O, Levav J (2008) Choice construction versus preference construction: the instability of preferences learned in context. J Market Res 45:145–158
Ariely D, Loewenstein G, Prelec D (2003) Coherent arbitrariness: stable demand curves without stable preferences. Q J Econ 118(1):73–105
Ariely D, Loewenstein G, Prelec D (2006) Tom Sawyer and the construction of value. J Econ Behav Organ 60:1–10
Axelrod R (1984) The evolution of cooperation. Basic, New York
Bettman JR, Luce MF, Payne JW (1998) Constructive consumer choice processes. J Consum Res 25(3):187–217
Binmore K (1999) Why experiment in economics? Econ J 109:F16–F24
Brandts J, Charness G (2009) The strategy method: a survey of experimental evidence. Mimeo, Institut d'Anàlisi Econòmica, CSIC, Bellaterra, Spain
Cubitt RP, Starmer C, Sugden R (2001) Discovered preferences and the experimental evidence of violations of expected utility theory. J Econ Methodol 8(3):385–414
Erev I, Ert E, Roth AE (2010) A choice prediction competition for market entry games: an introduction. Games 1(2):117–136
Friedman D, Rust J (eds) (1993) The double auction market. Westview, Boulder, CO
Grether DM, Plott CR (1979) Economic theory of choice and the preference reversal phenomenon. Am Econ Rev 69:623–638
Griffin D, Liu W, Khan U (2005) A new look at constructed choice processes. Market Lett 16(3/4):321–333
Hennig-Schmidt H (1999) Bargaining in a video experiment – determinants of boundedly rational behavior, vol 467, Lecture notes in economics and mathematical systems. Springer, Berlin
Hey JD (2005) Why we should not be silent about noise. Exp Econ 8:325–345
Hoeffler S, Ariely D, West P (2006) Path dependent preferences: the role of early experience and biased search in preference development. Organ Behav Hum Decis Process 101(2):215–229
Jacobsen E, Sadrieh A (1996) Experimental proof for the motivational importance of reciprocity. SFB-Discussion Paper B-386, University of Bonn
Kahneman D (1994) New challenges to the rationality assumption. J Inst Theor Econ 150:18–36
Kahneman D, Tversky A (1979) Prospect theory: an analysis of decision under risk. Econometrica 47(2):263–291
Kalai E, Stanford W (1988) Finite rationality and interpersonal complexity in repeated games. Econometrica 56(2):397–410
Keser C, Gardner R (1999) Strategic behavior of experienced subjects in a common pool resource game. Int J Game Theory 28(2):241–252
Loomes G, Sugden R (1995) Incorporating a stochastic element into decision theories. Eur Econ Rev 39:641–648
Loomes G, Sugden R (1998) Testing different stochastic specifications of risky choice. Economica 65:581–598
Loomes G, Moffatt PG, Sugden R (2002) A microeconometric test of alternative stochastic theories of risky choice. J Risk Uncertain 24:103–130
Machina M (1982) Expected utility analysis without the independence axiom. Econometrica 50:277–323
McFadden D (1999) Rationality for Economists? Journal of Risk and Uncertainty 19:73–105
Ockenfels A, Selten R (2005) Impulse balance equilibrium and feedback in first price auctions. Games Econ Behav 51:155–170
Payne JW, Bettman JR, Johnson EJ (1992) Behavioral decision research: a constructive processing perspective. Annu Rev Psychol 43(1):87–131
Plott CR (1996) Rational individual behavior in markets and social choice processes: the discovered preference hypothesis. In: Arrow KJ, Colombatto E, Perlman M, Schmidt C (eds) The rational foundations of economic behavior. Macmillan, Basingstoke, pp 225–250

Quiggin J (1982) A theory of anticipated utility. J Econ Behav Organ 3:323–343
Rockenbach B, Wolff I (2009) Institution design in social dilemmas: how to design if you must? Mimeo, University of Erfurt
Sadrieh A, Hassels M, Ridder D (2010) Heuristics for choosing from pairs of randomly generated lotteries. Mimeo, University of Magdeburg
Sauermann H, Selten R (1959) Ein Oligopolexperiment. Zeitschrift für die gesamte Staatswissenschaft 115:427–471
Selten R (1967) Die Strategiemethode zur Erforschung des eingeschränkt rationalen Verhaltens im Rahmen eines Oligopolexperiments. In: Sauermann H (ed) Beiträge zur experimentellen Wirtschaftsforschung. J.C.B. Mohr (Paul Siebeck), Tübingen, pp 136–168
Selten R (1978) The chain store paradox. Theory Decis 9(2):127–159
Selten R (1990) Bounded rationality. J Inst Theor Econ 146(4):649–658
Selten R (1991) Anticipatory learning in two-person games. In: Selten R (ed) Game equilibrium models I. Springer, Berlin, pp 98–153
Selten R (1993) In search of a better understanding of economic behavior. In: Heertje A (ed) The makers of modern economics, vol 1. Harvester Wheatsheaf, New York, NY, pp 115–139
Selten R (1994) New challenges to the rationality assumption: comment. J Inst Theor Econ 150:42–44
Selten R (1998a) Aspiration adaptation theory. J Math Psychol 42:191–214
Selten R (1998b) Features of experimentally observed bounded rationality. Eur Econ Rev 42:413–436
Selten R, Abbink K, Buchta J, Sadrieh A (2003) How to play (3×3)-games. A strategy method experiment. Games and Economic Behavior 45:19–37
Selten R, Abbink K, Cox R (2005) Learning direction theory and the winner's curse. Exp Econ 8:5–20
Selten R, Mitzkewitz M, Uhlich GR (1997) Duopoly Strategies Programmed by Experienced Players. Econometrica 65(3):517–555
Selten R, Sadrieh A, Abbink K (1999) Money Does not Induce Risk Neutral Behavior, but Binary Lotteries Do Even Worse. Theory and Decision 46(3):213–252
Simon H (1957) Models of man: social and rational. Wiley, New York, NY
Starmer C (2000) Developments in non-expected utility theory: the hunt for a descriptive theory of choice under risk. J Econ Lit 38(2):332–382
Tversky A, Kahneman DP (1974) Judgment under uncertainty: heuristics and biases. Science 27(185):1124–1131
Tversky A, Kahneman DP (1992) Advances in prospect theory: cumulative representation of uncertainty. J Risk Uncertain 5:297–323
von Neumann J, Morgenstern O (1947) Theory of games and economic behavior. Princeton University Press, Princeton, NJ
Wilcox NT (1993) Lottery choice: incentives, complexity and decision time. Econ J 103 (421):1397–1417

MIX
Papier aus verantwortungsvollen Quellen
Paper from responsible sources
FSC® C105338

If you have any concerns about our products,
you can contact us on
ProductSafety@springernature.com

In case Publisher is established outside the EU,
the EU authorized representative is:
Springer Nature Customer Service Center GmbH
Europaplatz 3, 69115 Heidelberg, Germany

Printed by Libri Plureos GmbH
in Hamburg, Germany